Springer Optimization and Its Applications

Volume 197

Aims and Scope
Optimization has continued to expand in all directions at an astonishing rate. New algorithmic and theoretical techniques are continually developing and the diffusion into other disciplines is proceeding at a rapid pace, with a spot light on machine learning, artificial intelligence, and quantum computing. Our knowledge of all aspects of the field has grown even more profound. At the same time, one of the most striking trends in optimization is the constantly increasing emphasis on the interdisciplinary nature of the field. Optimization has been a basic tool in areas not limited to applied mathematics, engineering, medicine, economics, computer science, operations research, and other sciences.

The series **Springer Optimization and Its Applications (SOIA)** aims to publish state-of-the-art expository works (monographs, contributed volumes, textbooks, handbooks) that focus on theory, methods, and applications of optimization. Topics covered include, but are not limited to, nonlinear optimization, combinatorial optimization, continuous optimization, stochastic optimization, Bayesian optimization, optimal control, discrete optimization, multi-objective optimization, and more. New to the series portfolio include Works at the intersection of optimization and machine learning, artificial intelligence, and quantum computing.

Volumes from this series are indexed by Web of Science, zbMATH, Mathematical Reviews, and SCOPUS.

Ioannis K. Kookos

Practical Chemical Process Optimization

With MATLAB® and GAMS®

 Springer

Ioannis K. Kookos
Department of Chemical Engineering
University of Patras
Patras, Greece

ISSN 1931-6828   ISSN 1931-6836   (electronic)
Springer Optimization and Its Applications
ISBN 978-3-031-11300-0   ISBN 978-3-031-11298-0   (eBook)
https://doi.org/10.1007/978-3-031-11298-0

This Springer imprint is published by the registered company Springer Nature Switzerland AG
The registered company address is: Gewerbestrasse 11, 6330 Cham, Switzerland

To my beloved children

Kostas and Georgia

*for the happiness you have offered
to me by simply watching you grow!*

Preface

This course book on chemical process optimization offers an undergraduate introduction to the systematic optimization of chemical processing systems. The aim of this book is to provide readers with a practical introduction to the terms, concepts, algorithms, and application of the optimization theory to systems that are of interest to chemical and bio-chemical engineering practice. Optimization theory is a vast subject with many excellent books available. However, the level of the mathematics involved makes the subject difficult for the average chemical engineering student while most examples involve purely mathematical problems.

After working in the field of process synthesis and optimization for almost 30 years, the author of the book has discovered that, although an understanding of the mathematical proofs is beneficial, it is of little (if any) help when it comes down to solving practical, engineering problems. It is particularly unlikely that a chemical engineer will develop algorithms for solving optimization problems in their professional life. As a result, there is limited potential in devoting a whole course on the mathematical properties and proofs related to the several classes of optimization problems. Having said that, it is also important to stress that chemical engineers must have a basic understanding of the philosophy and the limitations of the existing solution algorithms. They must be able to anticipate when a specific formulation will be easy to solve and must be able to comment on the results and "read" correctly the solution obtained using commercial software. Readers who are keen in developing their own algorithms can use this book as a "gentle" or "informal" introduction to optimization theory or as a source of interesting case studies.

The book is organized in seven chapters. The first chapter introduces the general form of optimization problems and then concentrates on single-variable, unconstrained optimization problems. Chapter 2 presents the generalization of the theory presented in Chap. 1 to the case of multivariable unconstrained optimization. Chapter 3 introduces the constrained, nonlinear optimization with both equality and inequality constraints. Chapter 4 presents the simplex methodology for solving linear programming problems. Chapter 5 introduces the modeling of decisions and logical conditions which gives rise to mixed-integer programming problems and introduces the Branch and Bound algorithm for solving complex problems. In Chap. 6, the GAMS® software is introduced through some interesting recreational applications. Chapter 7 includes several classical and interesting case studies from chemical and bio-chemical engineering.

The book is based heavily on the use of MATLAB® (www.mathworks.com) for solving the case studies presented in Chaps. 1, 2, 3, 4, and 5 and MATLAB Optimization Toolbox™ is introduced from the very first chapter of the book. Appendix A offers a gentle and short introduction to MATLAB. Although MATLAB's built-in optimization routines have been developed over the years, they are not as efficient as most of the specialized solvers available from other sources. To solve the more demanding case studies in Chaps. 6 and 7, use is made of GAMS software (www.gams.com).

Patras, Greece   Ioannis K. Kookos

Contents

Chapter 1
Preliminary Concepts and Definitions

1.1 An Introductory Example

Consider the situation where you have been asked to design a vertical, cylindrical storage tank with torispherical heads (see Fig. 1.1). The only information
that you have been provided with is the operating pressure of the tank (from
which you can deduce the thickness of the shell) and the volume of the tank.
The question is how to proceed and present a practical solution to this
seemingly simple problem. The first thing to realize is that the problem at
hand has two degrees of freedom which are the tank diameter d and the tank
height h. The meaning is that once you specify the diameter and the height of
the tank, then the design is practically finished. Any combination of d and h that
satisfy the volume constraint:

$$\left(\frac{\pi d^2}{4}\right)h = V \tag{1.1}$$

is a *feasible solution* to the problem. However, there are an infinite number of
combinations between d and h that satisfy the constraint on the capacity of
the tank. Which one should we select? Clearly to take a decision, we need a
performance criterion.

A performance criterion can be used to evaluate alternative solutions, and
based on this property of the performance criterion, we can compare and rank
alternative solutions. Furthermore, we can develop a solution strategy to locate
the best possible solution. So, the problem comes down to selecting a criterion
of performance.

Cost is of paramount importance when building or operating a chemical
plant. Let us assume that we have decided to design the tank that is cheaper to
construct (storage tanks normally do not have operating costs involved). The

© The Author(s), under exclusive license to Springer Nature Switzerland
AG 2022
I. K. Kookos, *Practical Chemical Process Optimization*, Springer Optimization
and Its Applications 197, https://doi.org/10.1007/978-3-031-11298-0_1

Fig. 1.1 A vertical,
cylindrical storage tank

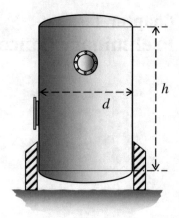

purchase cost of chemical process equipment can usually be expressed through
the following well-known formula:

$$\frac{C}{C_0} = \left(\frac{w}{w_0}\right)^n \tag{1.2}$$

where C is the purchase cost of our equipment, C_0 is the cost of equipment of
the same type (including design, material of construction, and operating conditions) but of different size, and w is the characteristic size of the particular
equipment which affects the cost (such as the heat transfer area for heat
exchangers, the weight for pressure vessels, the volume for storage tanks,
etc.). w_0 and C_0 refer to a particular case for which we know the characteristic
size of the equipment (w_0) and the cost (C_0). n obtains values which are close to
0.6 in most cases.

By getting the advise of any undergraduate book on chemical process
design, we quickly find out that the characteristic size of storage tanks is the
mass (or the weight) of the tank. We therefore have to express the mass of the
tank as a function of the two degrees of freedom. The mass is simply the
volume of the metal used times its density. So we need to determine the
volume of the metal. The volume is the surface area multiplied by the thickness
t. The mass of the cylindrical part is:

$$w_{\text{cylinrical}} = \text{area} \times \text{thickness} \times \text{density} = (\pi d \cdot h) \times t \times \rho_M \tag{1.3}$$

where ρ_M is the density of the metal used to construct the tank. For each head
we similarly have that:

$$w_{\text{head}} = \text{area} \times \text{thickness} \times \text{density} = \left(a\frac{\pi d^2}{4}\right) \times K \times t \times \rho_M \tag{1.4}$$

where a is the ratio between the area of the torispherical dished (ASME) end sections of the tank and the cross-sectional area of the cylindrical part of the tank (a circle with diameter d, in which case $a \approx 1.124$). K is the ratio between the thickness of the end sections and the cylindrical part of the tank (which, for dished heads, is approximately equal to 1.77). We therefore observe that $aK \approx 2$. The overall mass of the tank is therefore:

$$w = \pi dht\rho_M + 2\alpha K\left(\frac{\pi d^2}{4}\right)t\rho_M$$

or:

$$w = \pi\left(dh + d^2\right)t\rho_M \tag{1.5}$$

We may at this point summarize what we have found so far in a mathematical way as follows:

$$\text{minimize over } d,h,w \quad C_0\left(\frac{w}{w_0}\right)^n$$
$$\text{subject to} \tag{1.6}$$
$$w = \pi\left(dh + d^2\right)t\rho_M$$
$$V = \left(\frac{\pi d^2}{4}\right)h$$
$$0 \leq d$$
$$0 \leq h$$

We ask the question whether we know how to solve the mathematical problem (1.6) that summarizes, in a compact way, our physical problem. The answer is most probably negative (you need to study this book before you can give a positive answer).

We can express the problem in a more general, abstract (mathematical) form. We first observe that there are three variables appearing in our formulation: d, h, and w. We can arrange them as the elements of a vector, say vector $\mathbf{x}$:

$$\mathbf{x} = \begin{bmatrix} x_1 \\ x_2 \\ x_3 \end{bmatrix} = \begin{bmatrix} d \\ h \\ w \end{bmatrix} \tag{1.7}$$

We can proceed further and define the following functions of $\mathbf{x}$:

$$f(\mathbf{x}) : R^3 \mapsto R, \quad f(\mathbf{x}) = \frac{C_0}{w_0^n}x_3^n \tag{1.8}$$

$$h_1(\mathbf{x}) : R^3 \mapsto R, \ h_1(\mathbf{x}) = x_3 - \pi\left(x_1 x_2 + x_1^2\right) t \rho_M \tag{1.9}$$

$$h_2(\mathbf{x}) : R^3 \mapsto R, \ h_2(\mathbf{x}) = V - \frac{\pi}{4} x_1^2 x_2 \tag{1.10}$$

$$g_1(\mathbf{x}) : R^3 \mapsto R, \ g_1(\mathbf{x}) = -x_1 \tag{1.11}$$

$$g_2(\mathbf{x}) : R^3 \mapsto R, \ g_2(\mathbf{x}) = -x_2 \tag{1.12}$$

these are the *equality* and *inequality constraints* of the problem. Then we may write (1.6) as:

$$
\begin{aligned}
&\min_{x} \ \frac{C_0}{w_0^n} x_3^n \\
&\text{s.t.} \\
&h_1(\mathbf{x}) = x_3 - \pi\left(x_1 x_2 + x_1^2\right) t \rho_M = 0 \\
&h_2(\mathbf{x}) = 4V - \pi x_1^2 x_2 = 0 \\
&g_1(\mathbf{x}) = -x_1 \leq 0 \\
&g_2(\mathbf{x}) = -x_2 \leq 0
\end{aligned}
\tag{1.13}
$$

or, to use the standard notation in mathematical programming and process optimization:

$$
\begin{aligned}
&\min_{x} \ f(\mathbf{x}) \\
&\text{s.t.} \\
&\mathbf{h}(\mathbf{x}) = \mathbf{0} \\
&\mathbf{g}(\mathbf{x}) \leq \mathbf{0}
\end{aligned}
\tag{1.14}
$$

where $\mathbf{h} = [h_1 \ h_2]^T$ and $\mathbf{g} = [g_1 \ g_2]^T$. This is the general form of a nonlinear programming problem or simply NLP.

1.2 Commonly Encountered Problems in Optimization

When all functions appearing in (1.14) are linear, i.e., a constant times a variable, then one obtains the following problem:

$$
\begin{aligned}
&\min_{x} \ \mathbf{c}^T \mathbf{x} \\
&\text{s.t.} \\
&\mathbf{Ax} - \mathbf{b} = \mathbf{0} \\
&\mathbf{Bx} - \mathbf{d} \leq \mathbf{0} \\
&\mathbf{0} \leq \mathbf{x}
\end{aligned}
\tag{1.15}
$$

where **c**, **b**, and **d** are constant vectors and **A** and **B** are constant matrices. Formulation (1.15) is known as a linear programming problem (or simply LP) and is the class of optimization problems for which we do have today very efficient algorithms to attack. The requirement for the vector **x** to have only non-negative elements is traditional to the formulation of the LP problems. A related problem is that of the quadratic programming problem in which the objective function is quadratic:

$$\min_{x} \ c^T x + \frac{1}{2} x^T Q x$$
$$\text{s.t.} \qquad\qquad (1.16)$$
$$Ax - b = 0$$
$$Bx - d \leq 0$$
$$0 \leq x$$

where **Q** is a constant matrix. When we do not have inequality constraints, we end up with an *equality constrained problem*, while when we do not have equality constraints, we end up with an *inequality constrained problem*. When we only have *bound constraints* on x, then we have a bound constrained problem. When we do not have any constraints at all, we end up with an *unconstrained optimization* problem:

$$\min_{x} \ f(x) \qquad\qquad (1.17)$$

In the general case, $x \in R^n$ and $f{:}R^n \to R$. In the simplest possible case, x is scalar, i.e., $x \in R$ and $f{:}R \to R$ (i.e., f is a real valued function of a single variable).

$$\min_{x \in R} \ f(x) \qquad\qquad (1.18)$$

We will start from the study of problem (1.18), which you most probably have seen already, and then move backward to finally discuss problem (1.14).

1.3   Optimization of Functions of a Single Variable

We start our journey with the simplest possible problem in optimization, i.e., the minimization of a real valued function of a single variable:

$$\min_{x \in R} \ f(x) \qquad\qquad (1.19)$$

We will make use of the definition of the stationary point and the Taylor's approximation of a function around a point. We need to remind ourselves these definitions. A *stationary point* of $f(x)$ (a terminology that was borrowed from mechanics) is any point x_0 for which:

$$\left.\frac{df}{dx}\right|_{x_0} = 0 \qquad (1.20)$$

If $f(x)$ has continuous partial derivatives of all orders, then the value of $f(x)$ around x (at a distance δ from x) is given by the following Taylor series:

$$f(x+\delta) = f(x) + \left.\frac{df}{dx}\right|_x \delta + \frac{1}{2}\left.\frac{d^2f}{dx^2}\right|_x \delta^2 + o\left(|\delta|^2\right) \qquad (1.21)$$

where the $o(|\delta|^2)$ is the remainder and denotes that $o(\zeta)$ approaches zero faster than ζ as ζ approaches zero, i.e., $|o(\zeta)|/\zeta \to 0$ as $\zeta \to 0$. Starting from (1.21), we can define the following linear and quadratic approximations of $f(x)$ which are more accurate when $|\delta| \to 0$:

$$f(x+\delta) = f(x) + \left.\frac{df}{dx}\right|_x \delta \qquad (1.22a)$$

$$f(x+\delta) = f(x) + \left.\frac{df}{dx}\right|_x \delta + \frac{1}{2}\left.\frac{d^2f}{dx^2}\right|_x \delta^2 \qquad (1.22b)$$

We now come to the definition of the *minimum*: we consider a function $f{:}S \to R$. We give the following definitions:

- f has a *local minimum* value at a point x_0 if $f(x) \geq f(x_0)$ for all $x \neq x_0$ lying in an open interval around x_0.
- f has a *strict local minimum* value at a point x_0 if $f(x) > f(x_0)$ for all $x \neq x_0$ lying in an open interval around x_0.

If the conditions $f(x) \geq f(x_0)$ and $f(x) > f(x_0)$ hold for all $x \neq x_0$ in S, then x_0 is called a *global minimum* and a *strict global minimum*, correspondingly. In what follows, we will be using the idea of the local minimum to denote the "minimum" of a function. The distinction between local and global minimum, although important from the theoretical point of view, is of little practical importance as determining a local minimum is the best we can achieve in practice in realistic problems.

Let's now assume that $f(x)$ has a local minimum at x_0. Then the situation will be similar (at least locally) to the situation shown in Fig. 1.2. For all x close to x_0 with $x < x_0$, $f(x) > f(x_0)$, and $f(x)$ is decreasing. When $f(x)$ is decreasing, its derivative is negative (to the left of x_0). For all x close to x_0 with $x > x_0$, $f(x) > f(x_0)$, and $f(x)$ is increasing. When $f(x)$ is increasing, its derivative is positive (to the right of x_0). It then follows that the derivative must be zero at x_0, but this is only a necessary condition. However, we can conclude from

Fig. 1.2 $f(x)$ around a
local minimum

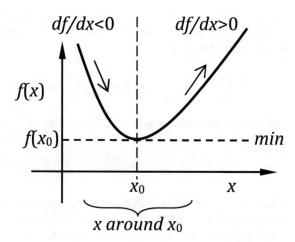

(1.22b) that when the first derivative is zero, the local behavior of $f(x)$ can be
determined from its second derivative.

More specifically, the second derivative has to be positive for $f(x)$ to increase
as we move away from x_0 and negative for $f(x)$ to decrease as we move away
from x_0. The following are sufficient conditions, i.e., conditions that guarantee
that x_0 is a local minimizer:

1. If $\left.\dfrac{df}{dx}\right|_{x_0} = 0$ & $\left.\dfrac{d^2f}{dx^2}\right|_{x_0} > 0$, then x_0 is a strict local minimum.   (1.23a)

2. If $\left.\dfrac{df}{dx}\right|_{x_0} = 0$ & $\left.\dfrac{d^2f}{dx^2}\right|_{x_0} < 0$, then x_0 is a strict local maximum.   (1.23b)

3. If $\left.\dfrac{df}{dx}\right|_{x_0} = 0$ & $\left.\dfrac{d^2f}{dx^2}\right|_{x_0} = 0$, then the criterion fails.   (1.23c)

In case 3, x_0 can be a local minimum, a local maximum, or a saddle point.
Consider, for instance, the functions $f_1 = x^3, f_2 = x^4$, and $f_3 = -x^4$. At $x_0 = 0$, we
have that the first and second derivatives are equal to zero in all cases.
However, f_1 has a saddle point at $x = x_0$, and f_2 has a minimum at $x = x_0$, but
f_3 has a maximum at $x = x_0$. Based on these three functions, we may also
demonstrate why the requirement that the second derivative is non-negative is
only a necessary (but not sufficient) condition for optimality. As the criterion
fails, we need to rely on the higher-order derivatives or on the study of the first
derivative around x_0 (the derivative needs to change sign for an extremum to
exist). The good news is that when the second derivative is positive at the
stationary point x_0, then we can conclude our investigation (i.e., we only need
to examine the second derivative at a single point and not in a region).

Let's now return to our introductory example of the storage tank. We may transform our problem to an unconstrained problem by first using (1.1) to eliminate h from (1.5):

$$w = \pi \left(\frac{1}{d} \frac{4V}{\pi} + d^2 \right) t \rho_M \qquad (1.24)$$

and finally substitute w into (1.2) to obtain:

$$\min_d \bar{f}(d) = \frac{C_0}{w_0^n} (\pi t \rho_M)^n \left[\left(\frac{4V}{\pi} \right) \frac{1}{d} + d^2 \right]^n = a \left(\beta \frac{1}{d} + d^2 \right)^n \qquad (1.25)$$

where $a = C_0 (\pi t \rho_M)^n / w_0^n$ and $\beta = 4\,V/\pi$ are constants. We note that we can proceed and search for the minimum without any information regarding the particular value of the constant a appearing in our model. We will therefore search for the optimal value of the function $f(d) = \frac{\bar{f}(d)}{a} = \left(\frac{\beta}{d} + d^2 \right)^n$. We will use MATLAB® to plot the function before getting into the algebra. We will first evaluate the value of the function for a single value of the tank diameter, and to simplify the algebra, we select $V = \pi/4$ (or $\beta = 1$):

```
>> clear            % remove all variables from the workspace
>> beta=1;          % assign a value to beta
>> n=0.6;           % assign a value to n
>> d=1;             % select a value for the diameter
>> f=(beta/d+d^2)^n % calculate f from Eq(1.25)


f = 1.5157
```

We see how easy it is to perform all the calculations in MATLAB. However, calculating the performance criterion or *objective function f* for a single value of the diameter is not very informative. Let's repeat the calculation with more values assigned to d (it is reminded that the dot symbol . used before the division and before raising to a power denotes that the intended calculation is performed on an element by element basis):

```
>> d=[0.1; 1; 5];        % select more value for the diameter
>> f=(beta./d+d.^2).^n;  % calculate f from Eq(1.25)
>> f


f = 3.9835
    1.5157
    3.8196
```

We now see that the cost function depends strongly on the value of the diameter. To investigate the complete variation of the objective function, we may plot the function for several values of d in the interval $[0.1, 3]$ and observe the results:

```
>> d=linspace(0.1,3,101);    % select 101 values uniformly
                                distributed
>> f=(beta./d+d.^2).^n;
>> plot(d,f)                  % plot the results
>> xlabel('Diameter (d)')
>> ylabel('Objective function f')
>> grid on
>> axis([0 3 1 4])
```

The results are presented in Fig. 1.3. Clearly the objective function has a minimum in the interval $0.1 < d < 3$. To find the minimum analytically, we set $x = d$ and apply (1.2):

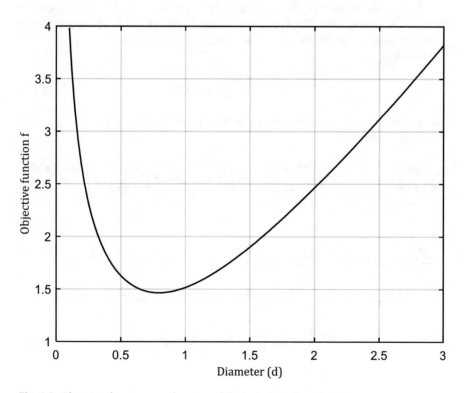

Fig. 1.3 Objective function as a function of the tank diameter ($\beta = 1$)

$$\frac{df}{dx} = \frac{d}{dx}\left[\left(\beta\frac{1}{x}+x^2\right)^n\right] = n\left(\beta\frac{1}{x}+x^2\right)^{n-1}\frac{d}{dx}\left[\left(\beta\frac{1}{x}+x^2\right)\right]$$

or:

$$\frac{df}{dx} = n\left(\beta\frac{1}{x}+x^2\right)^{n-1}\left(2x-\beta\frac{1}{x^2}\right) = 0 \tag{1.26}$$

MATLAB can also be used to determine the first derivative:

```
>> clear
>> syms d n beta f;      % define the symbolic variables d,n,β & f
>> f=(beta/d+d^2)^n;     % give function
>> diff(f,d)             % calculate first derivative

ans =

n*(2*d - beta/d^2)*(beta/d + d^2)^(n - 1)
```

Going back to solving the equation, we note that as d must always be positive, it follows that the first parenthesis in Eq. (1.26) is always positive. Then the expression in the second parenthesis in (1.26) must be zero, and it follows that:

$$x = d = \sqrt[3]{\beta/2} = 0.7937 \tag{1.27}$$

To rest our case, we need to determine the second derivative and to show that it is indeed positive:

$$\frac{d}{dx}\left(\frac{df}{dx}\right) = n\left(\beta\frac{1}{x}+x^2\right)^{n-1}\frac{d}{dx}\left(2x-\beta\frac{1}{x^2}\right) + n\left(2x-\beta\frac{1}{x^2}\right)\frac{d}{dx}\left[\left(\beta\frac{1}{x}+x^2\right)^{n-1}\right]$$

or:

$$\frac{d^2f}{dx^2} = n\left(\beta\frac{1}{x}+x^2\right)^{n-1}\left(2+2\beta\frac{1}{x^3}\right) + n(n-1)\left(2x-\beta\frac{1}{x^2}\right)^2\left(\beta\frac{1}{x}+x^2\right)^{n-2}$$

and, as the second term in the right hand side is zero, we finally obtain:

$$\frac{d^2f}{dx^2} = n\left(\beta\frac{1}{x}+x^2\right)^{n-1}\left(2+2\beta\frac{1}{x^3}\right) \tag{1.28}$$

We can easily conclude that the second derivative is always positive if d is positive. The second (as well as any other derivative) of the function can also be calculated in MATLAB:

```
>> diff(f,d,n)     % calculate n-th derivative
```

We have, therefore, located analytically the minimum that is shown in Fig. 1.3, and we have derived an analytical expression for the minimum. But we are not finished yet as we need to derive an expression for the height of the tank. To this end, we go back to Eq. (1.1) and substitute the optimal value of d:

$$h = \sqrt[3]{4\beta} \tag{1.29}$$

We can also determine the ratio of the height to the diameter:

$$\frac{h}{d} = \frac{\sqrt[3]{4\beta}}{\sqrt[3]{\beta/2}} = 2 \tag{1.30}$$

1.4 Convex Functions

We have so far discussed the existence and the calculation of the local minima of functions of a single variable. We need to emphasize the fact that the conditions for locating a minimum are only valid locally (in a region around a local minimum) and not global properties. This is depicted in Fig. 1.4 where a case of two local minima in S, x_1 and x_2, is shown. x_2, as can be seen by inspection, is also the global minimum in S. There is, however, a case in which a local minimum can be proven to be the global minimum, i.e., $f(x_0) < f(x)$ for every x in S. This can be guaranteed if $f(x)$ is a convex function.

Fig. 1.4 A case with two local minima (x_1 & x_2) – x_2 is also the global minimum

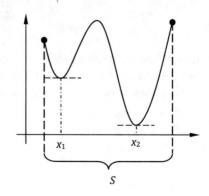

Fig. 1.5 The definition of
a convex or concave
upward function of a real
variable

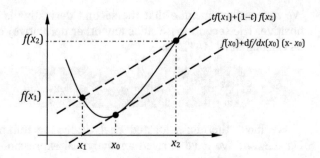

We start by presenting the definition of convexity for functions of a single
variable and then present a method for checking whether a function is convex
or not. We furthermore present a property of convex functions which is proven
to be of particular importance in the chapters that follow.

A real-valued function defined on an interval S is called convex (or concave
upward) if the line segment between any two points on the graph of the
function (say $(x_1, f(x_1))$ and $(x_2, f(x_2))$) lies above or on the graph as shown in
Fig. 1.5, or:

$$\forall x_1, x_2 \& t \in [0, 1], f(tx_1 + (1-t)x_2) \leq tf(x_1) + (1-t)f(x_2) \qquad (1.31)$$

A differentiable function of one variable is convex on an interval around x_0 if its
graph lies above all of its tangents, which has the following mathematical
interpretation:

$$f(x) \geq f(x_0) + \left.\frac{df}{dx}\right|_{x_0} (x - x_0) \qquad (1.32)$$

By comparing (1.32) with (1.22), it is easy to derive the local criterion for
convexity:

$$\left.\frac{d^2 f}{dx^2}\right|_{x_0} \geq 0 \qquad (1.33)$$

If (1.33) holds for all x in S (as it does for $f(x) = x^2$), then the function is convex
everywhere, and any local minima is also the global minimum.

Examples of convex functions are the following:

- e^{ax}, for any $a > 0$
- $e^{f(x)}$, if $f(x)$ is convex
- x^n, for $n < 0$ or $n > 1$, for positive x
- $ax + b$, for any a and b

Fig. 1.6 Structure of a crystalline solid (known as face-centered cubic structure or FCC)

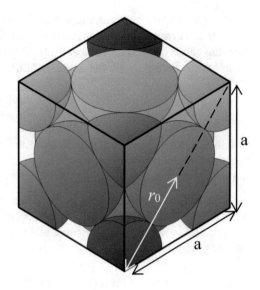

Proving that a function of a single variable is convex is usually done using (1.33). Despite the remarkable properties of convex functions, real problems are almost never convex, and convexity boils down to be a mathematical concept with little, if any, use in real problems (apart from the case of LP and QP problems).

1.5 Applications

Example 1.1 *Calculations in Crystalline Solids*
An impressive application of optimization theory in materials science is that of energy calculations in crystalline solids. A crystalline solid is a solid in which the atoms bond with each other in a regular pattern to form a periodic collection of atoms. In Fig. 1.6, the structure of the copper crystal is shown which is known to have a face-centered cubic (FCC) structure. The potential energy E per atom in a crystalline solid, as a function of the interatomic distance r, can be described by the following general equation:

$$E(r) = -\frac{A}{r^n} + \frac{B}{r^m} \tag{1.34}$$

where A and n are constants associated with the attractive forces developed, while B and m are constants associated with the repulsive forces. It is reminded that the force developed between the atoms is calculated with principles of classical mechanics and is the negative of the derivative of the potential energy:

$$F(r) = -\frac{dE(r)}{dr} \tag{1.35}$$

The interatomic distance at equilibrium is determined by the minimization of the potential energy which, according to (1.20) and (1.35), corresponds to zero net force between the atoms in the structure.

A particular example is that of Ne, which below 24.5 K is a crystalline solid with FCC structure (Kasap, S.O., *Principles of Electronic Materials and Devices*, 4th ed., McGraw Hill). Kasap proposes the following model for the (per atom) interatomic potential:

$$E(r) = -2\varepsilon\left[A\left(\frac{\sigma}{r}\right)^n - B\left(\frac{\sigma}{r}\right)^m\right] \tag{1.36}$$

with $n = 6$, $m = 12$, $A = 14.45$, $B = 12.13$, $\sigma = 0.274$ nm, and $\varepsilon = 0.003121$ eV (this is actually a general model first proposed in 1924 by John Lennard-Jones known as LJ-12-6 potential). $E(r)$ is expressed in eV/atom.

We will now show how we can calculate the equilibrium interatomic separation in the Ne crystal and how to estimate the bonding energy per atom in solid Ne. We will also calculate the density of solid Ne (atomic mass = 20.18).

We first use MATLAB to plot the potential:

```
clear
A=14.45;
B=12.13;
e=0.003121; % eV
sigma=0.274; % nm
r=linspace(0,3*sigma,1001);
E=-2*e*(A*(sigma./r).^6-B*(sigma./r).^12);
plot(r/sigma,E/(2*e),'LineWidth',1)
axis([0.9 3 -5 1])
grid on
xlabel('Intermolecular separation (r/\sigma)')
ylabel('potential (E/2\epsilon)')
```

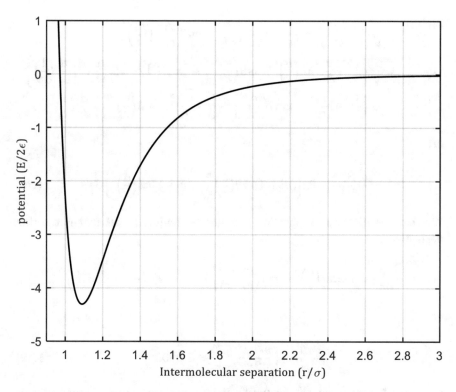

Fig. 1.7 Interatomic potential as a function of separation

The results are presented in Fig. 1.7, from which we immediately observe the existence of a minimum in potential energy. To derive the analytic expressions, we calculate the first and second derivative:

$$\frac{dE(r)}{dr} = -2\varepsilon\left[A\frac{d}{dr}\left(\frac{\sigma}{r}\right)^n - B\frac{d}{dr}\left(\frac{\sigma}{r}\right)^m\right]$$

$$= -2\varepsilon\left[nA\left(\frac{\sigma}{r}\right)^{n-1} - mB\left(\frac{\sigma}{r}\right)^{m-1}\right]\frac{d}{dr}\left(\frac{\sigma}{r}\right)$$

or:

$$\frac{dE(r)}{dr} = 2\varepsilon\left[nA\left(\frac{\sigma}{r}\right)^n - mB\left(\frac{\sigma}{r}\right)^m\right]\left(\frac{1}{r}\right) \tag{1.37}$$

and:

$$\frac{d^2E(r)}{dr^2} = \frac{d}{dr}\left(\frac{dE(r)}{dr}\right) = \frac{d}{dr}\left\{2\varepsilon\left[nA\left(\frac{\sigma}{r}\right)^n - mB\left(\frac{\sigma}{r}\right)^m\right]\left(\frac{1}{r}\right)\right\}$$

$$= 2\varepsilon\left\{\left(\frac{1}{r}\right)\frac{d}{dr}\left[nA\left(\frac{\sigma}{r}\right)^n - mB\left(\frac{\sigma}{r}\right)^m\right] + \left[nA\left(\frac{\sigma}{r}\right)^n - mB\left(\frac{\sigma}{r}\right)^m\right]\frac{d}{dr}\left(\frac{1}{r}\right)\right\}$$

$$= 2\varepsilon\left\{\left[n^2A\left(\frac{\sigma}{r}\right)^n - m^2B\left(\frac{\sigma}{r}\right)^m\right] + \left[nA\left(\frac{\sigma}{r}\right)^n - mB\left(\frac{\sigma}{r}\right)^m\right]\right\}\frac{d}{dr}\left(\frac{1}{r}\right)$$

or:

$$\frac{d^2E(r)}{dr^2} = 2\varepsilon\left[(m+1)mB\left(\frac{\sigma}{r}\right)^m - (n+1)nA\left(\frac{\sigma}{r}\right)^n\right]\frac{1}{r^2} \qquad (1.38)$$

We then calculate the separation distance by finding r that makes the first derivative equal to zero:

$$\frac{dE(r)}{dr} = 2\varepsilon\left[nA\left(\frac{\sigma}{r}\right)^n - mB\left(\frac{\sigma}{r}\right)^m\right]\left(\frac{1}{r}\right) = 0 \Rightarrow nA\left(\frac{\sigma}{r}\right)^n = mB\left(\frac{\sigma}{r}\right)^m$$

or:

$$\frac{r}{\sigma} = \left(\frac{nA}{mB}\right)^{1/(n-m)} = 1.0902 \qquad (1.39)$$

Using the fact that $nA\left(\frac{\sigma}{r}\right)^n = mB\left(\frac{\sigma}{r}\right)^m$, the second derivative at the optimum separation becomes:

$$\frac{d^2E(r)}{dr^2} = 2n(m-n)\frac{\varepsilon A}{\sigma^2}\left(\frac{\sigma}{r}\right)^{n+2} > 0 \qquad (1.40)$$

Things would have been simpler if we had used MATLAB to perform the algebra:

```
>> clear all
>> syms A B sigma e n m E r;
>> E=-2*e*(A*(sigma/r)^n-B*(sigma/r)^m);;
>> diff(E,r,1)

ans = -2*e*((B*m*sigma*(sigma/r)^(m - 1))/r^2 -
(A*n*sigma*(sigma/r)^(n - 1))/r^2)

>> simplify(diff(E,r,1))

ans = - (2*e*(B*m*(sigma/r)^m - A*n*(sigma/r)^n))/r
```

(continued)

```
>> simplify(diff(E,r,2))

ans = (2*e*(B*m*(sigma/r)^m - A*n*(sigma/r)^n + B*m^2*(sigma/r)
^m - A*n^2*(sigma/r)^n))/r^2
```

We will now use Fig. 1.6 to calculate the density of the crystal. From Fig. 1.6 and simple geometric arguments, we have that:

$$2a^2 = 4r_0^2 \Rightarrow a = r_0\sqrt{2} = 1.0902\sqrt{2}\sigma \tag{1.41}$$

The mass of an Ne atom is the molecular mass divided by the Avogadro number (N_A). From careful examination of Fig. 1.6, we conclude that there are four atoms in one cell. As the density is the mass per volume, we have that:

$$\rho_{Ne} = \frac{4mw_{Ne}}{N_A a^3}$$

$$= \frac{4 \cdot 20.18 \ \frac{g}{mol}}{6.02214076 \times 10^{23} \ \frac{1}{mol} \left(1.0902\sqrt{2} \cdot 0.274\right)^3 nm^3 \cdot \left(\frac{10^{-7} cm}{nm}\right)^3}$$

$$= 1.778 \ \frac{g}{cm^3}$$

Example 1.2 *Minimum Energy Consumption for a Natural Gas Compression System*

In this example, we consider the system of two compressors operating in series so as to compress natural gas (practically methane) from an initial pressure P_0 to a desired pressure P_f (see Fig. 1.8). After the first compressor, there is an intercooler located with the aim to lower the temperature to the initial temperature T_0.

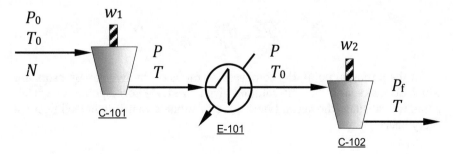

Fig. 1.8 A natural gas (methane) compression system

The power W consumed by a compressor to compress a gas from initial pressure P_{in} and absolute temperature T_{in} to a final pressure P_{out} is determined in undergraduate thermodynamics books and is given by:

$$W = N\frac{1}{e}\frac{RT_{in}}{a}\left[\left(\frac{P_{out}}{P_{in}}\right)^a - 1\right], \quad a = \frac{c_p/c_v - 1}{c_p/c_v} > 0 \qquad (1.42)$$

where N is the molar flow rate of the gas, R is the ideal gas constant, e (<1) is the fractional efficiency with respect to a reversible adiabatic compression, and c_P (c_v) is the constant pressure (volume) heat capacity. As the temperature in the suction of both compressors is the same, we have:

$$w = \frac{W}{\left(\frac{NRT_0}{e}\right)} = \frac{1}{a}\left[\left(\frac{P}{P_0}\right)^a - 1\right] + \frac{1}{a}\left[\left(\frac{P_f}{P}\right)^a - 1\right]$$

or:

$$w = \frac{1}{a}\left[\left(\frac{P}{P_0}\right)^a + \left(\frac{P_f}{P}\right)^a - 2\right] \qquad (1.43)$$

This is a function of one variable (the pressure P of the intermediate stream). To investigate the possibility of a minimum to exist, we calculate the first derivative:

$$\frac{dw}{dP} = \frac{1}{P}\left[\left(\frac{P}{P_0}\right)^a - \left(\frac{P_f}{P}\right)^a\right] \qquad (1.44)$$

We set the derivative equal to zero to find:

$$P = \sqrt{P_0 P_f} \qquad (1.45)$$

We may note that the ratio of the delivery to suction pressure (known as compression ratio) for the two compressors is:

$$\frac{P}{P_0} = \frac{\sqrt{P_0 P_f}}{P_0} = \sqrt{\frac{P_f}{P_0}} = \frac{P_f}{\sqrt{P_0 P_f}} = \frac{P_f}{P} \qquad (1.46)$$

i.e., at the optimum point, the compression ratios of the two compressors are equal. To make sure that the solution given by (1.45) corresponds to a minimum, we calculate the second derivative to validate that it is indeed positive everywhere:

$$\frac{d^2w}{dP^2} = \frac{2a}{P_f P_0}\sqrt{\frac{P_f}{P_0}} > 0 \qquad (1.47)$$

Fig. 1.9 Two CSTR in series system with first order reaction (isothermal conditions)

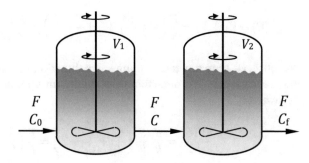

Example 1.3 *Minimum Reactor Volume for a First-Order Reaction*

In this application, we will consider a classical chemical engineering system. The first-order reaction R → P (R is the reactant and P the product) is taking place in two continuous stirred tank reactors (CSTR) with volumes V_1 and V_2, as shown in Fig. 1.9. The reaction is first order with respect to the reactant concentration with reaction rate constant k. The feed stream to the system has volumetric flow rate F, and the reactant concentration is C_0. The concentration of the reactant in the first reactor is C, and in the second is C_f. As we aim at achieving a specific conversion of the reactant, the concentration in the second reactor C_f is known and constant. Our aim is to determine the volumes of the two reactors and the concentration in the first one that minimize the overall reactor volume in the system.

In order to determine the combination of the two reactor volumes, we first write the material balance for the reactant in reactor 1:

$$FC_0 - FC - V_1 kC = 0 \tag{1.48}$$

and in reactor 2:

$$FC - FC_f - V_2 kC_f = 0 \tag{1.49}$$

We then solve both material balances for the volumes:

$$V_1 = \frac{F}{k}\left(\frac{C_0 - C}{C}\right) \tag{1.50}$$

$$V_2 = \frac{F}{k}\left(\frac{C - C_f}{C_f}\right) \tag{1.51}$$

The total volume is the sum of the volumes of the two reactors:

$$V(C) = V_1 + V_2 = \frac{F}{k}\left(\frac{C_0 - C}{C}\right) + \frac{F}{k}\left(\frac{C - C_f}{C_f}\right) = \frac{F}{k}\left(\frac{C_0}{C} + \frac{C}{C_f} - 2\right) \quad (1.52)$$

We observe that the total volume is a function of the concentration in the first reactor only. To determine the minimum volume, we calculate the first derivative:

$$\frac{dV(C)}{dC} = \frac{F}{k}\frac{d}{dC}\left(\frac{C_0}{C} + \frac{C}{C_f} - 2\right) = \frac{F}{k}\left(\frac{1}{C_f} - \frac{C_0}{C^2}\right) \quad (1.53)$$

If we set the first derivative equal to zero and solve for the concentration, we obtain:

$$C = \sqrt{C_0 C_f} \quad (1.54)$$

We then calculate the second derivative to check whether it is actually a minimum:

$$\frac{d^2V}{dC^2} = \frac{d}{dC}\left(\frac{dV}{dC}\right) = \frac{F}{k}\frac{d}{dC}\left(\frac{1}{C_f} - \frac{C_0}{C^2}\right) = 2\frac{F}{k}\frac{C_0}{C^3} > 0 \quad (1.55)$$

We could of course have performed all calculations in MATLAB:

```
>> clear
>> syms C F k C0 Cf V positive;
>> V=(F/k)*(C0/C+C/Cf-2);
>> diff(V,C)
ans = - (F*(C0/C^2 - 1/Cf))/k
>> solve(diff(V,C)==0,C)

ans = C0^(1/2)*Cf^(1/2)

>> diff(V,C,2)
ans = (2*C0*F)/(C^3*k)
```

As the second derivative is positive for all C in $[C_0, C_f]$, the solution determined by (1.54) corresponds to a global minimum.

We substitute the solution given by (1.54) to Eqs. (1.50) and (1.51) to determine the volumes:

$$V_1 = V_2 = \frac{F}{k}\left(\sqrt{\frac{C_0}{C_f}} - 1\right) \quad (1.56)$$

It follows that at the optimal solution, the two volumes are equal.

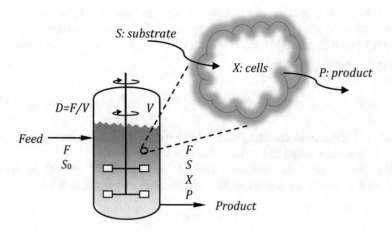

Fig. 1.10 A continuous bioreactor

Example 1.4 *Optimal Operation of a Continuous Bioreactor*
   In this example, we study a different type of a CSTR reactor. This time, we consider a bioreactor with constant volume V that is operated continuously. The bioreactor is shown in Fig. 1.10. The feed is sterile (free of microorganisms) and has volumetric flow rate F and substrate concentration S_0. The volume is constant, and under mild assumptions, this results to the conclusion that the product volumetric flow rate is equal to the feed volumetric flow rate. The biomass (cell mass) is denoted by X and the product concentration by P. The rate of production of biomass is denoted by r_X, the rate of production of the product by r_P, and the rate of consumption of substrate by r_S. At steady-state conditions, we may write the three material balances:

$$\text{Biomass}: \; -FX + Vr_X = 0 \tag{1.57}$$

$$\text{Substrate}: \; FS_0 - FS - Vr_S = 0 \tag{1.58}$$

$$\text{Product}: \; -FP + Vr_P = 0 \tag{1.59}$$

The production rate of the biomass is usually expressed through the definition of the specific production rate, denoted by μ:

$$r_X = \mu X \tag{1.60}$$

In many cases, as it is, for example, the case of bioethanol production, there is an inhibition effect caused by the product. The specific production rate is described by a classical Monod equation with an additional term to account for the inhibition:

$$\mu = \frac{r_X}{X} = \mu_{\max} \frac{S}{K_s + S} e^{-P/P_{\max}} \qquad (1.61)$$

$\mu_{\max}$ is the maximum specific production rate, K_s the saturation constant, and $P_{\max}$ a constant related to the inhibition by the product.

The rate of product formation can be considered, for growth-associated product formation, analogous to the rate of biomass production, i.e.:

$$r_P = a r_X = a\mu X \qquad (1.62)$$

When Eq. (1.60) is substituted into (1.57), an interesting result is obtained:

$$-FX + V\mu X = 0$$

or:

$$D = \frac{F}{V} = \mu \qquad (1.63)$$

where D is called the dilution rate. From (1.63), it follows that for a continuous bioreactor, the dilution rate is equal to the specific biomass production rate. Using this result, (1.58) becomes:

$$F(S_0 - S) - V \frac{\mu X}{Y_{X/S}} = 0$$

or:

$$S = S_0 - \frac{X}{Y_{X/S}} \qquad (1.64)$$

where we have used the yield coefficient $Y_{X/S}$ which can be defined as the ratio of r_X to r_S. If we now substitute (1.62) and (1.63) into (1.59), we also obtain:

$$-FP + Va\mu X = 0$$

or:

$$P = aX \tag{1.65}$$

where a is a constant. We finally substitute (1.63), (1.64), and (1.65) into (1.61):

$$D = \mu_{max} \frac{S_0 - \frac{X}{Y_{X/S}}}{K_S + S_0 - \frac{X}{Y_{X/S}}} e^{-aX/P_{max}} \tag{1.66}$$

We have achieved to eliminate all variables and express the dilution rate as a function of the biomass concentration.

The productivity of bioreactors is defined as the mass of the product produced per unit reactor volume and per unit time. In the case of a continuous bioreactor, the mass of the product produced is the flow rate of the product stream F times the concentration of the product. If we then divide by the reactor volume, we obtain the productivity Π:

$$\Pi = \frac{F \cdot P}{V} = D \cdot P \tag{1.67}$$

We then use Eqs. (1.65) and (1.66) to obtain:

$$\Pi = \mu_{max} \frac{S_0 - \frac{X}{Y_{X/S}}}{K_S + S_0 - \frac{X}{Y_{X/S}}} e^{-aX/P_{max}} aX \tag{1.68}$$

The productivity can also be expresses as a function of S (or P if necessary):

$$\Pi(S) = \xi \frac{S}{K_s + S} e^{-aY_{X/S}S/P_{max}} (S_0 - S) \tag{1.69}$$

where:

$$\xi = \mu_{max} e^{\frac{aY_{X/S}S_0}{P_{max}}} aY_{X/S} \tag{1.70}$$

Our aim is to maximize the productivity of the bioreactor Π. This is equivalent to minimizing $-\Pi$. Calculating the first derivative of $-\Pi$ with respect to X is lengthy, and we therefore present the result directly:

$$\frac{d(-\Pi)}{dS} = -\xi \left[e^{\frac{aY_{X/S}S}{P_{max}}} (S_0 - S) \frac{d}{dS} \left(\frac{S}{K_s + S} \right) + \frac{S(S_0 - S)}{K_s + S} \frac{d}{dS} \left(e^{\frac{aY_{X/S}S}{P_{max}}} \right) \right.$$

$$\left. + \frac{S}{K_s + S} e^{\frac{aY_{X/S}S}{P_{max}}} \frac{d(S_0 - S)}{dS} \right]$$

or:

$$\frac{d(-\Pi)}{dS} = -\xi \frac{S(S_0 - S)}{K_s + S} e^{\frac{aY_{X/S}S}{P_{max}}} \left[\frac{K_s}{S(K_s + S)} + \frac{aY_{X/S}}{P_{max}} - \frac{1}{(S_0 - S)} \right] \quad (1.71)$$

Setting the derivative equal to zero is equivalent to the expression inside the brackets to be equal to zero, from which we obtain (the algebra is again lengthy but involves only basic operations):

$$\alpha S^3 + \beta S^2 + \gamma S + \delta = 0 \quad (1.72)$$

where:

$$\alpha = \frac{aY_{X/S}}{P_{max}}, \quad \beta = 1 + \frac{aY_{X/S}}{P_{max}}(K_s - S_0),$$

$$\gamma = K_s \left(2 - \frac{aY_{X/S}}{P_{max}} S_0 \right), \quad \delta = -K_s S_0$$

Solving a third-order polynomial is not as simple as solving a second-order polynomial (the procedure is similar but messy). Deriving simple formulas for the optimum is therefore deemed impossible, and we need to apply the results on a case-by-case basis. To this end, we consider the following numerical example (numerical values correspond roughly to bioethanol production):

$$\mu_{max} = 0.41 \frac{1}{h}, \quad K_s = 0.4 \frac{g}{L}, \quad a = 2.44 \frac{g}{g}, \quad Y_{X/S} = 0.1 \frac{g}{g}, \quad S_0 = 100 \frac{g}{L},$$

$$P_{max} = 35 \frac{g}{L}$$

Let's perform the calculations in MATLAB and generate the plot of productivity as a function of the substrate concentration.

```
>> miumax=0.41;      % 1/h
>> Ks=0.40;          % g/L
>> a=2.44;           % g P /g S
>> Yxs=0.1;          % g X /g S
>> S0=100;           % g/L
>> Pmax=35;          % g/L
>> X=linspace(0.1,Yxs*S0,10001);
>> S=S0-X/Yxs;       % Eq. (1.64)
>> P=a*X;                                        % Eq. (1.65)
>> D=miumax*(S./(Ks+S)).*exp(-P/Pmax);  % Eq. (1.63) & (1.61)
>> figure(1)
>> plot(D,P,'k-.','Linewidth',1.5)
```

(continued)

```
>> hold on
>> plot(D,X,'k-','Linewidth',1.5)
>> plot(D,S,'k--','Linewidth',1.5)
>> axis([0 0.45 0 30])
>> figure(2)
>> plot(S,D.*P,'k-','Linewidth',1.5)
>> grid on
>> alpha=a*Yxs/Pmax;              % Eq. (1.72)
>> beta=1+alpha*(Ks-S0);
>> gamma=Ks*(2-alpha*S0);
>> delta=-Ks*S0;
>> roots([alpha beta gamma delta])
ans = -37.8671
      -15.6546
        9.6791
```

The results are presented in Figs. 1.11 and 1.12. The substrate concentration at the optimum is approximately 9.68 g/L. To determine the dilution rate, we can substitute into Eqs. (1.64), (1.65), and (1.66):

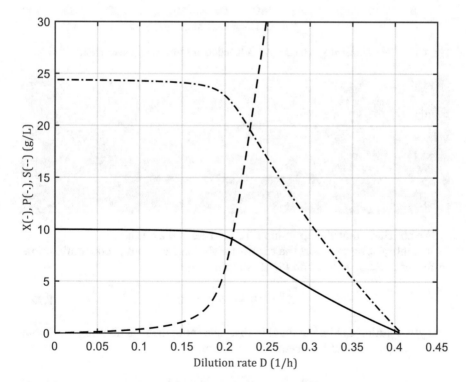

Fig. 1.11 Biomass, substrate, and product concentration as a function of dilution rate

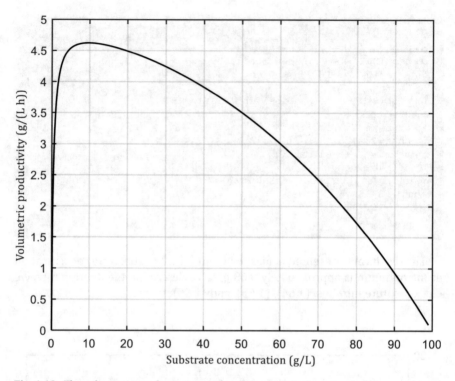

Fig. 1.12 The volumetric productivity as a function of substrate concentration

```
>> Sopt=max(ans)
Sopt =   9.6791
>> Xopt=Yxs*(S0-Sopt)
Xopt =   9.0321
>> Popt=a*Xopt
Popt =   22.0383
>> Dopt=miumax*Sopt*exp(-Popt/Pmax)/(Ks+Sopt)
Dopt =   0.2098
```

We conclude that the optimum dilution rate is 0.21 1/h.

It is interesting to report the results for the case without product inhibition. In this case $P_{\max} \to +\infty$ and (1.72) simplifies to:

$$S^2 + 2K_s S - K_s S_0 = 0 \qquad (1.72')$$

Analytic solution can be obtained in this case (note that S is non-negative), and the optimum substrate concentration is given by:

$$S = K_s \left(\sqrt{1 + S_0/K_s} - 1 \right) \qquad (1.73)$$

Example 1.5 *Optimal Operation of a Solid Oxide Fuel Cell*

In this example, we study a different type of reactor that produces electrical energy instead of chemicals called a fuel cell (see Fig. 1.13). A fuel, such as hydrogen, and an oxidizing agent, such as oxygen, are fed to the anode and the cathode of the cell, respectively, where they react under well-controlled conditions to "produce" water and electrical energy. The voltage E developed at an external load depends strongly on the operating current I of the cell. Electrochemists prefer to use the current density j (current per unit area of the cell, $I = jA$, where A is the area of the cell) when studying fuel cells. The voltage developed at almost zero current can be calculated from equilibrium thermodynamics and for the water formation reaction is around 1 V (depends on temperature). As current is produced from the cell, irreversibilities kick in, and the voltage drops. The observed voltage drop is abrupt at the beginning and linear thereafter. Finally, at large current densities, the voltage drops again abruptly as current approached the limiting current density j_L. The linear drop in the voltage is usually associated with Ohmic losses while the sudden drop in large currents with the depletion of the reactants close to the reaction sites.

In Fig. 1.14, a characteristic voltage-current density curve is shown in which the model predictions are compared to experimental values (Huang C., et al., *Journal of Power Sources*, 195, p. 5176, 2010). The mathematical model used is the following:

$$E(j) = E_0 - R_\Omega j + \frac{RT}{2F} \ln \left(1 - \frac{j}{j_L} \right) \tag{1.74}$$

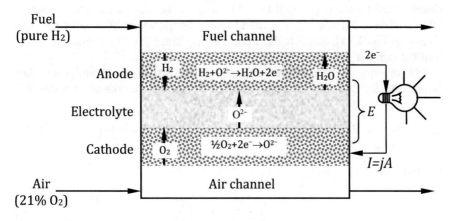

Fig. 1.13 The main parts of a fuel cell

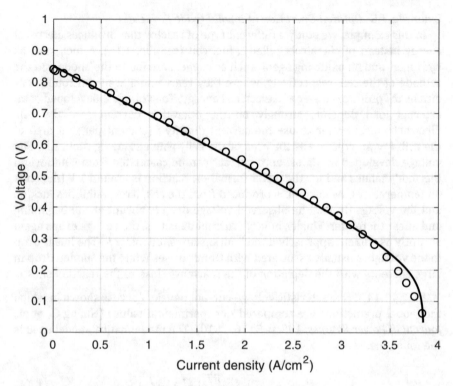

Fig. 1.14 Comparison of experimental results (o) with model predictions for a solid oxide fuel cell. (Experimental results from Huang et al., 2010)

where E_0 is the voltage at zero current ($E_0 = 0.85$ V), R_Ω is the (area specific) Ohmic resistance ($R_\Omega = 0.145$ $\Omega\cdot$cm^2), R is the ideal gas constant ($R = 8.3144$ V·C/(mol K)), T is the absolute temperature (873 K for the data shown in Fig. 1.14), and F is the Faraday's constant (96,485 C/mol). j_L is the limiting current density (3.85 V).

We consider the case where our goal is to extract the maximum power. The power density P (w/cm^2) is the product of the voltage and the current density:

$$P(j) = E(j) \cdot j = E_0 \cdot j - R_\Omega j^2 + \frac{RT}{2F} \ln\left(1 - \frac{j}{j_L}\right) \cdot j \qquad (1.75)$$

We note that:

$$\frac{d(-P)}{dj} = -E - j\frac{dE}{dj} \qquad (1.76)$$

and:

$$\frac{dE}{dj} = \frac{d}{dj}\left[E_0 - R_\Omega j + \frac{RT}{2F}\ln\left(1 - \frac{j}{j_L}\right)\right] = -R_\Omega - \frac{RT}{2F}\frac{1}{(j_L - j)} \qquad (1.77)$$

Putting everything together we obtain:

$$\frac{d(-P)}{dj} = -E_0 + 2R_\Omega j - f\ln\left(1 - \frac{j}{j_L}\right) + f\frac{j}{(j_L - j)} \qquad (1.78)$$

where $f = RT/(2F)$. Finding an analytic expression for the current density at which the first derivative is zero is impossible. We calculate the second derivative to check whether the extremum (if one exists) is actually a minimum:

$$\frac{d^2(-P)}{dj^2} = 2R_\Omega + f\frac{1}{(j_L - j)} + f\frac{j_L}{(j_L - j)^2} > 0 \qquad (1.79)$$

It follows that as $j < j_L$, the second derivative is always positive. Before closing this section, in Fig. 1.15, the variation of the power density as a function of the

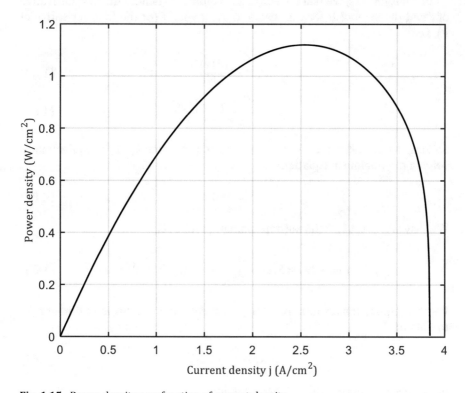

Fig. 1.15   Power density as a function of current density

current density is presented, from which we note that there is indeed a clear maximum (which, however, we cannot determine analytically).

1.6 The Numerical Solution of Single Variable Optimization Problems: Newton's Method

Let's summarize what we have learned from trying to solve some representative examples of single variable optimization. The optimization problem is transformed to a problem of solving a possibly nonlinear equation that is obtained when applying optimality conditions (1.23). We begin with the objective function $f(x)$ which we seek to minimize, and we end up with searching for the values of x that are roots of its first derivative. However, as we have seen, solving a nonlinear equation is far from easy, even when the equation involves a single variable only. We therefore have to rely on numerical techniques for solving the resulting nonlinear equation. We will therefore complete this chapter by reviewing the basics about the most well-known methodology for solving nonlinear equations which is the Newton's method.

In what follows, we will be using the symbol F to denote the first derivative of f and the symbol H for the second derivative of f (or the first derivative of F), i.e.:

$$F(x) = \frac{df(x)}{dx} \tag{1.80}$$

$$H(x) = \frac{d^2f(x)}{dx^2} = \frac{dF(x)}{dx} \tag{1.81}$$

Solving the problem of finding an extremum (or extrema) of f is equivalent to solving the nonlinear equation:

$$F(x) = 0 \tag{1.82}$$

We may write Taylor's linear approximation for $F(x)$:

$$F(x + \delta x) \approx F(x) + \left.\frac{dF}{dx}\right|_x \delta x = F(x) + H(x)\delta x \tag{1.83}$$

We then equate the linear approximation to zero and use the definition of $H(x)$ to obtain:

$$H(x)\delta x = -F(x) \tag{1.84}$$

or, as this is a simple scalar linear equation:

$$\delta x = -\frac{1}{H(x)} F(x) \tag{1.85}$$

The step calculated by (1.85) is called a Newton's step. It is called a step as Newton's method needs to be applied in an iterative fashion. We begin at iteration 0 by guessing a value for x. However, as it is extremely unlikely that we can guess the correct solution, we will have $F(x^0) = \varepsilon^0 \neq 0$ (superscript 0 denotes the iteration counter, and ε denotes the error). We calculate $H(x^0)$ and then the Newton step through (1.85) (for that to be possible, we must have $H(x) > 0$, as $H(x) = 0$ makes impossible to determine the step and $H(x) < 0$ will direct the algorithm toward a maximum). Then, we can determine our new estimate of the solution by adding the step to our current position:

$$x^1 = x^0 + \delta x^1 \tag{1.86}$$

We then proceed in this way and determine:

$$\delta x^{k+1} = -\frac{1}{H(x^k)} F(x^k) \tag{1.87}$$

and:

$$x^{k+1} = x^k + \delta x^{k+1} \tag{1.88}$$

until $F(x^k)$ becomes sufficiently close to zero, or:

$$\left| F(x^k) \right| < e_F \tag{1.89}$$

where e_F is a sufficiently small number (depends on the application and the size of x). Alternatively, we can stop when the step is small enough:

$$\left| \delta x^{k+1} \right| < e_x \tag{1.90}$$

This is a basic Newton's algorithm and is not very successful in practical applications unless we have a very good idea on the exact location of the solution. The problem lies in the unconditional acceptance of the Newton's step in (1.88). Do we have any guarantee that $f(x^{k+1}) < f(x^k)$? The answer is, unfortunately, negative unless our function has some very special characteristics. Exact conditions can be found in the suggested literature as we will try not to dive into the mathematical details. We will restrict ourselves into proposing a very rudimentary but practical technique, for modifying the Newton step if necessary. We may use the following simple idea:

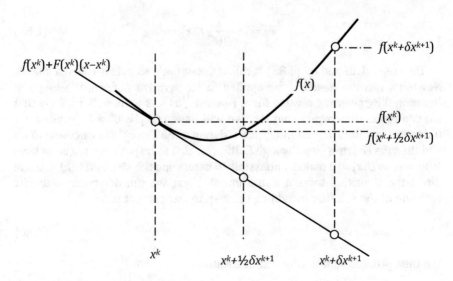

Fig. 1.16 Naïve backtracking algorithm (note that taking a full Newton step results in an increased value of $f(x)$, while one backtracking step results in a decreased value)

$$x^{k+1} = x^k + \left(\frac{1}{2}\right)^n \delta x^{k+1}, n \text{ is smallest } n : f\left(x^{k+1}\right) < f\left(x^k\right) \text{ is satisfied} \quad (1.91)$$

In other words, we start with $n = 0$ and check whether f has been decreased (see Fig. 1.16). If this turns out not to be the case, we set $n = n + 1$ and backtrack from our previous trial position. This is repeated until we discover the first point (smaller n) that results in a lower value for f relative to $f(x^k)$.

We can allow ourselves to backtrack a finite number of times; otherwise, as n increases $x^{k+1} \rightarrow x^k$, and the algorithm must terminate. When we are stack in our rudimentary algorithm, there is a way out: to try a different starting point x^0. The algorithms that handle the size of the step taken at each iteration in an efficient (robust and still fast) implementation of the Newton's method are rather sophisticated. We will return to this point when studying unconstrained optimization in many directions where the issue of backtracking is vital to the Newton's method. It is important to note that single-variable optimization is performed (without relying in most cases on derivative information) by using local interpolation with quadratic or cubic polynomials or the bisection algorithm.

Before closing this paragraph, it is important to note that there are two other options for setting the size of the Newton step. The first is the simplest and uses a constant factor (usually smaller than 1) that multiplies the full Newton step. This is very simple and can be efficient in some cases but needs a trial-and-error approach to find what is the appropriate factor in each case study. The other extreme is to handle the scale factor as the optimization variable and try to locate the exact optimal value. This is called exact line search and can be extremely costly to perform. The inexact line search that we describe in Eq. (1.91) can be equally efficient and much less costly (in terms of function evaluations).

We will now build our naïve implementation of Newton's method and try to solve the fuel cell problem that we have left open in the previous section. We will then use MATLAB to solve the same problem. We first define the function (Eq. 1.75) (a minus sign is added to account for the fact that we are solving a minimization problem), its first derivative (Eq. 1.78), and its second derivative (Eq. 1.79) as inline functions:

```
>>clear
>>f0=inline('-x*(0.85-0.145*x+0.0376*log(1-x/3.85))')
>>f1=inline('-0.85+2*0.145*x-0.0376*log(1-x/3.85)+0.0376*x/
(3.85-x)')
>>f2=inline('2*0.145+0.0376/(3.85-x)+0.0376*3.85/(3.85-x)
^2')
f0 = Inline function:
     f0(x) = -x*(0.85-0.145*x+0.0376*log(1-x/3.85))
f1 = Inline function:
     f1(x) = -0.85+2*0.145*x-0.0376*log(1-x/3.85)+0.0376*x/
     (3.85-x)
f2 = Inline function:
     f2(x) = 2*0.145+0.0376/(3.85-x)+0.0376*3.85/(3.85-x)^2
```

The symbol x is used instead of j for the current density as j is a reserved symbol in MATLAB ($j = (-1)^{1/2}$). We can now assign a value to x and then evaluate the function and its derivatives:

```
>> x=3.6;
>> f=feval(f0,x) ; % response f = -0.8107
>> F=feval(f1,x) ; % response F = 0.8383
>> H=feval(f2,x) ; % response H = 2.7566
>> NewtonStep=-F/H; % response NewtonStep = -0.3041
>> x=x+NewtonStep ; % response x = 3.2959
>> f=feval(f0,x) ; % response f = -0.9862
```

The tangent at $x = 3.6$ is shown in Fig. 1.17 and is given by the following equation ($k = 0$):

$$F\left(x^k + \delta x^{k+1}\right) = F\left(x^k\right) + H\left(x^k\right)\delta x^{k+1} = 0.8383 + 2.7566\,\delta x^{k+1} \qquad (1.92)$$

Equating this to zero and solving for δx^1 gives $\delta x^1 = -0.3041$, and finally the new estimate of the solution is $x^1 = x^0 + \delta x^1 = 3.6 - 0.3041 = 3.2959$ (point at which the tangent at x^0 becomes zero in Fig. 1.17). As the objective becomes

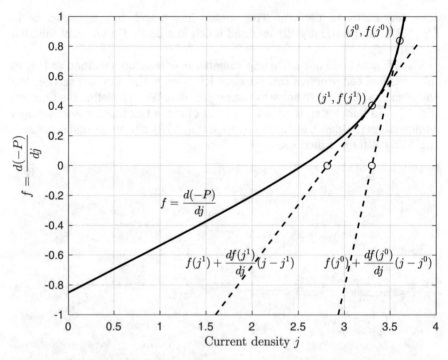

Fig. 1.17 Steps of the Newton methodology when applied to the fuel cell case study

smaller (more negative), the Newton step is accepted. We determine the new
tangent at this new point:

```
>> f=feval(f0,x)   ; % response f = -0.9862
>> F=feval(f1,x)   ; % response F =  0.4024
>> H=feval(f2,x)   ; % response H =  0.8294
>> NewtonStep=-F/H; % response NewtonStep =  -0.4851
>> x=x+NewtonStep ; % response x =  2.8108
>> f=feval(f0,x)   ; % response f = -1.1052
```

This concludes iteration 2, and the results obtained in the intermediate
iterations are summarized in Table 1.1. The method converges in the sixth
iteration where the derivative becomes close to 10^{-10}. The steps followed in
solving the fuel cell case study are organized in a general-purpose m-file which
is presented in Table 1.2. Understanding the details of the m-file is not particularly important, and you are advised against using this m-file for anything but
for validating your hand calculations or solving examples similar to the fuel cell
or the bioreactor.

In MATLAB®, the command `fminbnd` is available for performing single-
variable optimization. `fminbnd` is not based on derivative information (it is

Table 1.1 Results of applying Newton method to the fuel cell case study

k	$f(x^{k-1})$	$F(x^{k-1})$	$H(x^{k-1})$	δx^k	x^k
0					3.6000
1	−0.8107	+0.8383	+2.7566	−0.3041	+3.2959
2	−0.9862	+0.4024	+0.8294	−0.4851	+2.8108
3	−1.1052	+0.1161	+0.4602	−0.2522	+2.5586
4	−1.1205	+0.0076	+0.4059	−0.0186	+2.5400
5	−1.1206	$2.6826 \ 10^{-5}$	+0.4031	$-6.6558 \ 10^{-5}$	+2.5399
6	−1.1206	$3.3375 \cdot 10^{-10}$			

Table 1.2 Implementation of the Newton algorithm (to be used only for training)

```
function [x,fun,dfun,xk] = NaiveNewton(f,F,H,x0)
% f is the function to minimize (inline function)
% F is the 1st derivative of f  (inline function)
% H is the 2nd derivative of f  (inline function)
maxiter=100;     % do not allow more than maxiter iterations
maxbacktrack=10; % do not allow more than maxbacktrack steps
tol=1e-6;        % used in stopping criterion
k=0; x=x0; xk=x0;% initialization
fprintf('Iter            x             f(x)
||g(x)||\n')
while ( norm(feval(F,x))>tol & k<maxiter )
    fun   = feval(f,x);          % evaluate function
    dfun  = feval(F,x);          % evaluate 1st derivative
    ddfun = feval(H,x);          % evaluate 2nd derivative
    FullNewtonStep=-dfun/ddfun;  % Newton step

    n=0;                         % backtracking parameter
    newx=x+(1/2)^n*FullNewtonStep;  % new solution
    while ( feval(f,newx)>fun & n<maxbacktrack) % backtracking
        n=n+1;
        newx=x+(1/2)^n*FullNewtonStep;
        fprintf('Backtracking with n %4u \n',n)
    end
    if (n==maxbacktrack) ; k=maxiter; end
    x=newx;                      % update x
    k=k+1;                       % increase counter
    xk(:,k+1)=x;                 % store intermediate
points
    fprintf('%4u %14.8f   %14.8f %14.8f\n',k,x,fun,norm(dfun))
end

if (ddfun>eps)
    fprintf('Possible local minimum')
else
    fprintf('Newton has failed')
end
```

based on a combination of the Golden section method and the cubic interpolation method), and therefore only the function is required, which is a significant advantage. Its general syntax is as follows:

> xopt=**fminbnd**(fun,x1,x2,options)
>
> returns a value x that is a *local minimizer* of the *scalar valued function* that is described in fun in the interval x1 < x < x2.

We can solve the fuel cell case study in MATLAB as follows:

```
>> options=optimset('display','iter');
>> f0=inline('-x*(0.85-0.145*x+0.0376*log(1-x/3.85))');
>> xopt=fminbnd(f0,0.1,3.6,options)

Func-count    x              f(x)        Procedure
    1        1.43688      -0.896739       initial
    2        2.26312      -1.10559        golden
    3        2.77376      -1.10917        golden
    4        2.53754      -1.12056        parabolic
    5        2.53588      -1.12056        parabolic
    6        2.5398       -1.12056        parabolic
    7        2.53991      -1.12056        parabolic
    8        2.53987      -1.12056        parabolic
    9        2.53994      -1.12056        parabolic
Optimization terminated:
the current x satisfies the termination criteria using OPTIONS.TolX
of 1.000000e-04
xopt =    2.5399
```

The procedure takes nine iterations to locate the optimum which agrees with the one found using the Newton's method.

Learning Summary

The most general class of optimization with continuous variables is the NLP problem:

$$\min_{x} f(\boldsymbol{x})$$
$$\text{s.t.}$$
$$\boldsymbol{h}(\boldsymbol{x}) = \boldsymbol{0}$$
$$\boldsymbol{g}(\boldsymbol{x}) \leq \boldsymbol{0}$$

When all functions are linear, then the LP problem is obtained:

$$\min_{x} \; c^T x$$

$$\text{s.t.}$$

$$Ax - b = 0$$

$$Bx - d \le 0$$

$$0 \le x$$

When there are no (inequality or equality) constraints, we end up with an *unconstrained optimization* problem:

$$\min_{x} \; f(x)$$

where x is a vector. In the simplest possible case, x is scalar:

$$\min_{x \in R} \; f(x)$$

In this chapter, we show that the minimum of functions of a single variable is always a stationary point:

$$\left. \frac{df}{dx} \right|_{x_0} = 0$$

The Taylor's second-order approximation of a twice differentiable $f(x)$ is:

$$f(x) = f(x_0) + \left. \frac{df}{dx} \right|_{x_0} (x - x_0) + \frac{1}{2} \left. \frac{d^2 f}{dx^2} \right|_{x_0} (x - x_0)^2 + o\left(|x - x_0|^2 \right)$$

The definition of the local minimum: let D be the domain of f. f has a *strict local minimum* value at a point x_0 if $f(x) > f(x_0)$ for all x lying in an open interval around x_0. The conditions that need to be satisfied at an extremum are (sufficient conditions):

1. If $\left. \dfrac{df}{dx} \right|_{x_0} = 0 \,\&\, \left. \dfrac{d^2 f}{dx^2} \right|_{x_0} > 0,$ then x_0 is a strict local minimum.

2. If $\left. \dfrac{df}{dx} \right|_{x_0} = 0 \,\&\, \left. \dfrac{d^2 f}{dx^2} \right|_{x_0} < 0,$ then x_0 is a strict local maximum.

3. If $\left. \dfrac{df}{dx} \right|_{x_0} = 0 \,\&\, \left. \dfrac{d^2 f}{dx^2} \right|_{x_0} = 0,$ then the criterion fails.

Terms and Concepts

You must be able to discuss the concepts of:

Concave up or convex function
Extremum
Linear programming
Maximum (local)
Minimum (local)
Nonlinear programming problem
Quadratic programming
Stationary point
Taylor approximation
Unconstrained optimization and the optimality conditions for single variable
   cases

Problems

1.1   Consider the reaction scheme R → P → W which is realized in a constant
      volume V and constant temperature T well-mixed batch reactor. Initially
      only reactant R is loaded into the reactor with concentration $C_{R,0}$. The rate
      at which the reactant R is consumed is first order with respect to the
      reactant concentration:

$$r_1 = -k_1 C_R$$

where C_R is the reactant concentration and k_1 is the reaction rate constant. The desired product is consumed to produce an unwanted
by-product W with a rate that is first order with respect to the product
concentration:

$$r_2 = -k_2 C_P$$

where C_P is the desired product concentration and k_2 is the reaction rate
constant. Develop the material balances of the components R and P and
show that their solution is given by:

$$\frac{C_P(t)}{C_{R,0}} = \frac{k_1}{k_2 - k_1} \left(e^{-k_1 t} - e^{-k_2 t} \right)$$

Determine the time that the maximum concentration of the desired
product is achieved.

1.2 The potential energy is many ion pairs (such as Na^+-Cl^-) can be described by the following potential or bonding energy (in eV per pair)

$$E(r) = -\frac{\alpha}{r} + \frac{\beta}{r^\nu}$$

where α, β, and ν are constants ($\nu = 8$ most commonly) and r is the separation (center to center) distance. Determine an expression to calculate the separation distance and the bonding energy. Prove that this corresponds to a global minimum. Use the values $a = 4.03 \cdot 10^{-28}$, $\beta = 6.97 \cdot 10^{-96}$, and $\nu = 8$ (for which case, the separation distance is expressed in m and the bonding energy in J).

1.3 Consider the case of a cylindrical pipe used to transfer a fluid with temperature T_i. The pipe is insulated (see Fig. P1.3) with insulation material (with thermal conductivity k) with inside radius r_i and outside radius r_o (i is used to denote inside surface of insulating material and subscript o to denote the outside surface of the insulating material). The resistance to heat loss in the film inside the pipe and the metal wall is negligible compared to the resistance to the insulating material and outside film of air. The temperature of air is T_∞, and the outside heat transfer coefficient is h_o while the temperature of the outside surface of the insulating material is T_o.

The heat loss of the pipe is a function of the outside radius of the insulation:

$$\dot{q} = A_o U_o (T_i - T_\infty)$$

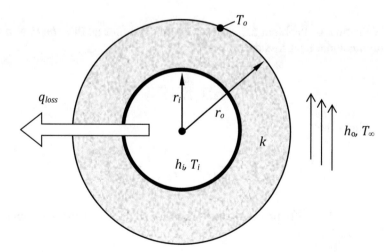

Fig. P1.3 An insulated pipe

where $A_o = 2\pi r_o L$ is the outside area available to heat transfer and U_o is the overall heat transfer coefficient with respect to the outside heat transfer area. Their product is given by:

$$\frac{1}{A_o U_o} \approx \ln\left(\frac{r_o}{r_i}\right) \cdot \frac{1}{2\pi Lk} + \frac{1}{2\pi r_o L} \cdot \frac{1}{h_o}$$

(a) Explain why increasing the outside radius (i.e., the thickness of the insulation) may result in an increase in the heat transfer losses.

(b) Calculate the critical thickness of the insulation by determining the conditions at which losses are maximized (prove that it is actually a maximum).

1.4 We continue with Problem 1.3. Consider now the case where instead of a cylindrical pipe you have a spherical storage tank. What are the conditions of the critical thickness of insulation? It is reminded that the heat losses in this case are given by:

$$\dot{q} = A_o U_o (T_i - T_\infty)$$

$$\frac{1}{A_o U_o} \approx \frac{r_o - r_i}{4\pi k r_o r_i} + \frac{1}{4\pi r_o^2} \cdot \frac{1}{h_o}$$

(a) Explain why increasing the outside radius (i.e., the thickness of the insulation) may result in an increase in the heat transfer losses.

(b) Calculate the critical thickness of the insulation by determining the conditions at which losses are maximized (prove that it is actually a maximum).

1.5 We return to Problem 1.3. Define the Biot number as $\text{Bi} = h r_i / k$ and the dimensional heat loss as:

$$Q = \frac{\dot{q}}{2\pi Lk(T_i - T_\infty)}$$

Show that

$$Q = \frac{1}{\ln\left(\frac{r_o}{r_i}\right) + \left(\frac{r_i}{r_o}\right) \cdot \frac{1}{\text{Bi}}}$$

Use MATLAB to produce Fig. P1.5. What conclusion can be drown from Fig. P1.5?

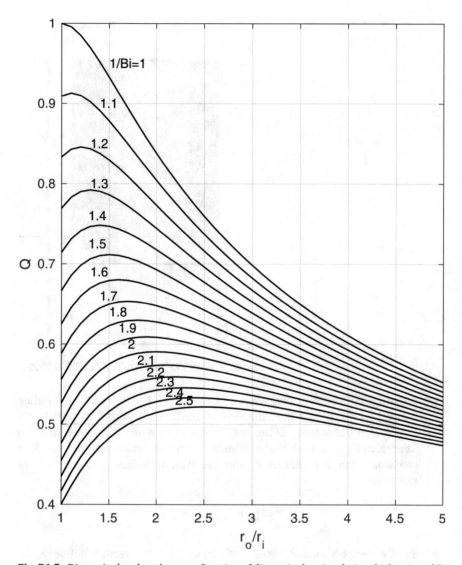

Fig. P1.5 Dimensionless heat loss as a function of dimensionless insulation thickness and Bi number

Use the data: $r_i = 3$ cm, $T_i = 80$ °C, $T_\infty = 20$ °C, $h = 3$ W/(m^2 K), $k = 0.18$ W/(m K). Determine what needs to be the thickness of the insulation so as to have a decrease in the heat loss (Ans. maximum heat loss at $r_o = 6$ cm or thickness of 3 cm, $r_o/r_i \approx 4.9$ for a decrease to be observed or thickness 11.7 cm!).

Fig. P1.6 A storage tank
with conical basis

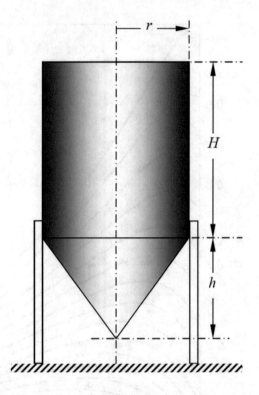

1.6 A storage tank with conical basis and a flat top (see Fig. P1.6) has a radius
   r equal to the height h of the conical basis. The cylindrical part has a
   height H. If the volume of the tank needs to be equal to $V = 9\pi$, find the
   dimensions of the tank that minimize the outer surface of the tank. It is
   reminded that the volume V_c and the lateral surface S_c of a cone are
   given by:

$$V_c = \frac{\pi}{3}r^2 h, \ S_c = \pi r \sqrt{r^2 + h^2}$$

1.7 Snell's law is a formula used to describe the relationship between the
   angles of incidence and refraction, when referring to light passing
   through a boundary between two different isotropic media, such as
   water and air. The refraction index, n, of a transparent medium is defined
   as the ratio of speed of light in vacuum to the speed of light in that
   medium u. If the speed of light in air is u_1 (refraction index $n_1 = c/u_1$)
   and in water is u_2, prove that:

$$n_1 \sin \theta_1 = n_2 \sin \theta_2$$

Fig. P1.7 Snell's law

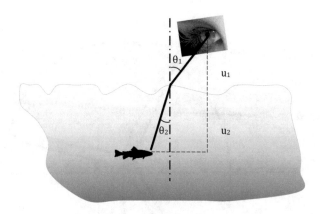

where the angles θ_1 and θ_2 are defined in Fig. P1.7. Use the idea that light will follow the path that minimizes the time needed to travel from the eye to the fish!

1.8 A wind turbine (WT) is shown in Fig. P1.8 which has a sweep area A_{WT}. A WT extracts energy by "slowing down" the wind (transforming the kinetic energy of the wind to a kinetic energy of the blades). A 100% efficient WT would need to stop completely the wind—but then the rotor would have to be a solid disk, and it would not turn. A wind turbine with just one rotor blade will not work either, as most of the wind passing through the area swept by the WT blade would miss the blade completely. Let's assume that the upstream and downstream velocities of the wind are u_0 and u_∞ and that the velocity of the wind at the turbine is the mean between these two: $u = (u_0 + u_\infty)/2$. Perform the following step to prove the Betz limit on the energy that can be extracted from the wind (Betz's law):

(a) Calculate the maximum energy that can be extracted from the wind if the wind "slows down" from u_0 to u_∞ (change in the kinetic energy ΔE_K).

(b) Calculate the mass flow rate of the wind through the sweep area of the WT using the velocity of the wind at the turbine.

(c) Substitute the mass flow rate into the equation for the change in the kinetic energy, and prove that the maximum change in the kinetic energy is realized when $u_\infty = u_0/3$.

(d) Substitute this result to show that the maximum fraction of the kinetic energy of the incoming wind that can be recovered is equal to 16/27 (or 59.26%).

1.9 The reaction A $\leftrightarrow$ B is taking place in two identical CSTRs in series that operate in the same temperature. The reaction rate is given by:

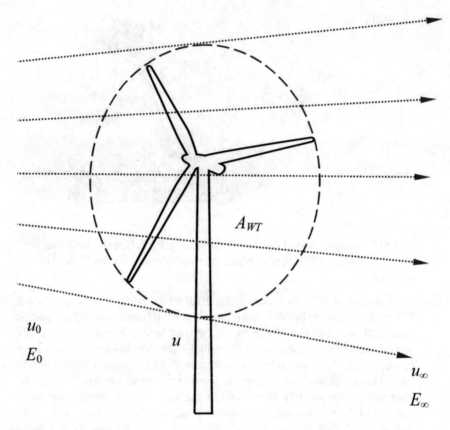

Fig. P1.8 A wind turbine

$$r = k\left(C_A - \frac{C_B}{K_E}\right)$$

where C_A is the concentration of the reactant A in kmol/m^3, C_B is the concentration of the product B in kmol/m^3, $k = \exp(17.2-5800/T)$ is the reaction rate constant in 1/min, $K_E = \exp(-24.7 + 9000/T)$ is the equilibrium constant, and T is the absolute temperature. The concentration of the feed is 10 kmol A/m^3, the volumetric flow rate of the feed is $F = 1$ m^3/min, and the conversion is 95%. Propose a mathematical formulation that can be used to determine the optimal temperature that minimizes the total volume of the two reactors. Use MATLAB to find the minimum.

1.10 In the process control course you have calculated the unit step response of a prototype second order system with transfer function:

$$G(s) = \frac{\omega_n^2}{s^2 + 2\zeta\omega_n s + \omega_n^2}$$

where ω_n is the natural frequency and ζ is the damping ratio. When $\zeta < 1$, it follows that the unit step response is:

$$y(t) = 1 - e_n^{-\zeta\omega_n t}\left[\cos(\omega_d t) + \frac{\zeta}{\sqrt{1-\zeta^2}}\sin(\omega_d t)\right]$$

where:

$$\omega_d = \omega_n\sqrt{1-\zeta^2}$$

The unit step response of a second-order underdamped system with $\zeta = 0.2$ is shown in Fig. P1.9. From the figure, we note that the unit step response exhibits local minima and maxima at $\omega_n t_1$, $\omega_n t_2$, $\omega_n t_3$, etc. Determine their exact location and the ratio $(y(\omega_n t_3) - 1)/(y(\omega_n t_1) - 1)$ (amplitude decay ratio DR). How would you define $t = t_3 - t_1$, and what is its physical meaning?

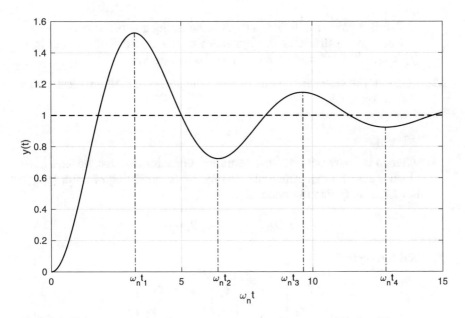

Fig. P1.9 Unit step response of a prototype second-order system with $\zeta = 0.2$

1.11 The impulse response of a linear system is the response to an impulse to the input of the system. As impulse is the derivative of the unit step input, the impulse response is the derivative to the unit step response. Therefore, you have already calculated the impulse response of a prototype second-order system with $\zeta < 1$ in Problem 1.10. Calculate the maxima and minima of the impulse response, and find their relationship with the characteristics of the unit step response.

1.12 Consider the first-order system:

$$\frac{dx(t)}{dt} + ax(t) = bu(t), \ x(0) = x_0 = 1$$

with $a, b > 0$ and the feedback control law $u(t) = -kx(t)$. Our aim is to find the value of k so that the following criterion is minimized:

$$J(k) = \int_0^\infty \left(x^2(t) + ru^2(t) \right) dt, \ r \geq 0$$

Prove that the minimum is achieved at the root of the following quadratic equation that guarantees stability:

$$rk^2 + 2\left(\frac{a}{b}\right) rk - 1 = 0$$

Validate your results in MATLAB by selecting $a = b = r = 1$.

Note: prove first that the dynamics of the closed loop system is $x(t) = x_0 e^{-(a + kb)t}$, and then perform the integration.

1.13 Consider the case where you are given the potential mathematical model of a process:

$$y = ax$$

where a is a parameter to be determined. Consider also that we have been given a set of n experimental measurements consisting of pairs (x_i, y_i), $i = 1, 2, \ldots, n$. Define the residuals:

$$r_i = ax_i - y_i, \quad i = 1, 2, \cdots, n$$

and the vectors:

$$R = \begin{bmatrix} r_1 \\ r_2 \\ \vdots \\ r_n \end{bmatrix}, \quad X = \begin{bmatrix} x_1 \\ x_2 \\ \vdots \\ x_n \end{bmatrix}, \quad Y = \begin{bmatrix} y_1 \\ y_2 \\ \vdots \\ y_n \end{bmatrix}$$

Prove that the problem of:

$$\min_{a} J(a) = \sum_{i=1}^{n} r_i^2 = \sum_{i=1}^{n} (ax_i - y_i)^2$$

has the solution:

$$X^T X a = X^T Y$$

1.14 A batch reactor of active volume $V = 1 \text{ m}^3$ is used to perform a first-order reaction A $\rightarrow$ B with reaction rate $r = kC$, where k is the rate constant and C is the concentration of the reactant A. The reactor is initially loaded with pure A and is operated isothermally, and the downtime for cleaning, loading, and uploading the reactor is 1 h. If the reaction rate constant is $k = 1 \ 1/\text{h}$ and the initial concentration $C_0 = 10 \text{ kmol/m}^3$, determine the final concentration and the batch time that maximizes the annual productivity of the product B.

Hint: show that the amount of product produced in 1 year is:

$$P_B = V(C_0 - C) \cdot \frac{t_y}{t_{\text{batch}} + t_{\text{downtime}}}$$

where t_{batch} is the time needed to achieve the final concentration C in the batch reactor, which is given by:

$$t_{\text{batch}} = \frac{1}{k} \ln\left(\frac{C_0}{C}\right)$$

t_y is the annual operating time in h.

1.15 The reaction A $\leftrightarrow$ B is performed under constant temperature in a plug flow reactor (PFR). The reactor feed is pure reactant A, and the reaction rate is given by:

$$r = k_f C_A - k_b C_B$$

where $k_f = 20\exp(-600/T)$ (in 1/min) is the reaction rate constant of the forward reaction, $k_b = 40\exp(-1500/T)$ (in 1/min) the reaction rate constant of the backward reaction, and T the absolute temperature. Show that the volumetric flow rate F of the feed in m^3/min and the reactor volume V in m^3 are related to the conversion achieved through the following equation:

$$\frac{V}{F} = \int\limits_{0}^{X} \frac{dX}{k_f - (k_f + k_b)X} = \frac{1}{k_f + k_b} \ln \left[\frac{k_f}{k_f - (k_f + k_b)X} \right]$$

For a given conversion $X = 0.8$, develop a methodology to calculate the temperature in the range 25–150 °C that minimizes the ratio V/F. Use MATLAB to find the temperature that will minimize the reactor volume.

1.16 There are many cases in optimization problems in chemical engineering in which, although they are practically single-variable optimization problems, they cannot be expressed as explicit functions of a single variable. A classical example is the design of binary distillation columns for constant relative volatility systems for which the Fenske, Underwood, and Gilliland (FUG) method can be used. In this method, the feed to a distillation is completely defined in terms of feed molar flow rate F, feed composition z, and feed thermal condition q. The top and bottom product compositions (x_D and x_B) are also given together with the (constant) relative volatility a of the binary mixture. Additional thermophysical properties include the heat of vaporization at the column and at the bottom of the column (Δh_D, Δh_B) and the vapor density ρ_v at the top of the column. The application of the FUG method is achieved in the following steps:

(a) Calculation of the minimum reflux ratio rr_{min} and the minimum number of equilibrium trays N_{min}:

$$rr_{min} = \left(\frac{1}{a-1} \right) \left[\frac{x_D}{z} - a \frac{1 - x_D}{1 - z} \right]$$

$$N_{min} = \frac{\ln \left[\left(\frac{x_D}{1 - x_D} \right) \cdot \left(\frac{1 - x_B}{x_B} \right) \right]}{\ln a}$$

(b) The only optimization variable λ, which is the ratio between the actual reflux ratio (rr) to the minimum reflux (rr_{min}), is selected, and the actual reflux ratio and the actual number of theoretical trays N_T are estimated:

$$\frac{N_T - N_{T,min}}{N_T + 1} = \frac{3}{4} \left[1 - \left(\frac{rr - rr_{min}}{rr + 1} \right)^{0.5668} \right]$$

(c) The height of the column is estimated based on an assumed tray efficiency E_o and tray spacing TS:

$$H = 1.2 \left(\frac{N_T - 1}{E_o} \right) TS$$

(d) The diameter of the column D_C is estimated (based on the vapor velocity u_V and the net tray area A_{net}):

$$u_V = \frac{2}{\sqrt{\rho_V}}$$

$$u_V \rho_V A_{net} = mw \cdot (1 + r) D_C$$

$$0.88 \frac{\pi D_C^2}{4} = A_{net}$$

(e) *The heat load of the reboiler Q_B and condenser Q_D are calculated followed by the calculation of the heat transfer area of the reboiler and the condenser (A_B, A_D), by assuming that the product of the heat transfer coefficient and the logarithmic mean temperature difference are approximately equal to 100 kW/m^2:*

$$Q_B = (1 + rr) D \cdot \Delta h_B$$

$$100 A_B = Q_B$$

$$Q_D = (1 + rr) D \cdot \Delta h_D$$

$$100 A_D = Q_D$$

(f) *The equipment cost is estimated:*

$$C_{EQ} = \frac{1}{3} \left(\underbrace{20,000\, H_C^{0.81}\, D_C^{1.05}}_{\text{shell}} + \underbrace{700\, H_C^{0.97}\, D_C^{1.45}}_{\text{trays}} + \underbrace{8000 A_D^{0.65}}_{\text{condenser}} + \underbrace{8000 A_B^{0.65}}_{\text{reboiler}} \right)$$

(g) The cost of the utilities (practically the heating steam) is estimated using the operating time t_y and the unitary utility cost c_{lps}:

$$C_{UT} \approx \frac{Q_B}{\Delta h_{lps}} t_y c_{lps}$$

(h) *Finally, the total, approximate annual cost (TAC) is estimated:*

$$TAC = C_{EQ} + C_{UT}$$

The model presented above is implemented in MATLAB as shown in Table P1.16. The m-file accepts the ratio λ of the actual to minimum reflux ration as input parameters and finally calculates the total annual cost. The reader is encouraged to understand all details of the calculations. Then, the program must be used to obtain the optimum value of λ for a number of values of the relative volatility and then develop an approximate equation for selecting λ as a function of a. A common rule of thumb is that $\lambda \approx 1.2$. Do you agree with that? When do you propose using a slightly larger of a slightly lower value? Is the optimum sensitive to the value of λ?

Table P1.16 Implementation of the FUG method for designing binary distillation columns for constant relative volatility systems

```
function [C,Ceq,Cut]=DistColumnFUG(lamda)
% binary column cost calculation using the FUG method
a=2;              % relative volatility           -
z=0.5;            % feed composition              mole fraction
F=100/3600;       % feed flowrate                 kmol/s
xD=0.99;          % distillate composition        mole faction
xB=0.01;          % bottoms composition           mole fraction
Eo=0.75;          % tray efficiency               -
TS=0.61;          % tray spacing                  m (24 inch)
rhoV=1;           % vapour density                kg/m^3
mw=50;            % molecular weight              kg/kmol
DhD=20000;        % heat of vaporization, distillate kJ/kmol
DhB=22000;        % heat of vaporization, bottoms     kJ/kmol
D=F*(z-xB)/(xD-xB);              % distillate in kmol/s
rrmin = (1/(a-1))*(xD/z - a*(1-xD)/(1-z));% minimum reflux
rr=lamda*rrmin;                 % actual reflux
Nmin=log((xD/(1-xD))*((1-xB)/xB))/log(a); % minimum # of trays
X=(rr-rrmin)/(1+rr);
Y=0.75*(1-X^0.5668);            % Eduljee correlation
NT=(Y+Nmin)/(1-Y);             % actual # of th. trays
N=(NT-1)/Eo;                   % actual trays
H=1.2*(N-1)*TS;                % column height in m
uV=2/sqrt(rhoV);% 2 in (m/s)(kg/m^3)^(1/2)% vapor vel. in m/s
V=(1+rr)*D ;                   % vapor flowrate kmol/s
Anet=V*mw/(uV*rhoV);           % net tray area in m^2
Dc=sqrt((4*Anet/pi)/0.88);     % diameter in m
QD=(1+rr)*D*DhD;               % cond. heat load in kW
AD=QD/100;                     % cond.heat tr.area m^2
QB=(1+rr)*D*DhB;               % reboil.heat load kW
AB=QB/100;                     % reboil.heat tr.area m^2
                               % equipment cost in $
Ceq=(2e4*Dc^1.05*H^0.81+700*Dc^1.45*H^0.97+8000*(AD^0.65+AB^0.
65))/3;
Cut=10*(QB/2000)*(3600*24*365)/1000;% utilities cost in $/y
C=Ceq+Cut;                     % total annual cost in $/y
```

Now you have become an expert in the field of optimizing distillation columns for separating ideal binary mixtures. Can you improve the model used by relating the thermophysical properties used to the actual binary pair used? Later in this book, we will see how we can design and optimize distillation columns using detailed models, but for the time being, this exercise is a useful introduction to the practice of process optimization!

Chapter 2
Multidimensional Unconstrained Optimization

2.1 From Single Variable to Multivariable Optimization

In the previous chapter, we discussed single-variable optimization problems and reviewed the conditions for solving problems of the general form:

$$\min_{x \in R} f(x) \tag{2.1}$$

while in this chapter, we will consider its generalization:

$$\min_{\mathbf{x} \in R^n} f(\mathbf{x}) \tag{2.2}$$

The similarity is striking, and as we will see shortly, this similarity will help us derive the optimality conditions and solution methodologies.

A key concept in developing the optimization theory for problem (2.1) was the first and second derivative and the (linear and quadratic) Taylor approximation of a function of a single variable. The immediate question to answer is: What is the generalization of these concepts to a function of several variables? Let's consider a function of two variables $f{:}R^2 \to R$, i.e. $f(x_1,x_2)$. What is the definition of the first derivative? As we have two variables, we need to consider two first derivatives, one with respect to x_1 and one with respect to x_2. This operation follows directly from the operation that we apply when we calculate the derivative of a function of a single variable. The only difference is that when we calculate the derivative with respect to x_1, we treat x_2 as constant, i.e., we perform the operation:

© The Author(s), under exclusive license to Springer Nature Switzerland
AG 2022
I. K. Kookos, *Practical Chemical Process Optimization*, Springer Optimization
and Its Applications 197, https://doi.org/10.1007/978-3-031-11298-0_2

$$\frac{d}{dx_1} f(x_1, \ x_2)|_{x_2 = \text{constant}} = \frac{\partial f}{\partial x_1} \qquad (2.3)$$

and we use the symbol ∂ to denote the operation of partial differentiation. Of course, we have as many partial derivatives as variables, and we stack them in a column vector called the gradient of f. We use the nabla symbol ∇ to denote the vector of partial derivatives:

$$\nabla f = \text{grad}(f) = \begin{bmatrix} \dfrac{\partial f}{\partial x_1} \\ \dfrac{\partial f}{\partial x_2} \end{bmatrix} \qquad (2.4)$$

In the case of n variables, we obtain:

$$\nabla f = \text{grad}(f) = \begin{bmatrix} \dfrac{\partial f}{\partial x_1} \\ \dfrac{\partial f}{\partial x_2} \\ \vdots \\ \dfrac{\partial f}{\partial x_n} \end{bmatrix} \qquad (2.5)$$

The generalization of the second derivative is the Hessian matrix which is a square and symmetric matrix of all combinations of second derivatives:

$$\mathbf{H} = \begin{bmatrix} \dfrac{\partial}{\partial x_1}\left(\dfrac{\partial f}{\partial x_1}\right) & \dfrac{\partial}{\partial x_1}\left(\dfrac{\partial f}{\partial x_2}\right) & \cdots & \dfrac{\partial}{\partial x_1}\left(\dfrac{\partial f}{\partial x_n}\right) \\ \dfrac{\partial}{\partial x_2}\left(\dfrac{\partial f}{\partial x_1}\right) & \dfrac{\partial}{\partial x_2}\left(\dfrac{\partial f}{\partial x_2}\right) & \cdots & \dfrac{\partial}{\partial x_2}\left(\dfrac{\partial f}{\partial x_n}\right) \\ \vdots & \vdots & \ddots & \vdots \\ \dfrac{\partial}{\partial x_n}\left(\dfrac{\partial f}{\partial x_1}\right) & \dfrac{\partial}{\partial x_n}\left(\dfrac{\partial f}{\partial x_2}\right) & \cdots & \dfrac{\partial}{\partial x_n}\left(\dfrac{\partial f}{\partial x_n}\right) \end{bmatrix} \qquad (2.6)$$

The next, and most important, step is to present the generalization of the Taylor approximation. Let's remind ourselves the form of the quadratic approximation around x_0 for the case of a function of a single variable:

$$f(x) = f(x_0) + \frac{df}{dx}\bigg|_{x_0} (x - x_0) + \frac{1}{2}(x - x_0)\frac{d^2 f}{dx^2}\bigg|_{x_0} (x - x_0) \qquad (2.7)$$

In the case of a multivariable function, the corresponding Taylor approximation follows directly from (2.7):

$$f(\mathbf{x}) = f(\mathbf{x}_0) + \nabla^T f(\mathbf{x}_0)(\mathbf{x} - \mathbf{x}_0) + \frac{1}{2}(\mathbf{x} - \mathbf{x}_0)^T \mathbf{H}(\mathbf{x}_0)(\mathbf{x} - \mathbf{x}_0) \qquad (2.8)$$

In the case of two variables, this simplifies to:

$$f(x_1, x_2) = f(\mathbf{x}_0) + \begin{bmatrix} \dfrac{\partial f}{\partial x_1} & \dfrac{\partial f}{\partial x_2} \end{bmatrix}_{\mathbf{x}_0} \begin{bmatrix} x_1 - x_{0,1} \\ x_2 - x_{0,2} \end{bmatrix}$$

$$+ \frac{1}{2} \begin{bmatrix} x_1 - x_{0,1} & x_2 - x_{0,2} \end{bmatrix} \begin{bmatrix} \dfrac{\partial}{\partial x_1}\left(\dfrac{\partial f}{\partial x_1}\right) & \dfrac{\partial}{\partial x_1}\left(\dfrac{\partial f}{\partial x_2}\right) \\ \dfrac{\partial}{\partial x_2}\left(\dfrac{\partial f}{\partial x_1}\right) & \dfrac{\partial}{\partial x_2}\left(\dfrac{\partial f}{\partial x_2}\right) \end{bmatrix}_{\mathbf{x}_0} \begin{bmatrix} x_1 - x_{0,1} \\ x_2 - x_{0,2} \end{bmatrix}$$

$$(2.9)$$

For convex functions, there is a useful property that follows directly from (1.32):

$$f(\mathbf{x}) \geq f(\mathbf{x}_0) + \nabla^T f(\mathbf{x}_0)(\mathbf{x} - \mathbf{x}_0) \qquad (2.10)$$

A *stationary point* of $f(\mathbf{x})$ is a point at which all partial derivatives become simultaneously equal to zero:

$$\nabla f(\mathbf{x}_0) = \mathbf{0} \qquad (2.11)$$

If we restrict attention to stationary points, then from (2.8), the local behavior of $f(\mathbf{x})$ is determined from the Hessian matrix:

$$f(\mathbf{x}) - f(\mathbf{x}_0) \approx \frac{1}{2}(\mathbf{x} - \mathbf{x}_0)^T \mathbf{H}(\mathbf{x}_0)(\mathbf{x} - \mathbf{x}_0) \qquad (2.12)$$

If $f(\mathbf{x}_0)$ is a (*strict*) *local minimum*, then it follows that $f(\mathbf{x}) - f(\mathbf{x}_0) > 0$ and therefore:

$$(\mathbf{x} - \mathbf{x}_0)^T \mathbf{H}(\mathbf{x}_0)(\mathbf{x} - \mathbf{x}_0) > 0 \qquad (2.13)$$

If this holds true, then $\mathbf{H}(\mathbf{x}_0)$ is called a *positive definite matrix* (p.d.). A matrix is positive definite if and only if all of its eigenvalues are positive. We may, therefore, calculate the eigenvalues of the Hessian matrix and determine whether it is a local minimum or not. If $f(\mathbf{x}_0)$ is a (*strict*) *local maximum*, then the Hessian matrix is negative definite (n.d., i.e., all its eigenvalues are negative). If the Hessian happens to be positive definite everywhere, then the results are global as the function is convex and (2.10) holds true. To summarize, the sufficient conditions for a local minimum in the case of a differentiable function of many variables are:

$$\nabla f(\mathbf{x}_0) = \mathbf{0} \qquad (2.14)$$

$$\&(\mathbf{x} - \mathbf{x}_0)^T \mathbf{H}(\mathbf{x}_0)(\mathbf{x} - \mathbf{x}_0) > 0 \text{ or } \mathbf{H}(\mathbf{x}_0) \text{ is p.d.} \qquad (2.15)$$

Let's now consider the following function of two variables:

$$f(x, y) = x^3 + y^3 + 2x^2 + 4y^2 + 6 \qquad (2.16)$$

$f(x,y):R^2 \rightarrow R$. The surface generated in MATLAB® is shown in Fig. 2.1. The following commands have been used:

```
>> clear
>> syms x y;
>> fmesh(x.^3+y.^3+2*x.^2+4*y.^2+6, [-1.5 0.5 -3.5 0.5])
```

In Fig. 2.2, the contour plots of the same function are given and have also been generated in MATLAB:

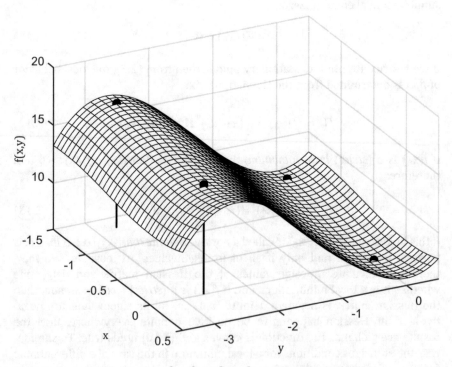

Fig. 2.1 Surface plot of $f(x,y) = x^3 + y^3 + 2x^2 + 4y^2 + 6$ generated in MATLAB

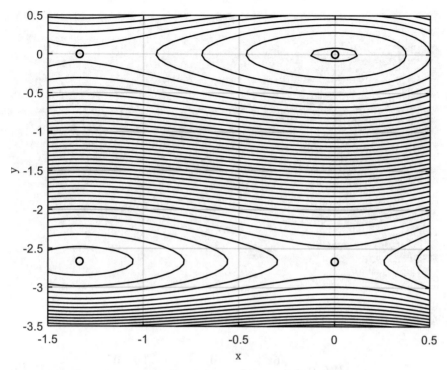

Fig. 2.2 Contour plot of $f(x,y) = x^3 + y^3 + 2x^2 + 4y^2 + 6$ generated in MATLAB

```
>>    fcontour (x.^3+y.^3+2*x.^2+4*y.^2+6,    [-1.5    0.5    -3.5
0.5],'LevelList',6.03:0.15:50/3)
```

From the surface and the contours, we note that there are four stationary points. To determine them, we calculate the gradient:

$$\nabla f(x, y) = \begin{bmatrix} x(3x + 4) \\ y(3y + 8) \end{bmatrix} \tag{2.17}$$

The stationary points are $(0,0)$, $(0,-8/3)$, $(-4/3,0)$, and $(-4/3,-8/3)$. We will study these points one by one, but we first calculate the Hessian:

$$\mathbf{H}(x, y) = \begin{bmatrix} 6x + 4 & 0 \\ 0 & 6y + 8 \end{bmatrix} \tag{2.18}$$

We could have performed all these algebraic calculations in MATLAB:

```
>> clear all
>> syms x y;
>> f=x^3+y^3+2*x^2+4*y^2+6

f =
x^3 + 2*x^2 + y^3 + 4*y^2 + 6

>> g=gradient(f)

g =
  3*x^2 + 4*x
  3*y^2 + 8*y

>> H=hessian(f)

H =
[ 6*x + 4,      0]
[    0, 6*y + 8]
```

A. Point (0,0). The gradient is zero while the hessian is:

$$\mathbf{H}(0,0) = \begin{bmatrix} 6x+4 & 0 \\ 0 & 6y+8 \end{bmatrix}_{(0,\,0)} = \begin{bmatrix} 4 & 0 \\ 0 & 8 \end{bmatrix} \tag{2.19}$$

To determine the eigenvalues, we need to solve the system of linear equations:

$$|\lambda \mathbf{I}_2 - \mathbf{H}(x,\ y)| = 0 \tag{2.20}$$

where $\mathbf{I}_2$ is the 2-by-2 unit matrix. However, as the hessian is diagonal, the solutions follow immediately, and the eigenvalues are $\lambda_1 = +4 > 0$ and $\lambda_2 = +8 > 0$. As both eigenvalues are positive, the hessian is positive definite, and point (0,0) is a local minimum. The corresponding value of the function is $f(0,0) = 6$. Locally, the function can be approximated by:

$$f(x,y) \approx f(0,0) + \frac{1}{2}[x \quad y]\mathbf{H}(0,0)\begin{bmatrix} x \\ y \end{bmatrix} = 6 + \frac{1}{2}\left(4x^2 + 8y^2\right) \tag{2.21}$$

and the contour lines $(f(x,y) = c)$ are:

$$\left(\frac{x}{2}\right)^2 + \left(\frac{y}{\sqrt{2}}\right)^2 = c' = \frac{c-6}{8} \tag{2.22}$$

(this is an equation of an ellipse with major axis parallel to the x-axis—when the eigenvalues are equal, it becomes the equation of a circle). This justifies the form of the contour lines around the point (0,0) shown in Fig. 2.2.

B. Point (0,−8/3). The gradient is zero while the hessian is:

$$\mathbf{H}(0, \,-8/3) = \begin{bmatrix} 6x+4 & 0 \\ 0 & 6y+8 \end{bmatrix}_{(0, \,-8/3)} = \begin{bmatrix} 4 & 0 \\ 0 & -8 \end{bmatrix} \tag{2.23}$$

As the hessian is again diagonal, the eigenvalues follow immediately and are $\lambda_1 = 4 > 0$ and $\lambda_2 = -8 < 0$. As one eigenvalues is positive and one is negative, the hessian is indefinite (a square matrix is indefinite if and only if it has at least one positive eigenvalue and at least one negative eigenvalue), and point (0,−8/3) is neither a local maximum or a local minimum. When a point is a stationary point and the hessian matrix is indefinite, then the point is a saddle point (the general form of the surface and the contour lines around a saddle point are shown in Figs. 2.3 and 2.4). Locally, the function can be approximated by:

$$f(x, y) \approx f(0, \,-8/3) + \frac{1}{2}\left[x \quad \left(y+\frac{8}{3}\right) \right] \mathbf{H}\left(0, \,-\frac{8}{3}\right) \begin{bmatrix} x \\ \left(y+\frac{8}{3}\right) \end{bmatrix}$$

$$= \frac{418}{27} + \frac{1}{2}\left[4x^2 - 8\left(y+\frac{8}{3}\right)^2 \right] \tag{2.24}$$

and the contour lines ($f(x,y) = c$) are:

$$\left(\frac{x}{2}\right)^2 - \left(\frac{y+\frac{8}{3}}{\sqrt{2}}\right)^2 = c' = \frac{c - \frac{418}{27}}{8} \tag{2.25}$$

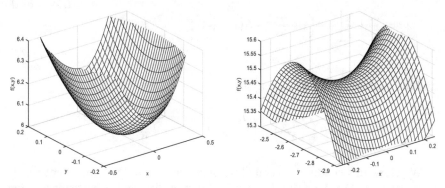

Fig. 2.3 Surface around the minimum (0,0) (left) and one saddle point (0, − 8/3) (right) of the function $f(x,y) = x^3 + y^3 + 2x^2 + 4y^2 + 6$

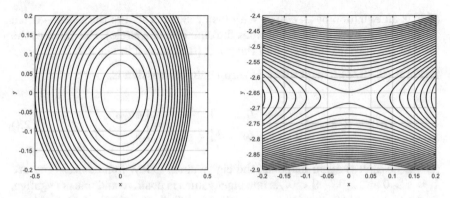

Fig. 2.4 Contours around the minimum (0,0) (left) and one saddle point (0, − 8/3) (right) of the function $f(x,y) = x^3 + y^3 + 2x^2 + 4y^2 + 6$

(which is the standard form of the equation of a hyperbola with center (0,–8/3)). This justifies the form of the contour lines around the point (0,–8/3) shown in Fig. 2.2 that resemble the case shown in Figs. 2.3 and 2.4. Point (−4/3,0) has the same characteristics as point (0,–8/3).

C. Point (−4/3,–8/3). The gradient is zero while the hessian is:

$$\mathbf{H}(-4/3, -8/3) = \begin{bmatrix} 6x + 4 & 0 \\ 0 & 6y + 8 \end{bmatrix}_{(-4/3, -8/3)} = \begin{bmatrix} -4 & 0 \\ 0 & -8 \end{bmatrix} \quad (2.26)$$

The hessian is again diagonal, and the eigenvalues follow immediately and are $\lambda_1 = -4 < 0$ and $\lambda_2 = -8 < 0$. As both eigenvalues are negative, point (−4/3,–8/3) is a local maximum. Locally, the function can be approximated by:

$$f(x, y) \approx f(-4/3, -8/3) + \frac{1}{2} \left[\left(x + \frac{4}{3}\right) \quad \left(y + \frac{8}{3}\right) \right] \mathbf{H}\left(-\frac{4}{3}, -\frac{8}{3}\right) \begin{bmatrix} \left(x + \frac{4}{3}\right) \\ \left(y + \frac{8}{3}\right) \end{bmatrix}$$

$$= \frac{50}{3} + \frac{1}{2} \left[-4\left(x + \frac{4}{3}\right)^2 - 8\left(y + \frac{8}{3}\right)^2 \right]$$

$$(2.27)$$

and the contour lines $(f(x,y) = c)$ are:

$$\left(\frac{y + \frac{4}{3}}{2}\right)^2 + \left(\frac{y + \frac{8}{3}}{\sqrt{2}}\right)^2 = c' = \frac{\frac{50}{3} - c}{8} \quad (2.28)$$

We close this section by reminding ourselves of another important result from calculus. Consider the differential function $f(\mathbf{x})$ and a particular $\mathbf{x}_0$ for

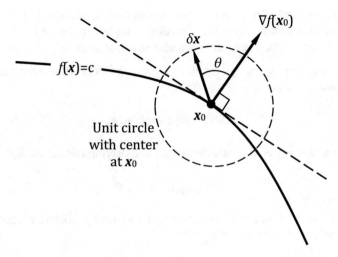

Fig. 2.5 Product of the gradient vector and a unit vector

which $f(\mathbf{x}_0) = c$. Obviously, the contour line $f(\mathbf{x}) = c$ passes through $\mathbf{x}_0$ as shown in Fig. 2.5. Consider now any unit vector $\delta\mathbf{x}$ emanating from $\mathbf{x}_0$. Any vector that satisfies the unit length criterion has its end point on the unit circle with center at $\mathbf{x}_0$. Let us now calculate the linear approximation of $f(\mathbf{x})$ around $\mathbf{x}_0$ as shown:

$$f(\mathbf{x}_0 + \delta\mathbf{x}) = f(\mathbf{x}_0) + \nabla^T f(\mathbf{x}_0) \cdot \delta\mathbf{x} \qquad (2.29)$$

The dot product of two vectors can be calculated as the product of their magnitudes and the cosine of their angle θ (see Fig. 2.5). We write (2.29) as follows:

$$f(\mathbf{x}_0 + \delta\mathbf{x}) - f(\mathbf{x}_0) = \nabla^T f(\mathbf{x}_0) \cdot \delta\mathbf{x} = \left\|\nabla^T f(\mathbf{x}_0)\right\| \cdot \|\delta\mathbf{x}\| \cos\theta \qquad (2.30)$$

However, $\delta\mathbf{x}$ is a unit vector and therefore (as the norm of any vector is a scalar):

$$\frac{f(\mathbf{x}_0 + \delta\mathbf{x}) - f(\mathbf{x}_0)}{\left\|\nabla^T f(\mathbf{x}_0)\right\|} = \cos\theta \qquad (2.31)$$

$\cos\theta$ is maximum when $\theta = 0$, i.e., when $\delta\mathbf{x}$ is collinear and has the same direction as the gradient vector, then the change in $f(\mathbf{x})$ is maximized. When $\theta = -\pi$, then $\cos\theta$ is minimum, i.e., when $\delta\mathbf{x}$ is parallel with the gradient and has direction opposite to the direction of the gradient vector, then the change in $f(\mathbf{x})$ is minimized. When $\theta = \pm\pi/2$, then the change in $f(\mathbf{x})$ is zero (i.e., we stay on the same contour at least locally). We therefore conclude that if we are at a point $\mathbf{x}_0$ in R^n, then:

- The direction of $+\nabla f(\mathbf{x}_0)$ is the direction of fastest increase in $f(\mathbf{x})$ at $\mathbf{x}_0$.
- The direction of $-\nabla f(\mathbf{x}_0)$ is the direction of fastest decrease in $f(\mathbf{x})$ at $\mathbf{x}_0$.
- $+\nabla f(\mathbf{x}_0)$ and $-\nabla f(\mathbf{x}_0)$ are perpendicular to the tangent of $f(\mathbf{x})$ at $\mathbf{x}_0$.

From the last one, we may also derive the conditions that must satisfy the tangent plane to a surface at a point $\mathbf{x}_0$:

$$\nabla^T f(\mathbf{x}_0)(\mathbf{x} - \mathbf{x}_0) = 0 \tag{2.32}$$

In addition, we may also conclude that all $\mathbf{x}$ satisfying the inequality:

$$\nabla^T f(\mathbf{x}_0)(\mathbf{x} - \mathbf{x}_0) > 0 \tag{2.33}$$

belong to the half space of directions of increasing objective function. In addition, all $\mathbf{x}$ satisfying the inequality:

$$\nabla^T f(\mathbf{x}_0)(\mathbf{x} - \mathbf{x}_0) < 0 \tag{2.34}$$

belong to the half space of directions of decreasing objective function.

2.2   Algorithms for Multivariable Unconstrained Optimization

Our objective is to devise methods for solving the optimization problem:

$$\min_{\mathbf{x} \in R^n} f(\mathbf{x}) \tag{2.35}$$

The optimality conditions are:

$$\nabla f(\mathbf{x}_0) = \mathbf{0} \tag{2.36}$$

$$\& \mathbf{H}(\mathbf{x}_0) \text{ is positive definite} \tag{2.37}$$

The method to locate any $\mathbf{x}$ that satisfies (2.36) will be of iterative nature. We start by writing the quadratic approximation of $f(\mathbf{x})$ around a point that we have arrived at iteration k:

$$f(\mathbf{x}^k + \delta\mathbf{x}^k) = f(\mathbf{x}^k) + \nabla^T f(\mathbf{x}^k)\delta\mathbf{x}^k + \frac{1}{2}(\delta\mathbf{x}^k)^T \mathbf{H}(\mathbf{x}^k)\delta\mathbf{x}^k \tag{2.38}$$

where $\delta\mathbf{x}^k = \mathbf{x} - \mathbf{x}^k$. We take the derivative with respect to $\delta\mathbf{x}^k$ to obtain an approximation for the gradient:

$$\nabla f\left(\mathbf{x}^k + \delta\mathbf{x}^k\right) = \nabla f\left(\mathbf{x}^k\right) + \mathbf{H}\left(\mathbf{x}^k\right)\delta\mathbf{x}^k \qquad (2.39)$$

Using (2.36) and (2.39), we obtain the following iterative methodology for calculating the step at iteration k:

$$\mathbf{H}\left(\mathbf{x}^k\right)\delta\mathbf{x}^k = -\nabla f\left(\mathbf{x}^k\right) \qquad (2.40)$$

This is a system of linear equations as $\mathbf{H}(\mathbf{x}^k)$ is a constant matrix, and $-\nabla f(\mathbf{x}^k)$ is a constant vector. To solve the system, the condition $\det(\mathbf{H}(\mathbf{x}^k)) \neq 0$ must be satisfied. In addition, for the algorithm to move toward a decreasing direction for the function, the hessian matrix must be positive definite. Based on Eq. (2.40), we can propose the following algorithm, which is the generalization of the Newton's method in multidimensional functions:

Step 0: Select an initial estimate of the solution $\mathbf{x}^0 \in R^n$, at iteration $k = 0$.
Step 1: Calculate $f(\mathbf{x}^k)$, $\nabla f(\mathbf{x}^k)$ and $\mathbf{H}(\mathbf{x}^k)$.
Step 2: Determine the Newton step $\delta\mathbf{x}^k$ by solving (2.40) to obtain
$\mathbf{x}^{k+1} = \mathbf{x}^k + \delta\mathbf{x}^k$.
Step 3: If $||\nabla f(\mathbf{x}^k)|| < \varepsilon$, then STOP; else set $k = k + 1$, and return to Step 1.

As we know from the previous chapter, this is the basic Newton's method which has no guarantee to locate the (local) optimal solution. However, before moving further, we will examine the general performance of the method when applied to representative functions.

We first examine the following function (see also Fig. 2.6):

$$f(\mathbf{x}) = 10\left(x_1^2 - x_2\right)^2 + (x_1 - 1)^2 + 4 \qquad (2.41a)$$

which has the following gradient and hessian matrix:

$$\nabla f(\mathbf{x}) = \begin{bmatrix} 40\left(x_1^2 - x_2\right)x_1 + 2x_1 - 2 \\ -20\left(x_1^2 - x_2\right) \end{bmatrix} \qquad (2.41b)$$

$$\mathbf{H}(x, y) = \begin{bmatrix} 120x_1^2 - 40x_2 + 2 & -40x_1 \\ -40x_1 & 20 \end{bmatrix} \qquad (2.41c)$$

The initial point chosen is $\mathbf{x}^0 = [-1\ 3]^T$. We calculate the gradient and the hessian at the initial point:

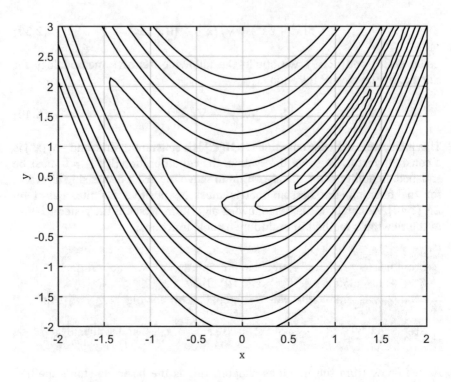

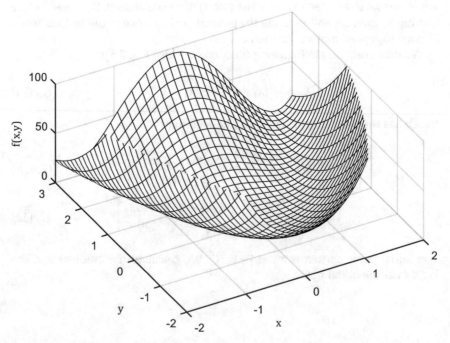

Fig. 2.6 Contour and surf plots generated in MATLAB for $f(x,y)$ given by (2.41)

$$\nabla f(\mathbf{x}^0) = \begin{bmatrix} 40(x_1^2 - x_2)x_1 + 2x_1 - 2 \\ -20(x_1^2 - x_2) \end{bmatrix} = \begin{bmatrix} 76 \\ 40 \end{bmatrix}$$

$$\mathbf{H}(\mathbf{x}^0) = \begin{bmatrix} 120x_1^2 - 40x_2 + 2 & -40x_1 \\ -40x_1 & 20 \end{bmatrix} = \begin{bmatrix} 2 & 40 \\ 40 & 20 \end{bmatrix}$$

We then calculate the full newton step:

$$\delta \mathbf{x}^0 = -\mathbf{H}^{-1}(\mathbf{x}^0)\nabla f(\mathbf{x}^0) = \frac{1}{39}\begin{bmatrix} 0.5 & -1 \\ -1 & 0.05 \end{bmatrix}\begin{bmatrix} 76 \\ 40 \end{bmatrix} = \frac{1}{39}\begin{bmatrix} -2 \\ -74 \end{bmatrix}$$

$$= -\begin{bmatrix} 0.0513 \\ 1.8974 \end{bmatrix}$$

and the new point:

$$\mathbf{x}^1 = \mathbf{x}^0 - \mathbf{H}^{-1}(\mathbf{x}^0)\nabla f(\mathbf{x}^0) = \begin{bmatrix} -1 \\ +3 \end{bmatrix} - \begin{bmatrix} 0.0513 \\ 1.8974 \end{bmatrix} = \begin{bmatrix} -1.0513 \\ +1.1026 \end{bmatrix}$$

The value of the objective function is decreased significantly as $f(\mathbf{x}^0) = 48$ and $f(\mathbf{x}^1) = 8.2078$. These steps can be performed efficiently in MATLAB:

```
>> clear all
>> f=inline('10*(x(1)^2-x(2))^2+(x(1)-1)^2+4');
>> g=inline('[40*(x(1)^2-x(2))*x(1)+2*x(1)-2; -20*(x(1)^2-x
(2))]');
>> H=inline('[120*x(1)^2-40*x(2)+2 -40*x(1);-40*x(1) 20]');
>> x=[-1,3]';
>> dx=-inv(H(x))*g(x)

dx = -0.0513
     -1.8974

>> x=x+dx

x = -1.0513
     1.1026

>> f(x)

ans = 8.2078
```

Repeating the same process for a second time, we obtain the following:

```
>> dx=-inv(H(x))*g(x)

dx =  1.9488
     -4.0948

>> x=x+dx

x =  0.8975
    -2.9922

>> f(x)

ans = 148.2397
```

We note that in the second iteration, $f(\mathbf{x}^2) = 148.2397 > f(\mathbf{x}^1) = 8.2078$! The reader is advised to use MATLAB to perform the remaining steps shown in Fig. 2.7. The unexpected behavior of the Newton's algorithm should be noted.

Let's now study an additional example:

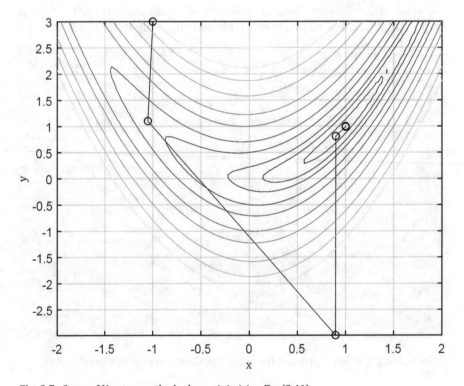

Fig. 2.7 Steps of Newton method when minimizing Eq. (2.41)

$$f(\mathbf{x}) = \sqrt{1 + x_1^2} + \sqrt{1 + x_2^2} \qquad (2.42a)$$

$$\nabla f(\mathbf{x}) = \begin{bmatrix} \dfrac{x_1}{\sqrt{1 + x_1^2}} \\ \dfrac{x_2}{\sqrt{1 + x_2^2}} \end{bmatrix} \qquad (2.42b)$$

$$\mathbf{H}(\mathbf{x}) = \begin{bmatrix} \dfrac{1}{\left(1 + x_1^2\right)^{3/2}} & 0 \\ 0 & \dfrac{1}{\left(1 + x_2^2\right)^{3/2}} \end{bmatrix} \qquad (2.42c)$$

The surface of the function is shown in Fig. 2.8. Let's start from $\mathbf{x}^0 = [5\ 5]^T$. We determine the value of the function, the gradient, and the hessian:

$$f(\mathbf{x}^0) = 2\sqrt{26}$$

$$\nabla f(\mathbf{x}^0) = \frac{5}{\sqrt{26}} \begin{bmatrix} 1 \\ 1 \end{bmatrix}$$

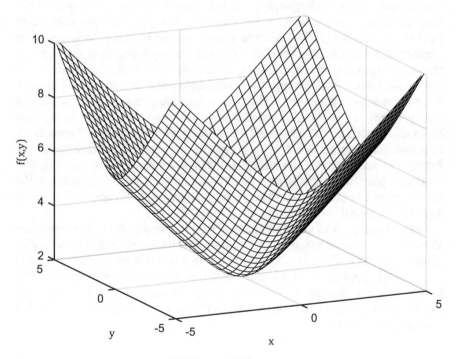

Fig. 2.8 Surface plot of $f(\mathbf{x}) = \sqrt{1 + x_1^2} + \sqrt{1 + x_2^2}$ generated in MATLAB

$$\mathbf{H}(\mathbf{x}^0) = \frac{1}{\left(\sqrt{26}\right)^3} \begin{bmatrix} 1 & 0 \\ 0 & 1 \end{bmatrix}$$

We then determine the full Newton's step:

$$\begin{bmatrix} \delta x_1^0 \\ \delta x_2^0 \end{bmatrix} = -\left(\sqrt{26}\right)^3 \begin{bmatrix} 1 & 0 \\ 0 & 1 \end{bmatrix}^{-1} \frac{5}{\sqrt{26}} \begin{bmatrix} 1 \\ 1 \end{bmatrix} = -130 \begin{bmatrix} 1 \\ 1 \end{bmatrix}$$

Then:

$$\begin{bmatrix} x_1^1 \\ x_2^1 \end{bmatrix} = \begin{bmatrix} x_1^0 \\ x_2^0 \end{bmatrix} + \begin{bmatrix} \delta x_1^0 \\ \delta x_2^0 \end{bmatrix} = \begin{bmatrix} 5 \\ 5 \end{bmatrix} - 130 \begin{bmatrix} 1 \\ 1 \end{bmatrix} = -\begin{bmatrix} 125 \\ 125 \end{bmatrix}!$$

The method diverges as in the second step, it moves further away from the origin (which is the unique global minimum). There was nothing to give us a warning that this could happen. Clearly, we need to devise a strategy to prevent Newton's method from accepting rather aggressive steps that could cause divergence.

We already know from the previous chapter that the "naïve" approach will fail in many cases and a backtracking scheme is necessary to make the algorithm robust. There is a huge amount of literature on this issue. There are three basic strategies that can be followed. The first one uses a constant scale factor (<1) that multiplies the Newton's step with the aim to make the method less aggressive. It has the advantage that it is easy to apply and does not increase the computational load as it does not involve any additional calculations of the function. It has the disadvantage that as the scale factor is constant, it will work in some cases but not in others. The other extreme is to use exact calculation of the optimum in the direction of the Newton's step. This turns to be very demanding in terms of additional evaluations of the function. There is an intermediate strategy which only seeks for an acceptable decrease in the function at every step of the algorithm. This is arguably the most successful practical approach as it achieves convergence with reasonable increase in the computational load. Here we will present a simple, yet efficient, backtracking algorithm that will suffice for the purposes of this book.

We will not get into the mathematical details, but we will only present the steps that need to be taken to apply a line search algorithm that is known as *Armijo's inexact line search*. The user needs to select two parameters $\alpha \in (0,1)$ and $\beta \in (0,1)$ (usually $\alpha = 0.2$ and $\beta = 1/2$ will work fine in most cases). Then in Step 2 of the Newton's algorithm, we do not accept immediately the full Newton's step, but we determine the minimum n, starting from $n = 0$, that satisfies the criterion.

$$f\left(\boldsymbol{x}^k + \beta^n \delta \boldsymbol{x}^k\right) < f\left(\boldsymbol{x}^k\right) + \alpha\beta^n \nabla f^{\mathrm{T}}\left(\boldsymbol{x}^k\right)\delta \boldsymbol{x}^k \tag{2.43}$$

When (2.43) is satisfied then, as it is proven in the literature, the decrease in the objective function is satisfactory so as to achieve convergence to a stationary point. With this modification, we may now construct out general-purpose Newton function that can handle efficiently most cases. The corresponding m-file is presented in Table 2.1. The reader is encouraged to study the lines of the file carefully before moving forward. We will now use this m-file to solve the peculiar problem that we studied before:

Table 2.1 Implementation of Newton's method with Armijo's line search

```
function x=myNewton(f,F,H,x0)
% Newton's method with Armijo's inexact line search
% f  is the obj. function             (inline function)
% F  is the gradient of the obj. function (inline function)
% H  is the hessian of the obj. function  (inline function)
% x0 is an initial guess of the solution
a = 0.2; beta=1/2; maxiter=100; tol=1e-6; % modify if
necessary

x=x0; fun =f(x) ; grad=F(x) ; hess=H(x) ; % STEP 1

NewtonStep = hess \ (-grad) ;                % STEP 2

iter=0;
fprintf('Iter      Objective        Gradient\n',iter,fun)
while ((norm(grad)>tol)&&(iter<maxiter))   % STEP 3
    iter=iter+1;

    n=0;
    % Backtrack if Eq (2.2.9) is not satisfied
    while( f(x+beta^n*NewtonStep) >
f(x)+a*beta^n*grad'*NewtonStep )
        n=n+1;
        fprintf('Backtracking with n=%4u \n',n)
    end

    x=x+beta^n*NewtonStep;                   % STEP 2
    fun=f(x); grad=F(x); hess=H(x); NewtonStep= hess \ (-grad);
    fprintf('%4u %14.8f %14.8f \n',iter,fun,norm(grad))
end
```

```
>> clear
>> f=@(x)sqrt(1+x(1)^2)+sqrt(1+x(2)^2);
>> F=@(x)[x(1)/sqrt(x(1)^2+1);x(2)/sqrt(x(2)^2+1)];
>> H=@(x)diag([1/(x(1)^2+1)^1.5,1/(x(2)^2+1)^1.5]);
>> x=myNewton(f,F,H,[5;5])
```

The response is interesting:

```
>> runmyNewton
Iter   Objective    Gradient
Backtracking with n= 1
Backtracking with n= 2
Backtracking with n= 3
Backtracking with n= 4
  1   6.56220237  1.34693115
Backtracking with n= 1
Backtracking with n= 2
Backtracking with n= 3
  2   2.94421208  1.03783719
Backtracking with n= 1
  3   2.00813014  0.12712853
  4   2.00000054  0.00103988
  5   2.00000000  0.00000000
```

```
x =
      1.0e-09 *

     -0.3976
     -0.3976
```

Observe that in the first iteration, the algorithm enters the backtracking and performs four backtracking steps, and the new point is:

$$\begin{bmatrix} x^1 \\ y^1 \end{bmatrix} = \begin{bmatrix} x^0 \\ y^0 \end{bmatrix} + \delta \begin{bmatrix} x^0 \\ y^0 \end{bmatrix} = \begin{bmatrix} 5 \\ 5 \end{bmatrix} - \left(\frac{1}{2}\right)^4 130 \begin{bmatrix} 1 \\ 1 \end{bmatrix} = -\begin{bmatrix} 3.125 \\ 3.125 \end{bmatrix}$$

and:

$$f(x^1, y^1) = 2\sqrt{1 + (-3.125)^2} = 6.5622 < f(x^0, y^0) = 10.198$$

Clearly backtracking has been successful in solving the problem with the aggressive nature of Newton's method (at least for this peculiar example).

2.3 Application Examples

Example 2.1 *Minimum Energy Consumption for a Natural Gas Compressor*
   In Chap. 1 and Example 1.2, we studied the case of natural gas compression using a two-stage compressor with intercooler. We now extend our analysis into multiple (more than two)-stage compression systems. Consider the three-compressor system as shown in Fig. 2.9. The total power consumed is:

$$f(P_1, P_2) = \frac{w_1 + w_2 + w_3}{N\frac{1}{e}\frac{RT_0}{a}} = \left(\frac{P_1}{P_0}\right)^a + \left(\frac{P_2}{P_1}\right)^a + \left(\frac{P_f}{P_2}\right)^a - 3 \qquad (2.44)$$

We calculate the gradient:

$$\nabla f = \begin{bmatrix} a\frac{1}{P_1}\left[\left(\frac{P_1}{P_0}\right)^a - \left(\frac{P_2}{P_1}\right)^a\right] \\ a\frac{1}{P_2}\left[\left(\frac{P_2}{P_1}\right)^a - \left(\frac{P_f}{P_2}\right)^a\right] \end{bmatrix} \qquad (2.45)$$

The calculation can also be performed in MATLAB:

```
>> syms P0 P1 P2 PF a;
>> simplify(gradient((P1/P0)^a+(P2/P1)^a+(PF/P2)^a-3,[P1 P2]))

ans =
 (a*((P1/P0)^a - (P2/P1)^a))/P1
 (a*((P2/P1)^a - (PF/P2)^a))/P2

>> simplify(hessian((P1/P0)^a+(P2/P1)^a+(PF/P2)^a-3,[P1 P2]));
```

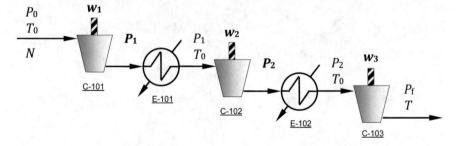

Fig. 2.9 A natural gas (methane) three-stage compression system

By setting the gradient equal to zero, we obtain the following result:

$$\lambda = \frac{P_1}{P_0} = \frac{P_2}{P_1} = \frac{P_f}{P_2} \qquad (2.46)$$

We note again that all compressors have the same compression ratio λ. To determine the exact value of the intermediate pressures, we make the following calculation:

$$P_n = \lambda P_{n-1} = \lambda^2 P_{n-2} = \ldots = \lambda^n P_0 \qquad (2.47)$$

or:

$$\lambda = \sqrt[n]{\frac{P_n}{P_0}} \qquad (2.48)$$

where n is the number of compression stages used. For the case of three compression stages that increase the pressure from 1 to 10 bar, for instance, we obtain:

$$\lambda = \sqrt[3]{\frac{10}{1}} = 2.15443469$$

and therefore the pressures are $P_1 = 2.15443469$ bar, $P_2 = 4.64158883$ bar, and $P_3 = P_f = 10$ bar.

We will use our implementation of Newton's method to solve the problem numerically:

```
clear
a=0.3;
P0=1;
Pf=10;

f=@(x) (x(1)/P0)^a + (x(2)/x(1))^a + (Pf/x(2))^a - 3;

F=@(x) [(a*((x(1)/P0 )^a - (x(2)/x(1))^a))/x(1)
    (a*((x(2)/x(1))^a - ( Pf/x(2))^a))/x(2)];

H=@(x) [ (a*((x(2)/x(1))^a-(x(1)/P0)^a+a*(x(1)/P0)^a +a*(x(2)/
x(1))^a))/x(1)^2, -(a^2*(x(2)/x(1))^a)/(x(1)*x(2))
    (a^2*(x(2)/x(1))^a)/(x(1)*x(2)), (a*((Pf/x(2))^a-(x(2)/x
(1))^a+a*(x(2)/x(1))^a+a*(Pf/x(2))^a))/x(2)^2];

x=myNewton(f,F,H,[2;4])
```

The response of the solver is:

```
>> runmyNewton
Iter    Objective     Gradient
   1    0.77718894    0.00276651
   2    0.77685083    0.00112495
   3    0.77678870    0.00045021
   4    0.77677826    0.00017999
   5    0.77677656    0.00007198
   6    0.77677629    0.00002879
   7    0.77677624    0.00001152
   8    0.77677624    0.00000461
   9    0.77677624    0.00000184
  10    0.77677624    0.00000074

x = 2.1544
    4.6415

>> eig(H(x))

ans = 0.0451
      0.0142
```

We will now consider the command available in MATLAB® for unconstrained optimization:

xopt=**fminunc**(fun,x0,options)
finds a local minimum of a function of several variables. Starts at x0 and attempts to find a local minimizer xopt of the function fun. fun (an inline function or an m-file) returns the value of the objective function. It may also return the gradient in which case one must use [f,g]=fun(x) and: options=optimoptions('fminunc','SpecifyObjectiveGradient',true); use also 'Diplay',Iter to display intermediate results.

A distinct advantage of the MATLAB implementation is that it does not require the hessian or the gradient as it builds them numerically. To solve the gas compression problem, we first note that the most practically useful way in solving medium scale applications is to supply the problem description through an m-file. For the example, the m-file accepts the two intermediate pressures as inputs and returns the value of the objective function:

```
function f=ThreeStageCompression(P)
P0 = 1; % initial pressure in bar
Pf = 10; % final pressure  in bar
a =0.3; % parameter of the model
f= (P(1)/P0)^a + (P(2)/P(1))^a + (Pf/P(2))^a - 3;
end
```

```
>> clear
>> options=optimoptions('fminunc','Display','iter');
>> x=fminunc(@ThreeStageCompression,[2;4],options)
                                                   First-order
 Iteration Func-count      f(x)        Step-size    optimality
     0          3       0.778671                      0.00639
     1          9       0.778294          10          0.0054
     2         12       0.776876           1          0.00289
     3         15       0.776777           1          0.00021
     4         18       0.776776           1          1.01e-05
     5         21       0.776776           1          1.4e-06
     6         24       0.776776           1          1.28e-08

Local minimum found.
Optimization completed because the size of the gradient is less
than
the default value of the optimality tolerance.

<stopping criteria details>

x = 2.1544
    4.6416
```

Two points that we will not be discussing in detail in this book (a lot of information can be found in the literature) are the numerical approximation of the first and second derivatives. The hessian matrix is usually calculated based on quasi-Newton methods that can guarantee that the hessian is positive definite (the BFGS method is one of the most popular members of this class). It is interesting to note that even when the hessian is available, using a quasi-Newton method is the preferred method to use. For approximating the gradient finite differences can be used. The forward finite differences are the preferred method as they only require one additional function evaluation:

$$\frac{df}{dx} \approx \frac{f(x + \delta x) - f(x)}{\delta x}, \ \delta x = \max\{1, |x|\} \cdot eps^{1/2} \qquad (2.49)$$

where eps is the accuracy of the computer (type $>>$ eps in MATLAB to get the accuracy of your computer system). Central finite differences are more accurate but require two function evaluations:

$$\frac{df}{dx} \approx \frac{f(x+\delta x) - f(x-\delta x)}{2\delta x}, \quad \delta x = \max\left\{1, |x|\right\} \cdot eps^{1/3} \tag{2.50}$$

Perform the following experiment in MATLAB to check the accuracy of the approximation of the first-order derivative of the function $f(x) = e^x$:

```
>> clear
>> format long
>> x=5;
>> exp(x)

ans = 1.484131591025766e+02

>> dx=max(1,abs(x))*sqrt(eps);
>> (exp(x+dx)-exp(x))/dx

ans = 1.484131645202637e+02

>> dx=max(1,abs(x))*eps^(1/3);
>> (exp(x+dx)-exp(x-dx))/(2*dx)

ans = 1.484131591241918e+02
```

Example 2.2 *Three CSTR in Series for a First-Order Reaction*

In Chap. 1 and Example 1.3, we studied the case of optimum design of two CSTR in series for the case of a first-order reaction. We now consider the case of three CSTR in series (see Fig. 2.10). For the case of a first-order reaction, the total volume of the reactors is given by:

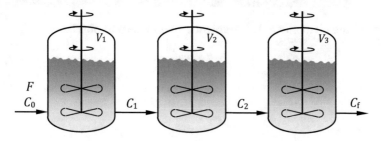

Fig. 2.10 Three CSTR in series system

$$V(C_1, C_2) = \frac{F}{k}\left[\left(\frac{C_0 - C_1}{C_1}\right) + \left(\frac{C_1 - C_2}{C_2}\right) + \left(\frac{C_2 - C_f}{C_f}\right)\right]$$
$$= \frac{F}{k}\left(\frac{C_0}{C_1} + \frac{C_1}{C_2} + \frac{C_2}{C_f} - 3\right)$$

(2.51)

We calculate the gradient and the hessian:

$$\nabla V = \frac{F}{k}\begin{bmatrix} -\dfrac{C_0}{C_1^2} + \dfrac{1}{C_2} \\[2ex] -\dfrac{C_1}{C_2^2} + \dfrac{1}{C_f} \end{bmatrix}$$

(2.52)

$$\mathbf{H} = \frac{F}{k}\begin{bmatrix} 2\dfrac{C_0}{C_1^3} & -\dfrac{1}{C_2^2} \\[2ex] -\dfrac{1}{C_2^2} & 2\dfrac{C_1}{C_2^3} \end{bmatrix}$$

(2.53)

The same calculations can be performed in MATLAB:

```
>> syms C0 C1 C2 Cf F k;
>> gradient((F/k)*(C0/C1+C1/C2+C2/Cf-3),[C1 C2])

ans = -(F*(C0/C1^2 - 1/C2))/k

       -(F*(C1/C2^2 - 1/Cf))/k

>> hessian((F/k)*(C0/C1+C1/C2+C2/Cf-3),[C1 C2])

ans = [ (2*C0*F)/(C1^3*k),      -F/(C2^2*k)]

      [     -F/(C2^2*k), (2*C1*F)/(C2^3*k)]
```

We set the elements of the gradient equal to zero to obtain:

$$C_1^2 = C_0 C_2$$

(2.54)

$$C_2^2 = C_1 C_f$$

(2.55)

or:

$$\frac{C_0}{C_1} = \frac{C_1}{C_2} = \left(\frac{C_0}{C_f}\right)^{1/3} \tag{2.56}$$

$$\frac{C_1}{C_2} = \frac{C_2}{C_f} = \left(\frac{C_0}{C_f}\right)^{1/3} \tag{2.57}$$

The volumes can be calculated:

$$V_1 = \frac{F}{k}\left[\left(\frac{C_0 - C_1}{C_1}\right)\right] = \frac{F}{k}\left[\left(\frac{C_0}{C_f}\right)^{1/3} - 1\right] \tag{2.58a}$$

$$V_2 = \frac{F}{k}\left[\left(\frac{C_1}{C_2}\right) - 1\right] = \frac{F}{k}\left[\left(\frac{C_0}{C_1}\right) - 1\right] = V_1 \tag{2.58b}$$

$$V_3 = \frac{F}{k}\left[\left(\frac{C_2}{C_f}\right) - 1\right] = \frac{F}{k}\left[\left(\frac{C_1}{C_2}\right) - 1\right] = V_2 = V_1 \tag{2.58c}$$

As in the case of the two CSTR in series, the volumes are equal.

Consider now the following numerical example: $F = 1$ m^3/h, $k = 1$ 1/h, $C_0 = 1$ kmol/m^3, and $C_f = 0.125$ kmol/m^3. It then follows from (2.56) and (2.57) that $C_1 = 0.5$ kmol/m^3 and $C_2 = 0.25$ kmol/m^3. The eigenvalues of the hessian are both positive, and the point determined is a local minimum.

The graphical representation of the optimal solution is shown in Fig. 2.11. It is reminded that the material balance of a single CSTR can be written as:

$$V = \frac{F}{r(C_{\text{out}})} \cdot [C_{\text{in}} - C_{\text{out}}] \tag{2.59}$$

where C_{in} is the concentration in the feed, C_{out} is the product concentration, and r is the rate of reaction. It then follows that if we plot F/r as a function of C, then the volume of each reactor is the area of the rectangle with height $F/r(C_{\text{out}})$ and width $C_{\text{in}} - C_{\text{out}}$.

From Fig. 2.11, it follows that if one CSTR is used, the necessary volume is $V = 8 \cdot (1 - 0.125) = 7$ m^3. The three reactors have a total volume of:

$$\begin{aligned}
V &= \frac{F}{kC_1}(C_0 - C_1) + \frac{F}{kC_2}(C_1 - C_2) + \frac{F}{kC_f}(C_2 - C_f) \\
&= \frac{1}{1 \cdot 0.5}(1 - 0.5) + \frac{1}{1 \cdot 0.25}(0.5 - 0.25) \\
&\quad + \frac{1}{1 \cdot 0.125}(0.25 - 0.125) = 1 + 1 + 1 = 3 \text{ m}^3
\end{aligned}$$

This is a significant decrease in the total volume (57% decrease) at the expense of having a more complex system to operate (three CSTR instead of one).

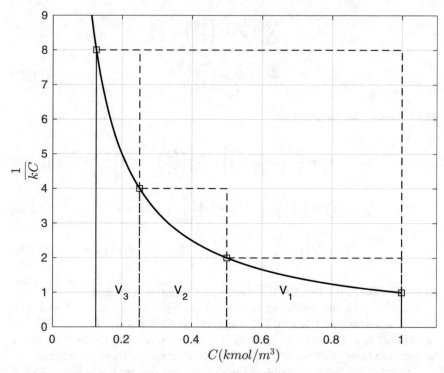

Fig. 2.11 Graphical representation of the optimal solution of the three CSTR in series with first-order reaction case study

The contour lines of the total reactor volume are shown in Fig. 2.12. Please note that not all points shown have a physical meaning. We use our implementation of Newton's method to find the minimum numerically:

```
>> clear
>> C0=1;
>> Cf=0.125;
>> f=@(C) (C0-C(1))/C(1) + (C(1)-C(2))/C(2) + (C(2)-Cf)/Cf;
>> F=@(C) [ -C0/C(1)^2+1/C(2); -C(1)/C(2)^2+1/Cf];
>> H=@(C) [2*C0/C(1)^3, -1/C(2)^2; -1/C(2)^2, 2*C(1)/C(2)^3];
>> Copt=myNewton(f,F,H,[0.6;0.3])
Iter   Objective    Gradient
  1    3.00984448   0.53867447
  2    3.00012125   0.05270747
  3    3.00000003   0.00084742
  4    3.00000000   0.00000021
Copt = 0.5000
        0.2500
>> eig(H(Copt))
ans = 11.1556
       68.8444
```

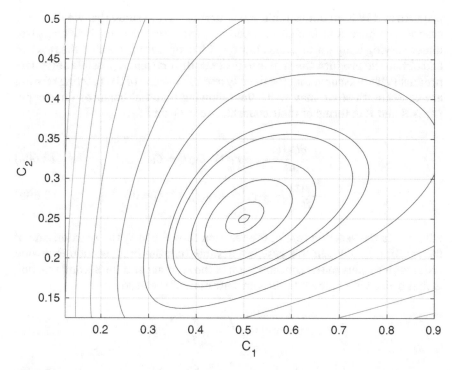

Fig. 2.12 Contour lines for the three CSTR case study with first-order reaction (for a point to have physical meaning, C_1 must be greater than C_2)

However, as the reader can verify, the algorithm diverges easily if we change the initial point (try (0.8,0.3), for instance). This is also true even for the MATLAB command `fminunc`, which seems not to be useful for this relatively simple example. The mathematical formulation is, unfortunately, incomplete as the constraints $0.125 < C_2 < C_1 < 1$ need to be incorporated into the formulation to make the algorithm robust and practically useful. This must wait for the discussion in the next chapter.

Example 2.3 *Optimal Operation of a Batch Reactor*

Describing the operation of a batch reactor involves solving differential equations as the system changes continuously over time. Furthermore, the optimization variables are not single variables but functions of time (i.e., of infinite dimension). As a result, optimization of dynamic systems, such as batch reactors, is particularly challenging to perform. To clarify these points, we consider the reaction scheme R → P → W, where R is the reactant, P is the product, and W is an undesirable waste product. We assume that we study the case where pure R is loaded in a batch reactor; the reaction is taking place under varying temperature, and after $t_f = 1$ h of operation, we stop the reaction and obtain the product. Our aim is to produce the maximum possible amount of product P. If we assume that the density and the volume of the reacting mixture are constant, then we may write the following material balances for components R and P in terms of their concentrations C_R and C_P:

$$\frac{dC_R(t)}{dt} = -r_1(t), \ C_R(t) = C_0 \tag{2.60a}$$

$$\frac{dC_P(t)}{dt} = +r_1(t) - r_2(t), \ C_P(t) = 0 \tag{2.60b}$$

where r_1 is the reaction rate of the reaction R $\rightarrow$ P and r_2 the reaction rate of the reaction P $\rightarrow$ W. In this case, we assume that the first reaction is second order with respect to the concentration of the reactant and the second reaction is first order with respect to the concentration of the product:

$$r_1(t) = k_{0,1} \exp\left(-\frac{E_1}{R_g T(t)}\right) C_R^2(t) \tag{2.61a}$$

$$r_2(t) = k_{0,2} \exp\left(-\frac{E_2}{R_g T(t)}\right) C_P(t) \tag{2.61b}$$

where $k_{0,1}$ and $k_{0,2}$ are the reaction rate constants, E_1 and E_2 the activation energies, and R_g the ideal gas constant. Temperature $T(t)$ is an unknown function of time, and we need to optimize its variation over time to maximize the production of the product:

$$\min_{T(t)} -C_P(t_f) \tag{2.62}$$

It is important to note at this point that the variable over which we solve the problem is defined over an infinite number of points as we need to determine the values of the temperature $T(t)$, for every $\in [0, t_f]$, where t_f is the final time. To alleviate this difficulty, we may define temperature to be a specific function of time such as a polynomial or an exponential function:

$$T(t) = \alpha_0 + \alpha_1 t + \alpha_2 t^2 + \cdots + \alpha_n t^n \qquad (2.63)$$

or:

$$T(t) = \alpha_0 + \alpha_1 \exp(\alpha_3 t) \qquad (2.64)$$

where α_i are parameters to be determined. In both cases, we have transformed the initial problem, which needs an infinite number of parameters to be solved, to a problem of finite number of parameters.

The other issue is that the model consists of differential equations that do not have an analytic solution. Apart from the problems involved in solving numerically a problem that involves differential equations, there are also problems associated with the calculation of the first- and second-order derivatives.

Consider the following numerical parameters: $C_0 = 1$ kmol/m^3; $k_{0,1} = 4000$ (m^3/kmol)/h; $k_{0,2} = 620,000$ 1/h; $E_1/R = 2500$ K; $E_2/R = 5000$ K; and linear temperature profile:

$$T(t) = T_0 + (T_1 - T_0)t \qquad (2.65)$$

We can summarize the overall problem:

$$\min_{T_0, T_1} - C_P(1)$$

$$\text{s.t.}$$

$$\frac{dC_R}{dt} = -4000 \exp\left(-\frac{2500}{T_0 + (T_1 - T_0)t}\right) C_R^2$$

$$\frac{dC_P}{dt} = +4000 \exp\left(\frac{-2500}{T_0 + (T_1 - T_0)t}\right) C_R^2 - 620,000 \exp\left(\frac{-5000}{T_0 + (T_1 - T_0)t}\right) C_P$$

$$t \in [0, 1], \ C_R(0) = 1, \ C_P(0) = 0$$

$$(2.66)$$

To solve this system of ordinary differential equations (ODE), we may use one of the several ode solvers available in MATLAB. Their general syntax is as follows:ode_solver_name(@fun_name, [initial_time final_time], initial_x) where the symbols used are self-explanatory. The solver ode113 (use the >> help ode113 command to obtain more information) is a very efficient general-purpose ode solver in MATLAB and is usually our first choice. The following m-file implements the model of the batch reactor:

```
function Cdot=RayReactor(t,C,T0,T1)
T=T0+(T1-T0)*t;
Cdot(1) = -4000*exp(-2500/T)*C(1)^2;
Cdot(2) = +4000*exp(-2500/T)*C(1)^2 -620000*exp(-5000/T)*C(2);
Cdot=Cdot(:);
```

To obtain the solution for $T_0 = 400$ K and $T_1 = 300$ K, we use the following command:

```
[t,C]=ode113(@RayReactor,[0 1],[1 0],[],400,300);
```

The solution obtained can be seen in Fig. 2.13. The value of the objective function is:

```
>> -C(end,2)

ans =  -0.5278
```

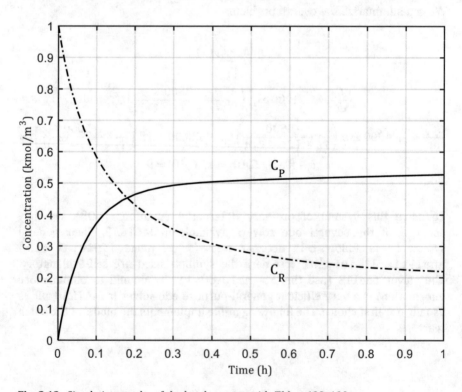

Fig. 2.13 Simulation results of the batch reactor with $T(t) = 400-100\,t$

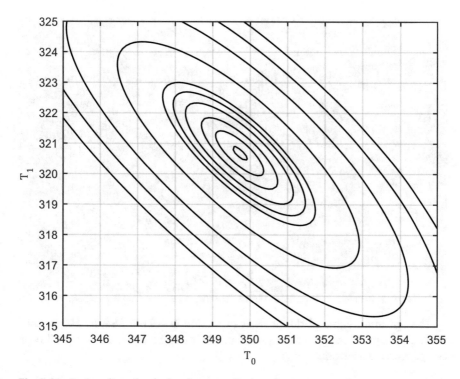

Fig. 2.14 Contour lines for the batch reactor final product concentration

We may repeat the procedure just presented and plot the results to obtain the contour plot shown in Fig. 2.14. We note that the case study has a clear minimum which can be determined using the MATLAB command fminunc:

```
function obj=RayReactorObj(T)
[t,C]=ode113(@RayReactor,[0 1],[1 0],[],T(1),T(2));
obj=-C(end,2);
```

```
>> clear
>> x0=[400;300];
>> options=optimoptions('fminunc','Display','Iter');
>> Topt=fminunc(@RayReactorObj,x0,options)
```

MATLAB response is as follows:

Iteration	Func-count	f(x)	Step-size	First-order optimality
0	3	-0.52778		0.00423
1	15	-0.543452	820	0.00378
2	18	-0.601359	1	0.000423
3	24	-0.606268	7.24839	0.000588
4	27	-0.608215	1	0.000311
5	30	-0.60893	1	2.02e-05
6	33	-0.608949	1	2.36e-05
7	36	-0.608954	1	1.11e-06
8	39	-0.608954	1	5.16e-09

Local minimum found.

Optimization completed because the size of the gradient is less than
the default value of the optimality tolerance.

ans = 349.7223
      320.6829

The optimum value of the objective function is -0.608954, or 60.9% con-version of the reactant to the desired product can be achieved using a linear temperature variation.

2.4   Parameter Estimation: Nonlinear Least Squares

In this chapter, we have studied methods and examples for the general prob-lem of unconstrained optimization problem with many variables:

$$\min_{x \in R^n} f(x) \tag{2.67}$$

This is not a very useful category of optimization problems in terms of practical applications. Two are the reasons why we have devoted a whole chapter to this subject. The first one is that unconstrained optimization helps introduce the algorithms and methods for solving constrained optimization problems. The second is that the problem of parameter estimation in nonlinear models is usually expressed as an unconstrained optimization problem with many vari-ables. This problem is known to most engineers as least-squares optimization or parameter estimation problem. The name is due to the objective function chosen in these problems, which is the sum of the squares of the deviations between the predictions of a theoretical model and the experimental data. This is neither the only nor the most appropriate objective function for estimating the parameters in a mathematical model, but it is the one most used.

To give a specific example of parameter estimation problem, let us consider the fuel cell system that we have studied in Chap. 1 and more specifically concentrate on Fig. 1.14. In this figure, the experimental results obtained from a particular system are shown. To obtain these results, the engineers varied the current density of the cell and observed the resulting variation of the voltage developed by the cell. The researchers had control over the equipment and changed in a predetermined (and in many cases well-thought and well-planned) way several parameters which are available to them to manipulate. We will use the symbol $\mathbf{x}$ to denote these parameters which are varied purposefully by the researcher. In the case of the fuel cell, the researchers only varied the current density ($\mathbf{x} = j$) and kept all other parameters constant. They obtained experimental observations of important outputs or responses of the system which we will denote by the symbol $\mathbf{y}$. In the case of the fuel cell, the response of the system is evaluated by recording the voltage developed by the cell, $\mathbf{y} = E$.

Based on scientific knowledge and experimental observations, the researchers usually develop mathematical models to validate their hypothesis about the systems they study. These models contain the controlled variables (such as the current density), the system responses (such as the cell voltage), and several parameters that must be estimated. We will denote these parameters by the symbol $\mathbf{p}$. The mathematical model φ is, therefore, of the following form:

$$y = \varphi(\mathbf{x}, \mathbf{p}) \tag{2.68}$$

In the case of the fuel cell, there is only one controlled variable and one response, and Eq. (2.68) can be simplified:

$$y = \varphi(x, \mathbf{p}) \tag{2.68'}$$

The experiments take place at discrete values of the controlled variable x_i, $i = 1, 2, \ldots, n$, and experimental observations are obtained $y^{\exp}(x_i)$. The discrepancies between the experimentally observed values of the response and the ones predicted by the model are defined by the residual function $r(\mathbf{p})$:

$$r_i(x_i, \mathbf{p}) = \varphi(x_i, \mathbf{p}) - y_i^{\exp} \tag{2.69}$$

The aim of a parameter estimation methodology is to develop an efficient way to determine the parameter vector $\mathbf{p}$ that minimizes a measure of disagreement between the experimental values and the model predictions. The most popular parameter estimation method, known as the nonlinear least squares method (NLLS), is based on minimizing the sum of the squares of the residuals:

$$\min_{\boldsymbol{p}} f(\boldsymbol{p}) = \frac{1}{2}\sum_{i=1}^{n}r_i^2(x_i, \boldsymbol{p}) = \frac{1}{2}\sum_{i=1}^{n}(\varphi(x_i, \boldsymbol{p}) - y_i^{\exp})^2 \qquad (2.70)$$

As residuals can be positive or negative, we need to use the squares of the residuals to avoid a situation where the model fails miserably to describe the experimental observations, but the residuals cancel each other. An undesirable characteristic of the sum of the squares is that large residuals affect strongly the value of the criterion.

One method to solve NLLS problems is the Newton's method. In applying Newton's method, we need to calculate the gradient and the hessian. We first calculate the gradient:

$$\frac{\partial f(\boldsymbol{p})}{\partial p_k} = \frac{1}{2}\sum_{i=1}^{n}\frac{\partial r_i^2(x_i; \boldsymbol{p})}{\partial p_k} = \frac{1}{2}\sum_{i=1}^{n}2r_i\frac{\partial r_i(x_i; \boldsymbol{p})}{\partial p_k}$$

$$= \sum_{i=1}^{n}r_i\frac{\partial(\varphi(x_i, \boldsymbol{p}) - y_i^{\exp})}{\partial p_k}$$

or:

$$\frac{\partial f(\boldsymbol{p})}{\partial p_k} = \sum_{i=1}^{n}r_i(x_i, \boldsymbol{p})\frac{\partial\varphi(x_i, \boldsymbol{p})}{\partial p_k} \qquad (2.71)$$

The last equation can also be written in matrix form:

$$\begin{bmatrix} \dfrac{\partial f(\boldsymbol{p})}{\partial p_1} \\ \dfrac{\partial f(\boldsymbol{p})}{\partial p_2} \\ \vdots \\ \dfrac{\partial f(\boldsymbol{p})}{\partial p_m} \end{bmatrix} = \begin{bmatrix} \dfrac{\partial\varphi(x_1, \boldsymbol{p})}{\partial p_1} & \dfrac{\partial\varphi(x_2, \boldsymbol{p})}{\partial p_1} & \cdots & \dfrac{\partial\varphi(x_n, \boldsymbol{p})}{\partial p_1} \\ \dfrac{\partial\varphi(x_1, \boldsymbol{p})}{\partial p_2} & \dfrac{\partial\varphi(x_2, \boldsymbol{p})}{\partial p_2} & \cdots & \dfrac{\partial\varphi(x_n, \boldsymbol{p})}{\partial p_2} \\ \vdots & \vdots & \ddots & \vdots \\ \dfrac{\partial\varphi(x_1, \boldsymbol{p})}{\partial p_m} & \dfrac{\partial\varphi(x_2, \boldsymbol{p})}{\partial p_m} & \cdots & \dfrac{\partial\varphi(x_n, \boldsymbol{p})}{\partial p_m} \end{bmatrix} \begin{bmatrix} r_1(\boldsymbol{p}) \\ r_2(\boldsymbol{p}) \\ \vdots \\ r_n(\boldsymbol{p}) \end{bmatrix}$$

or:

$$\nabla f(\boldsymbol{p}) = \boldsymbol{J}^T(\boldsymbol{p})\boldsymbol{r}(\boldsymbol{p}) \qquad (2.72)$$

where:

$$\nabla f(\boldsymbol{p}) = \begin{bmatrix} \dfrac{\partial f(\boldsymbol{p})}{\partial p_1} \\[2mm] \dfrac{\partial f(\boldsymbol{p})}{\partial p_2} \\[2mm] \vdots \\[2mm] \dfrac{\partial f(\boldsymbol{p})}{\partial p_m} \end{bmatrix}_{m \times 1}, \quad \boldsymbol{r}(\boldsymbol{p}) = \begin{bmatrix} r_1(\boldsymbol{p}) \\[1mm] r_2(\boldsymbol{p}) \\[1mm] \vdots \\[1mm] r_n(\boldsymbol{p}) \end{bmatrix}_{n \times 1} \tag{2.73}$$

and $\boldsymbol{J}(\boldsymbol{p})$ is the Jacobian matrix:

$$\boldsymbol{J}(\boldsymbol{p}) = \left[\frac{\partial \varphi(x_i, \, \boldsymbol{p})}{\partial p_j} \right] = \begin{bmatrix} \dfrac{\partial \varphi(x_1, \, \boldsymbol{p})}{\partial p_1} & \dfrac{\partial \varphi(x_1, \, \boldsymbol{p})}{\partial p_2} & \cdots & \dfrac{\partial \varphi(x_1, \, \boldsymbol{p})}{\partial p_m} \\[3mm] \dfrac{\partial \varphi(x_2, \, \boldsymbol{p})}{\partial p_1} & \dfrac{\partial \varphi(x_2, \, \boldsymbol{p})}{\partial p_2} & \cdots & \dfrac{\partial \varphi(x_2, \, \boldsymbol{p})}{\partial p_m} \\[3mm] \vdots & \vdots & \ddots & \vdots \\[3mm] \dfrac{\partial \varphi(x_n, \, \boldsymbol{p})}{\partial p_1} & \dfrac{\partial \varphi(x_n, \, \boldsymbol{p})}{\partial p_2} & \cdots & \dfrac{\partial \varphi(x_n, \, \boldsymbol{p})}{\partial p_m} \end{bmatrix} \tag{2.74}$$

It should be noted that m is the number of parameters and n is the number of experimental observations. We also calculate the hessian:

$$\frac{\partial}{\partial p_\ell} \left(\frac{\partial f(\boldsymbol{p})}{\partial p_k} \right) = \frac{\partial}{\partial p_\ell} \left(\sum_{i=1}^{n} r_i(x_i, \, \boldsymbol{p}) \frac{\partial \varphi(x_i, \, \boldsymbol{p})}{\partial p_k} \right) = \sum_{i=1}^{n} \frac{\partial \varphi(x_i, \boldsymbol{p})}{\partial p_\ell} \cdot \frac{\partial \varphi(x_i, \boldsymbol{p})}{\partial p_k}$$
$$+ \sum_{i=1}^{n} r_i(x_i, \boldsymbol{p}) \frac{\partial^2 \varphi(x_i, \boldsymbol{p})}{\partial p_\ell \partial p_k}$$

or by ignoring the second term in the right-hand side (as the residuals become small close to the solution):

$$\nabla^2 f(\boldsymbol{p}) \approx \boldsymbol{J}^T(\boldsymbol{p}) \boldsymbol{J}(\boldsymbol{p}) \tag{2.75}$$

When Newton's step is calculated by applying Eq. (2.40), we obtain:

$$\boldsymbol{J}^T(\boldsymbol{p}^k) \boldsymbol{J}(\boldsymbol{p}^k) \cdot \delta \boldsymbol{p}^k = -\boldsymbol{J}^T(\boldsymbol{p}^k) \boldsymbol{r}(\boldsymbol{p}^k) \tag{2.76}$$

which requires only the first derivatives of the mathematical model with respect to the parameters.

We have presented a great deal of theory which we will now try to elucidate through the fuel cell example. The experimental data available are pairs of applied current density and measured cell voltage $(x_i = j_i, \, y_i = E(j_i))$. The mathematical model is given by Eq. (1.74):

$$y_i(x_i; E_0, R_\Omega, j_L) = E_0 - R_\Omega x_i + \frac{RT}{2F} \ln\left(1 - \frac{x_i}{j_L}\right) \tag{2.77}$$

We define the vector of the three parameters:

$$\boldsymbol{p} = \begin{bmatrix} E_0 \\ R_\Omega \\ j_L \end{bmatrix} \tag{2.78}$$

and (2.78) becomes:

$$y_i(x_i; \boldsymbol{p}) = p_1 - p_2 x_i + \frac{RT}{2F} \ln\left(1 - \frac{x_i}{p_3}\right) \tag{2.79}$$

Experimental data for a particular solid oxide fuel cell (SOFC) that operates at a relatively low temperature (873 K) are presented in Table 2.2. We make use of these data to determine the parameters of the model (2.79). To this end, we build the following m-file that calculates the residuals:

```
function [r,f,jexp,Eexp]=SSR_FuelCell(p)
% Sum of Squares of Residuals for the SOFC model
jexp=[.020 .109 .600 1.087 1.588 2.089 2.583 3.083 3.462 3.835]';
Eexp=[.835 .827 .764 .690 .609 .533 .447 .346 .242 0.062]';
r=(p(1)-p(2)*jexp+0.0376*log(1-jexp/p(3))) - Eexp;
```

We then use the implementation of the NLLS algorithm in MATLAB (solver lsqnonlin: `lsqnonlin(@myfun,x0)`). The model predictions and the experimental data are compared in Fig. 2.15:

Table 2.2 Experimental data for the fuel cell application

i	j (A/cm^2)	E^{exp} (V)
1	0.020	0.835
2	0.109	0.827
3	0.600	0.764
4	1.087	0.690
5	1.588	0.609
6	2.089	0.533
7	2.583	0.447
8	3.083	0.346
9	3.462	0.242
10	3.835	0.062

Data have been extracted from figures appearing in Huang C., et al., *Journal of Power Sources*, 195, p. 5176, 2010

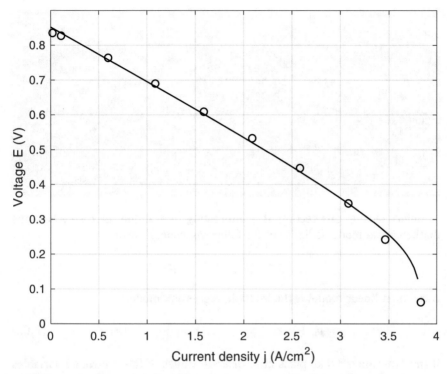

Fig. 2.15 Comparison of experimental results (o) and model predictions (-) for the fuel cell case study

```
>> clear
>> p=[0.9;0.2;4]
>> opts = optimoptions(@lsqnonlin,'Display','Iter');
>> po=lsqnonlin(@SSR_FuelCell,p,[],[],opts)
>> [r,f,jexp,Eexp]=SSR_FuelCell(po);
>> j=linspace(0,po(3),101);
>> plot(j,po(1)-po(2)*j+0.0376*log(1-j/po(3))),'k-
','LineWidth',1)
>> hold on
>> grid on
>> xlabel('Current density j (A/cm^{2})')
>> ylabel('Voltage E (V)')
>> plot(jexp,Eexp,'ko','LineWidth',1)
```

Iteration	Func-count	f(x)	Norm of step	First-order optimality
0	4	0.0538428		1.44
1	8	0.0538428	0.650709	1.44
2	12	0.00968978	0.162677	0.376
3	16	0.00116089	0.0459602	0.00237
4	20	0.00116068	8.36595e-05	9.67e-06

Local minimum possible.

po = 0.8517
     0.1445
     3.8422

Before closing this section, it is interesting to examine the case where the mathematical model is linear of the following general form:

$$y = \varphi(x, p) = A(x) \cdot p \tag{2.80}$$

A common linear model is the multiple regression model:

$$y = p_0 + p_1 x_1 + p_2 x_2 + \ldots p_q x_q + \cdots + p_n x_n \tag{2.81}$$

If the experiments take place at m discrete values of the controlled variables vector x_i, $i = 1,2,\ldots,m$, and experimental observations are obtained $y^{\exp}(x_i)$, then y, A, and p are defined as follows:

$$
y = \begin{bmatrix} y_1 \\ y_2 \\ \vdots \\ y_i \\ \vdots \\ y_m \end{bmatrix}, \quad
A(x) = \begin{bmatrix}
1 & x_{1,1} & x_{2,1} & \cdots & x_{q,1} & \cdots & x_{n,1} \\
1 & x_{1,2} & x_{2,2} & \cdots & x_{q,2} & \cdots & x_{n,2} \\
\vdots & \vdots & \vdots & \ddots & \vdots & & \vdots \\
1 & x_{1,i} & x_{2,i} & & x_{q,i} & & x_{n,i} \\
\vdots & \vdots & \vdots & & \vdots & \ddots & \vdots \\
1 & x_{1,m} & x_{2,m} & \cdots & x_{q,m} & \cdots & x_{n,m}
\end{bmatrix},
$$

$$
p = \begin{bmatrix} p_0 \\ p_1 \\ \vdots \\ p_q \\ \vdots \\ p_n \end{bmatrix} \tag{2.82}
$$

The discrepancies between the experimentally observed values and the ones predicted by the model are defined by the vector of residuals $r(\mathbf{p})$ with elements:

$$r_i(\mathbf{x}_i, \mathbf{p}) = y_i(\mathbf{x}_i, \mathbf{p}) - y_k^{\exp} \tag{2.83}$$

The linear least squares method (LLS) is based on minimizing the sum of the squares of the residuals:

$$
\begin{aligned}
\min_{\mathbf{p}} f(\mathbf{p}) &= \frac{1}{2} \sum_{i=1}^{n} r_i^2(\mathbf{x}_i, \mathbf{p}) = \frac{1}{2} \sum_{i=1}^{n} (y_i(\mathbf{x}_i, \mathbf{p}) - y_i^{\exp})^2 \\
&= \frac{1}{2} (A\mathbf{p} - \mathbf{y}^{\exp})^T (A\mathbf{p} - \mathbf{y}^{\exp}) \\
&= \frac{1}{2} (\mathbf{p}^T A^T A\mathbf{p} - \mathbf{p}^T A^T \mathbf{y}^{\exp} - \mathbf{y}^{\exp\,T} A\mathbf{p} + \mathbf{y}^{\exp\,T} \mathbf{y}^{\exp})
\end{aligned} \tag{2.84}
$$

or:

$$\min_{\mathbf{p}} f(\mathbf{p}) = \frac{1}{2} (\mathbf{p}^T A^T A\mathbf{p} - 2\mathbf{y}^{\exp\,T} A\mathbf{p} + \mathbf{y}^{\exp\,T} \mathbf{y}^{\exp}) \tag{2.85}$$

We then calculate the gradient of $f(\mathbf{p})$:

$$
\begin{aligned}
\frac{\partial f(\mathbf{p})}{\partial \mathbf{p}} &= \frac{1}{2} \frac{\partial}{\partial \mathbf{p}} (\mathbf{p}^T A^T A\mathbf{p} - 2\mathbf{y}^{\exp\,T} A\mathbf{p} + \mathbf{y}^{\exp\,T} \mathbf{y}^{\exp}) \\
&= \frac{1}{2} \frac{\partial}{\partial \mathbf{p}} (\mathbf{p}^T A^T A\mathbf{p}) - \frac{\partial}{\partial \mathbf{p}} (\mathbf{y}^{\exp\,T} A\mathbf{p}) = A^T A\mathbf{p} - A^T \mathbf{y}^{\exp}
\end{aligned}
$$

where we have used the identities:

$$\frac{\partial (\mathbf{z}^T Q\mathbf{z})}{\partial \mathbf{z}} = 2Q\mathbf{z}, \quad \frac{\partial (\mathbf{c}^T \mathbf{z})}{\partial \mathbf{z}} = \mathbf{c}$$

or, by setting the gradient equal to the zero:

$$[A^T A] \cdot \mathbf{p} = A^T \mathbf{y}^{\exp} \tag{2.86}$$

The last equation is a linear system of equations whose solution provides an estimate for $\mathbf{p}$.

Learning Summary

The conditions for a local minimum in the case of differentiable functions of many variables:

$$\min_{\boldsymbol{x} \in R^n} f(\boldsymbol{x})$$

are:

$$\nabla f(\boldsymbol{x}_0) = \mathbf{0}$$

$$(\boldsymbol{x} - \boldsymbol{x}_0)^T \boldsymbol{H}(\boldsymbol{x}_0)(\boldsymbol{x} - \boldsymbol{x}_0) > 0 \text{ or } \boldsymbol{H}(\boldsymbol{x}_0) \text{ is p.d.}$$

If we are at a point $\mathbf{x}_0$ in R^n then:

- The direction of $+\nabla f(\mathbf{x}_0)$ is the direction of fastest increase in $f(\mathbf{x})$ at $\mathbf{x}_0$.
- The direction of $-\nabla f(\mathbf{x}_0)$ is the direction of fastest decrease in $f(\mathbf{x})$ at $\mathbf{x}_0$.
- $+\nabla f(\mathbf{x}_0)$ and $-\nabla f(\mathbf{x}_0)$ are perpendicular to the tangent of $f(\mathbf{x})$ at $\mathbf{x}_0$.

All $\mathbf{x}$ satisfying the inequality:

$$\nabla^T f(\boldsymbol{x}_0)(\boldsymbol{x} - \boldsymbol{x}_0) > 0$$

belong to the half space of directions of increasing objective function.

All $\mathbf{x}$ satisfying the inequality:

$$\nabla^T f(\boldsymbol{x}_0)(\boldsymbol{x} - \boldsymbol{x}_0) < 0$$

belong to the half space of directions of decreasing objective function.

The generalization of the Newton's method in multidimensional functions is the following:

Step 0: Select an initial estimate of the solution $\mathbf{x}^0 \in \mathbb{R}^n$, at iteration $k = 0$.
Step 1: Calculate $f(\mathbf{x}^k)$, $\nabla(\mathbf{x}^k)$ and $\mathbf{H}(\mathbf{x}^k)$.
Step 2: Determine the Newton step $\delta\mathbf{x}^k$ by solving (2.40) to obtain $\mathbf{x}^{k+1} = \mathbf{x}^k + \delta\mathbf{x}^k$.
Step 3: If $||\nabla f(\mathbf{x}^k)|| < \varepsilon$, then STOP; else set $k = k + 1$, and return to Step 1.

In Step 2 of the Newton algorithm, we do not accept immediately the full Newton's step, but we determine the minimum n, starting from $n = 0$, that satisfies the criterion:

$$f\left(\mathbf{x}^k + \beta^n \delta\mathbf{x}^k\right) < f\left(\mathbf{x}^k\right) + \alpha\beta^n \nabla f^T\left(\mathbf{x}^k\right)\delta\mathbf{x}^k$$

When it is satisfied then, as it is proven in the literature, the decrease in the objective function is satisfactory to achieve convergence to a stationary point.

The most popular parameter estimation method, known as the nonlinear least squares method (NLLS), is based on minimizing the sum of the squares of the residuals between a proposed model $y = \varphi(x,\mathbf{p})$ and experimental data y^{exp}:

$$\min_{p} f(p) = \frac{1}{2}\sum_{i=1}^{n} r_i^2(x_i, p) = \frac{1}{2}\sum_{i=1}^{n}(\varphi(x_i, p) - y_i^{exp})^2$$

The NLLS parameter estimation is the most important (if not the only) application of unconstrained optimization.

Terms and Concepts

You must be able to discuss the concepts of:

Concave up or convex function
Hessian matrix
Gradient vector and its properties
Least squares parameter estimation
Newton method when applied to functions of many variables (describe the steps)
Positive definite hessian matrix
Stationary point of functions of many variables
Taylor approximation of functions of many variables
Unconstrained optimization and the optimality conditions for functions of many variable

Problems

2.1 Polarographic probes are used routinely to perform measurement of diluted oxygen in bioreactors. The experimental data obtained when performing a step test to a small bioreactor are presented in Table P2.1. Nitrogen is first fed to the bioreactor to remove all oxygen. Then air is fed, and the probe recordings are obtained. The mathematical analysis of the problem involves solving some complex partial differential equations that describe oxygen diffusion in the membrane of the polarographic probe. The result is summarized in the following equation:

$$y(t) = 1 + 2\sum_{n=1}^{\infty}(-1)^n \exp\left(-n^2\pi^2\frac{t}{\tau}\right)$$

Propose a strategy and a mathematical formulation to estimate the value of the model parameter τ using only one term approximation ($n = 1$ only) and using possibly some graphical means. After obtaining an initial estimate for τ, plot the sum of squares of the residuals for values of τ around your initial estimation. Finally, use nonlinear least squares to estimate τ.

Table P2.1 Experimental data from a step experiment in a bioreactor

Time (s)	$i(t)/i_\infty$
0	0
5	0.0217
10	0.1797
15	0.3523
20	0.4987
25	0.6247
30	0.7073
35	0.7677
40	0.8137
45	0.8523
50	0.8813
60	0.9220
70	0.9460
80	0.9623
90	0.9730
100	0.9803
150	0.9967
200	0.9993

Fig. P2.2 A simple holding tank

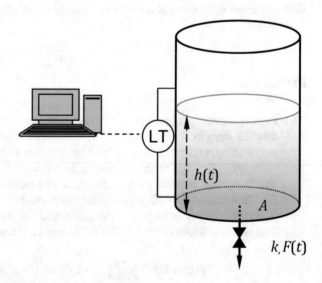

2.2 The tank shown in Fig. P2.2 is initially filled with water up to the level of $h = 68$ cm. The valve is opened at $t = 0$, and the level of the liquid in the tank is measured through a computer system every 10 s. The equation for the volumetric flow rate of the liquid is the following:

Table P2.2 Experimental data from the liquid storage tank shown in Fig. P2.2

	t (s)	h (cm)
1	0	63.8
2	10	58.1
3	20	48.6
4	30	42.5
5	40	36.6
6	50	32.3
7	60	25.7
8	70	21.9
9	80	14.9
10	90	12.9
11	100	9.1

$$\dot{Q}(t) = -k\sqrt{h(t)}$$

The cross-sectional area of the tank is $A = 154$ cm^2. Propose a method to determine the constant k in the model of the volumetric flow rate of the outgoing liquid from the tan k. Determine the constant k, and show graphically the adequacy of your estimation (using the data in Table P2.2).

2.3 Hill and Root (*Introduction to Chemical Engineering Kinetics and Reactor Design*, 2nd ed., Wiley, 2014) report experimental data on the bromination of meta-xylene at 17 °C. Small quantities of bromine are added into a xylene solution, and the rate of disappearance of bromine was measured (see Table P2.3). The rate expression is of the form:

$$\frac{dC}{dt} = -kC^n$$

where C is the bromine concentration and k and n are constants in the rate model. Prove that the solution of the rate model above is:

Table P2.3 Experimental data for meta-xylene bromination

i	Time (min)	$c_{Br2}(t)$ (mol/L)	i	Time (min)	$c_{Br2}(t)$ (mol/L)
1	0.00	0.3335	11	19.60	0.1429
2	2.25	0.2965	12	27.00	0.1160
3	4.50	0.2660	13	30.00	0.1053
4	6.33	0.2450	14	38.00	0.0830
5	8.00	0.2255	15	41.00	0.0767
6	10.25	0.2050	16	45.00	0.0705
7	12.00	0.1910	17	47.00	0.0678
8	13.50	0.1794	18	57.00	0.0553
9	15.60	0.1632	19	63.00	0.0482
10	17.85	0.1500			

$$\frac{1}{C^{n-1}(t)} - \frac{1}{C_0^{n-1}} = (n-1)kt$$

Use the solution just derived to propose a methodology for estimating the constants of the model using the data in Table P2.3. Plot the residuals around the values $\nu = 1.5$ and $k = 0.1$ (L/mol)$^{\nu-1}$/min. Use the least squares solver in MATLAB to determine the optimum values of the parameters.

2.4 An important problem in process control is model reduction of process models. A high-order model is usually obtained by linearizing a relative complex model to obtain a transfer function of the following form:

$$G(s) = \frac{b_0 + b_1 s + b_2 s^2 + \ldots + b_{n-1} s^{n-1}}{a_0 + a_1 s + a_2 s^2 + \ldots + a_n s^n} e^{-ds}$$

When our aim is to synthesize a low-order, fixed structure controller such as a proportional-integral (PI) controller, a simpler (reduced) model is usually more appropriate:

$$g(s) = \frac{\beta_0 + \beta_1 s + \beta_2 s^2 + \ldots + \beta_{m-1} s^{m-1}}{\alpha_0 + \alpha_1 s + \alpha_2 s^2 + \ldots + \alpha_m s^m} e^{-\theta s}, \quad m < < n$$

The reduced model can be obtained by several methods, and arguably, the simplest possible is by "matching" the step response or the frequency response of the two models. If, for instance, our aim is to match the frequency response of the two models, then we normally have available pairs of amplitude ratio AR and phase angle φ of the process at several frequencies:

$$\{\omega_i, AR_i, \varphi_i\}$$

The parameters of the reduced model $\mathbf{p} = [\beta_0 \; \beta_1 \ldots \beta_m \; \alpha_0 \; \alpha_1 \ldots \alpha_m \; \theta]^T$ are determined to minimize the sum of the squared deviations between the amplitude ratio and the phase angle of the process and the reduced model:

$$WSSE(\mathbf{p}) = \sum_i \left(AR_{G,i} - AR_{g,i}(\mathbf{p}) \right)^2 + \gamma \sum_i \left(\varphi_{G,i} - \varphi_{g,i}(\mathbf{p}) \right)^2$$

As an example, consider the process:

$$G(s) = \frac{1}{s^5 + 5s^4 + 10s^3 + 10s^2 + 5s + 1}$$

Using the following commands in MATLAB:

```
G=tf(1,[1 5 10 10 10 5 1]);
w=logspace(-2,0.5,10);
[AR,phi]=bode(G,w);
AR_G=squeeze(AR)
phi_G=squeeze(phi)/(180/pi)
```

we generate the following "experimental data":

ω_i (rad/s)	$AR_{G,i}$	$\varphi_{G,i}$ (rad)
0.0100	0.9998	−0.0500
0.0190	0.9991	−0.0948
0.0359	0.9968	−0.1796
0.0681	0.9885	−0.3401
0.1292	0.9595	−0.6422
0.2448	0.8645	−1.2006
0.4642	0.6140	−2.1728
0.8799	0.2385	−3.8081
1.6681	0.0359	−5.1538
3.1623	0.0025	−6.3226

Select the first order plus delay time model ($\mathbf{p} = [k \; \tau \; \theta]^T$), and determine its parameters:

$$g(s) = \frac{k}{\tau s + 1} e^{-\theta s}$$

where:

$$AR_g(\omega) = \frac{k}{\sqrt{(\omega \tau)^2 + 1}}$$

$$\varphi_g(\omega) = \arctan(-\omega \tau) - \omega \theta$$

2.5 Table P2.5 contains data for the saturation pressure P^s of water as a function of the absolute temperature T. Estimate the parameters of the following forms of the Antoine equation:

$$\ln P^s = A - \frac{B}{T}$$

$$\ln P^s = A - \frac{B}{T + C}$$

Table P2.5 Data for the saturation pressure of water	i	T [K]	P^s [mmHg]
	1	283.15	9.16
	2	293.15	17.47
	3	303.15	31.74
	4	313.15	55.19
	5	323.15	92.30
	6	333.15	149.04
	7	343.15	233.17
	8	353.15	354.53
	9	363.15	525.27
	10	373.15	760.00

2.6 Integral reactor data on the pyrolytic dehydrogenation of benzene (Ben: C_6H_6) to diphenyl (DPH: $C_{12}H_{10}$) and triphecyl (TPH: $C_{18}H_{14}$):

$$2\text{Ben} \Leftrightarrow \text{DPH} + H_2$$

$$\text{Ben} + \text{DPH} \Leftrightarrow \text{TPH} + H_2$$

are reported by Seinfield and Gavalas (*AIChE J.*, 16, p. 644, 1970). The authors also present a mathematical model consisting of two ordinary differential equations that involve two unknown parameters. Study the paper, and use the data to estimate the model parameters. Compare your results with the results presented in the paper. What are your conclusions?

2.7 In Fig. P2.7 a closed loop system is shown. A process with first order dynamics:

$$\tau_p \frac{dy(t)}{dt} + y(t) = u(t) + d(t)$$

is to be controlled with a PI-Controller with the following dynamics:

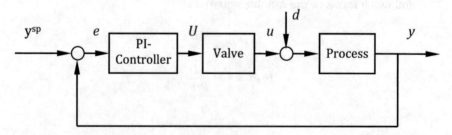

Fig. P2.7 A closed loop system with PI control

$$e(t) = y^{\text{sp}} - y(t)$$

$$\frac{dI(t)}{dt} = e(t)$$

$$U(t) = k_c \left(e(t) + \frac{I(t)}{\tau_I} \right)$$

where:

    y is the process output.

    y^{sp} is the set point (the desired value of the output).

    e is the error (deviation) signal.

    U is the controller output.

    d is a disturbance to the system.

    u is the valve input to the system.

    I is the integral of the error.

    τ_p is the time constant of the process ($\tau_p = 1$ s).

    k_c is the controller gain.

    τ_I is the integral time of the controller.

The controller output U drives a valve which also exhibits first-order system dynamics:

$$\tau_v \frac{du(t)}{dt} + u(t) = U(t)$$

where $\tau_v = 0.1$ s is the valve time constant.

You have been asked to design the PI-controller, i.e., select the controller gain and the controller integral time, for $y^{\text{sp}} = 0$ and $d = 1$ (unit step change in the load), so as to minimize the following performance index:

$$\min_{k_c, \tau_I} J = \int_0^{100} \left(|e(t)| + 100|1 + U(t)| \right) dt$$

Hint: build an m-file that solves the differential equations and also transforms the objective function to a differential equation for a given pair of k_c and τ_I. Create a contour plot for $k_c \in [5,12]$ and $\tau_I \in [0.5,1,5]$.

Chapter 3
Constrained Optimization

3.1 Introduction to Constrained Optimization

In the previous chapter, we discussed unconstrained optimization problems with many optimization variables:

$$\min_{x \in R^n} f(\mathbf{x}) \tag{3.1}$$

In this chapter, we will be studying problems with equality and inequality constraints which, when present, result in the most general form of nonlinear programming problems (NLP) with continuous variables:

$$\min_{\mathbf{x}} f(\mathbf{x})$$
$$\text{s.t.} \tag{3.2}$$
$$\mathbf{h}(\mathbf{x}) = \mathbf{0}$$
$$\mathbf{g}(\mathbf{x}) \leq \mathbf{0}$$
$$\mathbf{x}_L \leq \mathbf{x} \leq \mathbf{x}_U$$

where $\mathbf{x}$ is a n-th dimensional vector, $\mathbf{h}$ is vector valued function ($\mathbf{h} \colon R^n \to R^m$), $\mathbf{g}$ is also a vector valued function ($\mathbf{g} \colon R^n \to R^p$), and f is a real valued function $f \colon R^n \to R$. All vectors $\mathbf{x}$ that satisfy both the equality and inequality constraints define the *feasible space* of the problem. An important property of a feasible space is that of convexity. A great deal of space is devoted in almost all optimization books to define *convex sets* and convex functions. Generally speaking, a set S is convex if for any $\mathbf{x}_1 \in$ S and $\mathbf{x}_2 \in$ S, all points $\mathbf{x}$ in the line segment connecting $\mathbf{x}_1$ and $\mathbf{x}_2$ also belong to S, i.e., $\mathbf{x} \in$ S. However, this is a property that is never satisfied by nonlinear equality constraints. Therefore, convexity cannot be assumed for mainstream chemical engineering problems,

© The Author(s), under exclusive license to Springer Nature Switzerland AG 2022
I. K. Kookos, *Practical Chemical Process Optimization*, Springer Optimization and Its Applications 197, https://doi.org/10.1007/978-3-031-11298-0_3

and the extra effort and space needed to cover its mathematical details cannot be justified. We refer the interested reader into classical NLP theory books. As we have mentioned in Chap. 1, linear programming (LP) problems are indeed convex problems, and we will present the basic steps in solving LP problems in the next chapter without explicit reference to the general theory of convex sets and function (but we will present the basic elements necessary to establish whether a point is actually the solution to an LP problem or not).

3.2   Equality Constrained Problems

We start by examining the simplest case which is that of equality constrained problems:

$$\min_{\mathbf{x}} f(\mathbf{x})$$
$$\text{s.t.} \tag{3.3}$$
$$\mathbf{h}(\mathbf{x}) = 0$$

To derive the optimality conditions, we will define the Lagrangian function $\mathcal{L}$:

$$\mathcal{L}(\mathbf{x}, \lambda) = f(\mathbf{x}) + \lambda^T \mathbf{h}(\mathbf{x}) \tag{3.4}$$

where λ is a vector with as many elements as there are equality constrains (i.e., m). The elements of λ, known as Lagrange multipliers, are to be determined together with the elements of the vector $\mathbf{x}$. Note that $\mathcal{L}:R^{n+m} \to R$, i.e., the Lagrangian is a real valued function. We can also write (3.4) in the following form to facilitate understanding:

$$\mathcal{L}(\mathbf{x}, \lambda) = f(\mathbf{x}) + \sum_{i=1}^{m} \lambda_i h_i(\mathbf{x}) \tag{3.5}$$

Assume that we have managed to develop a methodology that determines the solution to the problem of minimizing the Lagrangian function which is achieved at point $\mathbf{x}_0$. Then, if $\mathbf{h}(\mathbf{x}_0) = \mathbf{0}$, then $\mathbf{x}_0$ is also the solution to problem (3.3) as the second term in the right-hand side of (3.5) is equal to zero. To develop such a methodology, we only need to observe that the problem of minimizing the Lagrangian function over $\mathbf{x}$ and λ is an unconstrained minimization problem with stationary points:

$$\nabla_{\mathbf{x}} \mathcal{L}(\mathbf{x}, \lambda) = \nabla f(\mathbf{x}) + \sum_{i=1}^{m} \lambda_i \nabla h_i(\mathbf{x}) = \mathbf{0} \tag{3.6}$$

$$\nabla_\lambda \mathcal{L}(\mathbf{x}, \lambda) = \mathbf{h}(\mathbf{x}) = \mathbf{0} \tag{3.7}$$

where $\nabla_\mathbf{x}$ denotes the gradient with respect to the vector $\mathbf{x}$ and ∇_λ denotes the gradient with respect to the vector λ. These results follow directly by noticing that:

$$\frac{\partial}{\partial x_j}\left(f(\mathbf{x}) + \sum_{i=1}^{m} \lambda_i h_i(\mathbf{x})\right) = \frac{\partial f(\mathbf{x})}{\partial x_j} + \sum_{i=1}^{m} \lambda_i \frac{\partial h_i(\mathbf{x})}{\partial x_j} \tag{3.8}$$

$$\frac{\partial}{\partial \lambda_k}\left(f(\mathbf{x}) + \sum_{i=1}^{m} \lambda_i h_i(\mathbf{x})\right) = \frac{\partial}{\partial \lambda_k}\left(\sum_{i=1}^{m} \lambda_i h_i(\mathbf{x})\right) = h_k(\mathbf{x}) \tag{3.9}$$

Equations (3.6) and (3.7) constitute a system of $n+m$ equations in $n+m$ unknowns (i.e., $\mathbf{x}$ and λ). We have, therefore, transformed the initial constrained problem to an unconstrained problem, which is then transformed (through the application of the optimality conditions) to a system of nonlinear equations. As satisfaction of the equality constraints (see Eq. (3.7)) is part of the final problem formulation, it follows that minimizing the Lagrangian is equivalent to minimizing the initial objective function.

In developing the algorithms for solving equality constrained problems, it will be useful to express Eq. (3.6) in matrix form. To achieve that, we rewrite Eq. (3.8) in the following form (we show in detail the intermediate steps as this can be confusing):

$$\frac{\partial}{\partial x_1}\left(f(x) + \sum_{i=1}^{m} \lambda_i h_i(\mathbf{x})\right) = \frac{\partial f(\mathbf{x})}{\partial x_1} + \lambda_1 \frac{\partial h_1(\mathbf{x})}{\partial x_1} + \lambda_2 \frac{\partial h_2(\mathbf{x})}{\partial x_1} + \ldots + \lambda_m \frac{\partial h_m(\mathbf{x})}{\partial x_1}$$

$$\frac{\partial}{\partial x_2}\left(f(x) + \sum_{i=1}^{m} \lambda_i h_i(\mathbf{x})\right) = \frac{\partial f(\mathbf{x})}{\partial x_2} + \lambda_1 \frac{\partial h_1(\mathbf{x})}{\partial x_2} + \lambda_2 \frac{\partial h_2(\mathbf{x})}{\partial x_2} + \ldots + \lambda_m \frac{\partial h_m(\mathbf{x})}{\partial x_2}$$

$$\vdots$$

$$\frac{\partial}{\partial x_n}\left(f(x) + \sum_{i=1}^{m} \lambda_i h_i(\mathbf{x})\right) = \frac{\partial f(\mathbf{x})}{\partial x_n} + \lambda_1 \frac{\partial h_1(\mathbf{x})}{\partial x_n} + \lambda_2 \frac{\partial h_2(\mathbf{x})}{\partial x_n} + \ldots + \lambda_m \frac{\partial h_m(\mathbf{x})}{\partial x_n}$$

or

$$\nabla_\mathbf{x}\mathcal{L}(\mathbf{x}, \lambda) = \begin{bmatrix} \dfrac{\partial f(\mathbf{x})}{\partial x_1} \\[6pt] \dfrac{\partial f(\mathbf{x})}{\partial x_2} \\[6pt] \vdots \\[6pt] \dfrac{\partial f(\mathbf{x})}{\partial x_n} \end{bmatrix} + \begin{bmatrix} \dfrac{\partial h_1(\mathbf{x})}{\partial x_1} & \dfrac{\partial h_2(\mathbf{x})}{\partial x_1} & \cdots & \dfrac{\partial h_m(\mathbf{x})}{\partial x_1} \\[6pt] \dfrac{\partial h_1(\mathbf{x})}{\partial x_2} & \dfrac{\partial h_2(\mathbf{x})}{\partial x_2} & \cdots & \dfrac{\partial h_m(\mathbf{x})}{\partial x_2} \\[6pt] \vdots & \vdots & \ddots & \vdots \\[6pt] \dfrac{\partial h_1(\mathbf{x})}{\partial x_n} & \dfrac{\partial h_2(\mathbf{x})}{\partial x_n} & \cdots & \dfrac{\partial h_m(\mathbf{x})}{\partial x_n} \end{bmatrix} \begin{bmatrix} \lambda_1 \\ \lambda_2 \\ \vdots \\ \lambda_m \end{bmatrix}$$

and finally:

$$\nabla_{\mathbf{x}}\mathcal{L}(\mathbf{x}, \lambda) = \nabla f(\mathbf{x}) + \nabla^T \mathbf{h}(\mathbf{x})\lambda \tag{3.10}$$

where the m-by-n Jacobian matrix of the equality constraints is defined by:

$$\nabla \mathbf{h}(\mathbf{x}) = \frac{\partial h_i(\mathbf{x})}{\partial x_j} \begin{bmatrix} \dfrac{\partial h_1(\mathbf{x})}{\partial x_1} & \dfrac{\partial h_1(\mathbf{x})}{\partial x_2} & \cdots & \dfrac{\partial h_1(\mathbf{x})}{\partial x_n} \\ \dfrac{\partial h_2(\mathbf{x})}{\partial x_1} & \dfrac{\partial h_2(\mathbf{x})}{\partial x_2} & \cdots & \dfrac{\partial h_2(\mathbf{x})}{\partial x_n} \\ \vdots & \vdots & \ddots & \vdots \\ \dfrac{\partial h_m(\mathbf{x})}{\partial x_1} & \dfrac{\partial h_m(\mathbf{x})}{\partial x_2} & \cdots & \dfrac{\partial h_m(\mathbf{x})}{\partial x_n} \end{bmatrix} \tag{3.11}$$

It is given as an exercise to the reader to use the following result:

$$\frac{\partial^2}{\partial x_\ell \partial x_j}\left(f(\mathbf{x}) + \sum_{i=1}^{m}\lambda_i h_i(\mathbf{x})\right) = \frac{\partial}{\partial x_\ell}\left(\frac{\partial f(\mathbf{x})}{\partial x_j} + \sum_{i=1}^{m}\lambda_i \frac{\partial h_i(\mathbf{x})}{\partial x_j}\right)$$
$$= \frac{\partial^2 f(\mathbf{x})}{\partial x_\ell \partial x_j} + \sum_{i=1}^{m}\lambda_i \frac{\partial^2 h_i(\mathbf{x})}{\partial x_\ell \partial x_j} \tag{3.12}$$

and to prove that the second derivative of the Lagrangian function is given by:

$$\nabla^2\mathcal{L} = \begin{bmatrix} \nabla^2_{\mathbf{xx}}\mathcal{L} & \nabla^2_{\mathbf{x\lambda}}\mathcal{L} \\ \nabla^2_{\lambda \mathbf{x}}\mathcal{L} & \nabla^2_{\lambda\lambda}\mathcal{L} \end{bmatrix} = \begin{bmatrix} \nabla^2 f(\mathbf{x}) + \sum_{i=1}^{m}\lambda_i \nabla^2 h_i(\mathbf{x}) & \nabla^T \mathbf{h}(\mathbf{x}) \\ \nabla \mathbf{h}(\mathbf{x}) & 0 \end{bmatrix} \tag{3.13}$$

Consider the following simple numerical example where we seek for the minimum distance from the origin in the x_1, x_2 space that satisfies a linear equality constraint:

$$\min_{x} x_1^2 + x_2^2$$
$$\text{s.t.} \tag{3.14}$$
$$x_1 + x_2 = 1$$

The graphical solution of the problem is shown in Fig. 3.1. The Lagrangian function is:

$$\mathcal{L}(\mathbf{x}, \lambda) = x_1^2 + x_2^2 + \lambda(x_1 + x_2 - 1) \tag{3.15}$$

Applying the optimality conditions (3.6) and (3.7) gives:

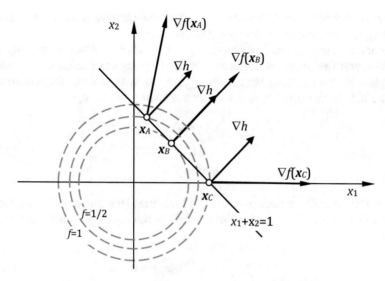

Fig. 3.1 Graphical solution to the equality constrained problem given by Eq. (3.14)

$$2x_1 + \lambda = 0 \tag{3.16}$$

$$2x_2 + \lambda = 0 \tag{3.17}$$

$$x_1 + x_2 - 1 = 0 \tag{3.18}$$

Equations (3.16, 3.17, and 3.18) consist a system of linear equations with solution $x_1 = x_2 = 1/2$, and $\lambda = -1$. The optimal point is shown as point $\mathbf{x}_B$ in Fig. 3.1. At this point, the gradients of the objective function and of the equality constraint are:

$$\nabla f = \begin{bmatrix} 2x_1 \\ 2x_2 \end{bmatrix} = \begin{bmatrix} 1 \\ 1 \end{bmatrix}, \quad \nabla h = \begin{bmatrix} 1 \\ 1 \end{bmatrix} \tag{3.19}$$

Apparently, the gradient of the objective function belongs to the same line as the gradient of the equality constraint. This only happens at point $\mathbf{x}_B$. There is no other feasible point (a point that satisfies the equality) where the two gradients belong to the same line. At point $\mathbf{x}_C$, for instance, the gradient of the objective function and the gradient of the equality constraint are:

$$\nabla f = \begin{bmatrix} 2x_1 \\ 2x_2 \end{bmatrix} = \begin{bmatrix} 2 \\ 0 \end{bmatrix}, \quad \nabla h = \begin{bmatrix} 1 \\ 1 \end{bmatrix} \tag{3.20}$$

and it is impossible to write the gradient of the objective function as a multiple of the gradient of the constraint (this is also clear from Fig. 3.1 as the two vectors do not belong to the same line). The same observation can be made for point $\mathbf{x}_A$ and any other feasible point apart from $\mathbf{x}_B$.

Before moving forward, there is a point to clarify: What is the physical meaning and the practical usefulness of the Lagrange multipliers? To answer this question, we consider the following problem with one equality constraint, but the RHS of the equality is equal to an arbitrary number β:

$$\min_{\mathbf{x}} f(\mathbf{x})$$
$$\text{s.t.} \tag{3.21}$$
$$h(\mathbf{x}) = \beta$$

If $\mathbf{x}_\beta$ is the optimum solution, then it follows that (the subscript β is used to denote that the result is valid when the RHS of the equality constraint is equal to β):

$$\nabla f\left(\mathbf{x}_\beta\right) + \lambda_\beta \nabla h\left(\mathbf{x}_\beta\right) = \mathbf{0} \tag{3.22}$$

We use the linear approximation of the objective function close to the solution:

$$f(\mathbf{x}) = f\left(\mathbf{x}_\beta\right) + \nabla^T f\left(\mathbf{x}_\beta\right)\left(\mathbf{x} - \mathbf{x}_\beta\right) \tag{3.23}$$

We now make use of the optimality conditions (3.22) to obtain:

$$f(\mathbf{x}) - f\left(\mathbf{x}_\beta\right) = -\lambda_\beta \nabla^T h\left(\mathbf{x}_\beta\right)\left(\mathbf{x} - \mathbf{x}_\beta\right) \tag{3.24}$$

We also perform linearization of the equality constraint:

$$h(\mathbf{x}) = h\left(\mathbf{x}_\beta\right) + \nabla^T h\left(\mathbf{x}_\beta\right)\left(\mathbf{x} - \mathbf{x}_\beta\right) \tag{3.25}$$

From (3.24) and (3.25) we finally obtain:

$$f(\mathbf{x}) - f\left(\mathbf{x}_\beta\right) = -\lambda_\beta\left(h(\mathbf{x}) - h\left(\mathbf{x}_\beta\right)\right) \tag{3.26}$$

Or, as both $f(\mathbf{x})$ and $h(\mathbf{x})$ are real valued functions and $h(\mathbf{x}) - h(\mathbf{x}_\beta) = (\beta + \Delta\beta) - \beta = \Delta\beta$:

$$\lambda_\beta = -\frac{\Delta f\left(\mathbf{x}_\beta\right)}{\Delta\beta} \tag{3.27}$$

We therefore conclude that the Lagrange multiplier is the ratio of the variation of the objective function at the optimum to the variation of the RHS of the equality constraint when written in the form (3.21). As the ratio can be both negative and positive (we have no reason to believe that in the general case, an increase in β will always result in an increase or always in a decrease in the objective function), then we conclude that the Lagrange multiplier of equality constrains can be negative, zero, or positive (any real number). To validate our findings, let's consider again Problem (3.14), which when $\beta = 1$ has the solution $x_1 = x_2 = 0.5$, $\lambda = -1$, and $f = 0.5$. If we increase β to $\beta = 1.1$, we obtain:

$$f(\mathbf{x}) \approx f(\mathbf{x}_\beta) - \lambda_\beta(h(\mathbf{x}) - h(\mathbf{x}_\beta)) = 0.5 - (-1)(1.1 - 1) = 0.6$$

Solving the exact problem, we obtain $x_1 = x_2 = 0.55$, $\lambda = -1.1$, and $f = 0.605$. We therefore conclude that (3.27) is an acceptable approximation to the sensitivity of the objective function to variations in the RHS of the equality constraints.

3.3   Application Examples

Example 3.1 *Minimum Reactor Volume for First-Order Reaction*
  We consider again Example 1.3 of Chap. 1 where two CSTR in series are studied, and the aim is to calculate the minimum total volume necessary to achieve a specific conversion. When a first-order reaction is assumed the optimization, problem obtained is a nonlinearly equality constrained optimization problem (see details presented in Example 1.3):

$$\min_{V_1, V_2, C} V_1 + V_2$$
$$\text{s.t.} \tag{3.28}$$
$$FC_0 - FC - V_1 kC = 0$$
$$FC_0 - FC_f - V_2 kC_f = 0$$

We define the Lagrangian function:

$$L = (V_1 + V_2) + \lambda_1(FC_0 - FC - V_1 kC) + \lambda_2(FC - FC_f - V_2 kC_f) \tag{3.29}$$

We calculate the stationary point of the Lagrangian by calculating the partial derivatives with respect to V_1, V_2, C, λ_1, and λ_2:

$$1 - \lambda_1 kC = 0 \tag{3.30a}$$

$$1 - \lambda_2 k C_f = 0 \tag{3.30b}$$

$$-\lambda_1 (F + V_1 k) + \lambda_2 F = 0 \tag{3.30c}$$

$$FC_0 - FC - V_1 k C = 0 \tag{3.30d}$$

$$FC - FC_f - V_2 k C_f = 0 \tag{3.30e}$$

We substitute (3.30a) and (3.30b) into (3.30c) and solve for V_1 to obtain:

$$V_1 = \frac{F}{k} \left(\frac{C}{C_f} - 1 \right) \tag{3.31}$$

We then solve (3.30d) for V_1:

$$V_1 = \frac{F}{k} \left(\frac{C_0}{C} - 1 \right) \tag{3.32}$$

From (3.31) and (3.32), we obtain:

$$C^2 = C_0 C_f \tag{3.33}$$

Finally, by substituting (3.33) into (3.30a) and (3.30b):

$$\lambda_1 = \frac{1}{k \sqrt{C_0 C_f}} \tag{3.34}$$

$$\lambda_2 = \frac{1}{k C_f} \tag{3.35}$$

From (3.30e) when solved for V_2, it follows that the two reactors have equal volumes. These results agree with the results obtained in Example 1.3 by variable elimination.

Example 3.2 *Optimal Operation of a Solid Oxide Fuel Cell*
In this example, we will consider again the fuel cell optimization studied in Example 1.5 of Chap. 1. As it is presented in detail in Chap. 1, the fuel cell model consists of the relationship between the voltage and the current density, and

the aim is to determine the current density at which the power density is maximum:

$$\min_{j,E} - E(j) \cdot j$$

$$\text{s.t.} \tag{3.36}$$

$$E - E_0 + R_\Omega j - \frac{RT}{2F} \ln\left(1 - \frac{j}{j_L}\right) = 0$$

where E_0 is the voltage at zero current ($E_0 = 0.85$ V), R_Ω is the (area specific) Ohmic resistance ($R_\Omega = 0.145$ $\Omega \cdot$cm^2), R is the ideal gas constant ($R = 8.3144$ V·C/(mol K)), T is the absolute temperature (873 K for the data shown in Fig. 1.14), and F is the Faraday's constant (96,485 C/mol). j_L is the limiting current density (3.85 V). We define the Lagrangian function:

$$\mathcal{L}(E, j, \lambda) = - Ej + \lambda\left[E - E_0 + R_\Omega j - \frac{RT}{2F} \ln\left(1 - \frac{j}{j_L}\right)\right] \tag{3.37}$$

We determine the partial derivative of the Lagrangian with respect to E, j, and λ:

$$-j + \lambda = 0 \tag{3.38a}$$

$$- E + \lambda R_\Omega + \lambda \frac{RT}{2F}\left(\frac{1}{j - j_L}\right) = 0 \tag{3.38b}$$

$$E - E_0 + R_\Omega j - \frac{RT}{2F} \ln\left(1 - \frac{j}{j_L}\right) = 0 \tag{3.38c}$$

This system of nonlinear equation cannot be solved analytically to determine the optimal point of operation, and we need to rely on a numerical solution.

Example 3.3 *Optimal Operation of Power Systems*
In this example, we will consider the controlled operation of power generation systems. All chemical plants operate power plants that consume fuels such as coal, diesel, natural gas, or LPG to produce electrical power and/or thermal power. These systems have typical input-output characteristics (fuel-consumption-power generation) that are similar to the one shown in Fig. 3.2.

They are described, in their most basic form, from a second-order polynomial function:

$$OC = a + \beta P + \gamma P^2 \tag{3.39}$$

where OC is the operating costs expressed in monetary units per kW or per operating hour and P is the power output in kW. The characteristic shown in Fig. 3.2 is idealized in that it is presented as a smooth, convex curve. Primary data obtained for actual power plant operation usually do not fall on a smooth curve. These units also have a minimum (P_{min}) and a maximum power output (P_{max}). Minimum load limitations are generally caused by fuel combustion stability and other inherent design constraints.

Consider the case where two power generation units are available with the following input-output characteristics (also shown in Fig. 3.2):

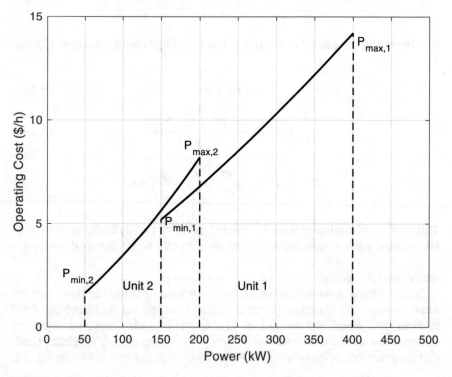

Fig. 3.2 Input-output characteristic curves for the two power plants considered

$$OC_1 \left[\frac{\$}{h}\right] = 1.0 + 0.025 \, P_1 + 0.000020 \, P_1^2 \tag{3.40}$$

$$OC_2 \left[\frac{\$}{h}\right] = 0.2 + 0.025 \, P_2 + 0.000\,075 \, P_2^2 \tag{3.41}$$

where P_1 and P_2 are the power outputs in kW. The aim is to determine P_1 and P_2 to satisfy the power demand constraint:

$$P_1 + P_2 = P_D \tag{3.42}$$

The overall problem is an equality constrained optimization (where we neglect at the moment the inequality constraint at the minimum and maximum power):

$$\min_{OC_1, OC_2, P_1, P_2} \quad OC_1 + OC_2$$
$$\text{s.t.} \tag{3.43}$$
$$OC_1 - \left(1.0 + 0.025 \; P_1 + 0.000\,020 \; P_1^2\right) = 0$$
$$OC_2 - \left(0.2 + 0.025 \; P_2 + 0.000\,075 \; P_2^2\right) = 0$$
$$P_1 + P_2 = P_D$$

To facilitate clarity of presentation is advantageous to substitute the input-output characteristic curves into the objective function to simplify the problem representation:

$$\min_{P_1, \; P_2} \quad \left(1.0 + 0.025 \; P_1 + 0.000020 \; P_1^2\right)$$
$$+ \left(0.2 + 0.025 \; P_2 + 0.000075 \; P_2^2\right)$$
$$\text{s.t.} \tag{3.44}$$
$$P_1 + P_2 = P_D$$

In the general case where n Power plants are available, the final formulation is:

$$\min_{P_i} \sum_{i=1}^{n} \left(\alpha_i + \beta_i P_i + \gamma_i P_i^2\right)$$
$$\text{s.t.} \tag{3.45}$$
$$\sum_{i=1}^{n} P_i = P_D$$

We can therefore define the following Lagrangian function:

$$\mathcal{L}(P_i, \lambda) = \sum_{i=1}^{n}\left(a_i + \beta_i P_i + \gamma_i P_i^2\right) + \lambda\left(\sum_{i=1}^{n}P_i - P_D\right) \tag{3.46}$$

The partial derivative of the Lagrangian with respect to P_i is:

$$\frac{d\mathcal{L}}{dP_i} = \beta_i + 2\gamma_i P_i + \lambda \tag{3.47}$$

These conditions together with the demand satisfaction constraint form a linear system of $n+1$ equation in $n+1$ unknowns:

$$\begin{bmatrix} 2\gamma_1 & 0 & \cdots & 0 & 1 \\ 0 & 2\gamma_2 & \cdots & 0 & 1 \\ \vdots & \vdots & \ddots & \vdots & \vdots \\ 0 & 0 & \cdots & 2\gamma_n & 1 \\ 1 & 1 & \cdots & 1 & 0 \end{bmatrix} \cdot \begin{bmatrix} P_1 \\ P_2 \\ \vdots \\ P_n \\ \lambda \end{bmatrix} = \begin{bmatrix} -\beta_1 \\ -\beta_2 \\ \vdots \\ -\beta_n \\ P_D \end{bmatrix} \tag{3.48}$$

For the case of two power production units, we may observe that Eq. (3.47) can be written as:

$$P_i + \frac{1}{2\gamma_i}\lambda = -\frac{\beta_i}{2\gamma_i} \tag{3.47}$$

and the system of linear equations becomes:

$$\begin{bmatrix} 1 & 0 & \dfrac{1}{2\gamma_1} \\ 0 & 1 & \dfrac{1}{2\gamma_2} \\ 1 & 1 & 0 \end{bmatrix} \begin{bmatrix} P_1 \\ P_2 \\ \lambda \end{bmatrix} = \begin{bmatrix} -\dfrac{\beta_1}{2\gamma_1} \\ -\dfrac{\beta_2}{2\gamma_2} \\ P_D \end{bmatrix} \tag{3.49}$$

We subtract rows 1 and 2 from row 3:

$$\begin{bmatrix} 1 & 0 & \dfrac{1}{2\gamma_1} \\ 0 & 1 & \dfrac{1}{2\gamma_2} \\ 0 & 0 & -\left(\dfrac{1}{2\gamma_1}+\dfrac{1}{2\gamma_2}\right) \end{bmatrix} \begin{bmatrix} P_1 \\ P_2 \\ \lambda \end{bmatrix} = \begin{bmatrix} -\dfrac{\beta_1}{2\gamma_1} \\ -\dfrac{\beta_2}{2\gamma_2} \\ P_D + \dfrac{\beta_1}{2\gamma_1}+\dfrac{\beta_2}{2\gamma_2} \end{bmatrix} \tag{3.50}$$

It then follows that for the general case, the solution can be obtained by the following equation (where we have assumed that $P_D = 500$ kW):

$$\lambda = -\frac{P_D + \sum_{i=1}^{n} \frac{\beta_i}{2\gamma_i}}{\sum_{i=1}^{n} \frac{1}{2\gamma_i}} \tag{3.51}$$

For the case study under investigation:

$$\lambda = -\frac{500 + \frac{0.025}{2 \cdot 0.000020} + \frac{0.025}{2 \cdot 0.000075}}{\frac{1}{2 \cdot 0.000020} + \frac{1}{2 \cdot 0.000075}} = -0.040789479$$

We then use (3.47) to obtain $P_1 = 394.7368$ kW and $P_2 = 105.2632$ kW.

Example 3.4 *Chemical Reaction Equilibrium*

In this example, we will consider one of the most impressive applications of equality constrained optimization in chemical engineering. The final composition of gas reacting mixtures can be predicted with relative accuracy when the reacting conditions and potential catalysts used facilitate equilibrium. At equilibrium, the Gibbs energy of the mixture attains its minimum, as it is explained in most chemical engineering thermodynamics textbooks. The equality constraints express the fact that atoms of each element are conserved as they can be combined and form different chemical species. The resulting mathematical problem can be used to predict the final composition of a reacting gas mixture with known initial composition. The idea is to search for the composition that minimizes the total Gibbs energy of the mixture. The impressive element in this approach is that the specific reactions or reaction networks that are present in the system need not to be known. Only the potential chemical species that can be detected in the final system need to be determined. The only additional information necessary is the thermophysical properties that can be calculated using tools from classical thermodynamics. This is an amazing and unique element of this methodology as the design of an industrial plant can easily be performed using the equilibrium assumption.

We begin our presentation by reminding that the Gibbs energy of an ideal gas mixture of NC components (each component denoted by i) is given by (the exact derivation can be found in the textbook by Elliott and Lira, *Introductory Chemical Engineering Thermodynamics*, Prentice Hall, NJ, 1999 or any other similar textbook):

$$G(T, n_i) = \sum_{i=1}^{NC} n_i \left[\left(\frac{\Delta G_{f,i}^0}{RT} \right) + \ln n_i - \ln \sum_{i=1}^{NC} n_i \right] \qquad (3.52)$$

where n_i are the mole of chemical species i, $\Delta G_{f,i}^0$ is the standard state Gibbs free energy of formation of chemical species i at the temperature of the reaction T, and R is the ideal gas constant. Let use a_{ik} to denote the number of atoms of element k in chemical species i. If we use ethane (C_2H_6) as an example, then $a_{C2H6,H} = 6$, and $a_{C2H6,C} = 2$. If the initial number of atoms of element k is denoted by a_k^0, then we can write the following material balance constraints:

$$\sum_{i=1}^{NC} n_i \, a_{i,k} = a_k^0, \quad k = 1, 2, \cdots, NE \qquad (3.53)$$

where NE denotes the number of elements present in the system. We can now define the following Lagrangian function:

$$\begin{aligned} L(n_i, \lambda_k) = &\sum_{i=1}^{NC} n_i \left[\left(\frac{\Delta G_{f,i}^0}{RT} \right) + \ln n_i - \ln \sum_{i=1}^{NC} n_i \right] \\ &+ \sum_{k=1}^{NE} \lambda_k \left(\sum_{i=1}^{NC} n_i \, a_{i,k} - a_k^0 \right) \end{aligned} \qquad (3.54)$$

This case study will be the subject of a detailed investigation latter in this chapter.

3.4   Inequality Constrained Problems

We now turn our attention to inequality constrained optimization problems of the following general form:

$$\min_{\mathbf{x}} f(\mathbf{x})$$
$$\text{s.t.} \qquad\qquad (3.55)$$
$$\mathbf{g}(\mathbf{x}) \leq \mathbf{0}$$

where $\mathbf{x}$ is a n-th dimensional vector and $\mathbf{g}$ is also a vector valued function ($\mathbf{g}\colon R^n \to R^p$) (as always f is a real valued function $f\colon R^n \to R$). All vectors $\mathbf{x}$ that satisfy all inequality constraints define the *feasible space* of the problem.

An important concept in the case of inequality constraints is that of an *active constraint*. An inequality constraint is active at point $\mathbf{x}$ if it is satisfied as equality, i.e., $g_i(\mathbf{x}) = 0$. It then follows that if we knew in advance which inequality constraints are active at the optimum, we could keep all active constraints and drop all inactive ones and solve the problem as an equality constrained problem. The only difference is that while $h(\mathbf{x}) = 0$ and $-h(\mathbf{x}) = 0$ are exactly the same equality constraint ($x - y = 0$ is exactly the same with $-(x - y)=0$ as both can be written as $x = y$), the same is not true for $g(\mathbf{x}) \leq 0$ and $-g(\mathbf{x}) \leq 0$ ($x - y \leq 0$ is different from $-(x - y) \leq 0$ as the first one implies that $x \leq y$ and the second one $y \leq x$). The result is that while the Lagrange multipliers of the equality constraint can be any real number, the Lagrange multiplier of an inequality constraint, denoted by μ, is always a non-negative number ($\mu \geq 0$).

In the case where the set of active constraints at the feasible point $\mathbf{x}$, denoted by $I = \{i:g_i(\mathbf{x}) = 0\}$, is known in advance, then the optimality conditions for problem (3.55) can be stated as (see also (3.6) and (3.7)):

$$\nabla f(\mathbf{x}) + \sum_{i\in I(\mathbf{x})}\mu_i\nabla g_i(\mathbf{x}) = \mathbf{0} \tag{3.56}$$

$$\left.\begin{array}{c} g_i(\mathbf{x}) = 0 \\ \mu_i \geq 0 \end{array}\right\}, \quad \forall i \in I(\mathbf{x}) \tag{3.57}$$

The Lagrange multipliers of inactive constraints are all equal to zero (i.e., either $g_i(\mathbf{x}) = 0$ or $\mu_i = 0$ always holds true). It takes a lot of careful thinking to realize that the following conditions are indeed equivalent to conditions (3.56) and (3.57) but require no prior knowledge of the active constraints at the feasible point $\mathbf{x}_0$:

$$\nabla f(\mathbf{x}) + \sum_{i=1}^{p}\mu_i\nabla g_i(\mathbf{x}) = \mathbf{0} \tag{3.58}$$

$$\mu_i g_i(\mathbf{x}) = 0, \quad \forall i \tag{3.59}$$

$$\mu_i \geq 0, \quad \forall i \tag{3.60}$$

The requirement that $\mathbf{x}$ is a feasible point is called the *primal feasibility* (PF) condition, whereas conditions (3.58) together with (3.60) are referred to as the *dual feasibility* (DF) conditions. The restriction (3.59) is referred to as the *complementary slackness* (CS) condition, and the three conditions together (PF, DF, and CS) are called the *Karush-Kuhn-Tucker* (or simply the KKT) conditions. Any point which satisfies the KKT conditions and the gradients of the active constraints are linearly independent (called also a *regular point*) is called a KKT point. Any KKT point is candidate solution to Problem (3.55).

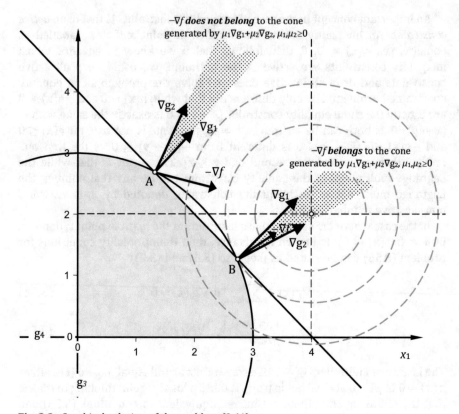

Fig. 3.3 Graphical solution of the problem (3.61)

Before moving further, consider a numerical example with a simple graphical solution (see Fig. 3.3):

$$\min_{x_1,x_2} f(x_1, x_2) = (x_1 - 4)^2 + (x_2 - 2)^2$$

$$\text{s.t.} \tag{3.61}$$

$$g_1(x_1, x_2) = x_1 + x_2 - 4 \leq 0$$
$$g_2(x_1, x_2) = x_1^2 + x_2^2 - 3^2 \leq 0$$
$$g_3(x_1, x_2) = -x_1 \leq 0$$
$$g_4(x_1, x_2) = -x_2 \leq 0$$

We consider the case where the information that the optimum lies at the intersection of inequality constraints g_1 and g_2 is given (i.e., points A and B in Fig. 3.3):

$$\begin{cases} x_1 + x_2 - 4 = 0 \\ x_1^2 + x_2^2 - 3^2 = 0 \end{cases} \Rightarrow \begin{cases} x_1 = 1.2929, \quad x_2 = 2.7071 \quad \text{point A} \\ x_1 = 2.7071, \quad x_2 = 1.2929 \quad \text{point B} \end{cases}$$

We first check whether point A is a KKT point. Point A is a feasible point (PF is satisfied). We calculate the gradients of the two constraints:

$$\nabla g_1 = \begin{bmatrix} 1 \\ 1 \end{bmatrix}, \quad \nabla g_2 = 2 \begin{bmatrix} x_1 \\ x_2 \end{bmatrix} = \begin{bmatrix} 2.5858 \\ 5.4142 \end{bmatrix}$$

We determine the gradient of the objective function: $\nabla f = 2 \begin{bmatrix} x_1 - 4 \\ x_2 - 2 \end{bmatrix} = \begin{bmatrix} -5.4142 \\ +1.4142 \end{bmatrix}$ and we apply (3.58):

$$\begin{bmatrix} -5.4142 \\ -0.5858 \end{bmatrix} + \begin{bmatrix} 1 & 2.5858 \\ 1 & 5.4142 \end{bmatrix} \begin{bmatrix} \mu_1 \\ \mu_2 \end{bmatrix} = \begin{bmatrix} 0 \\ 0 \end{bmatrix} \Rightarrow \begin{bmatrix} \mu_1 \\ \mu_2 \end{bmatrix} = \begin{bmatrix} 11.6569 \\ -2.4142 \end{bmatrix}$$

As $\mu_2 < 0$, the dual feasibility conditions are not satisfied, and point A is not a KKT point.

At point B, in Fig. 3.3, we have:

$$\nabla g_1 = \begin{bmatrix} 1 \\ 1 \end{bmatrix}, \quad \nabla g_2 = 2 \begin{bmatrix} x_1 \\ x_2 \end{bmatrix} = \begin{bmatrix} 5.4142 \\ 2.5858 \end{bmatrix}$$

We determine the gradient of the objective function:

$$\nabla f = 2 \begin{bmatrix} x_1 - 4 \\ x_2 - 2 \end{bmatrix} = \begin{bmatrix} -2.5858 \\ -1.4142 \end{bmatrix}$$

And:

$$\begin{bmatrix} -2.5858 \\ -1.4142 \end{bmatrix} + \begin{bmatrix} 1 & 5.4142 \\ 1 & 2.5858 \end{bmatrix} \begin{bmatrix} \mu_1 \\ \mu_2 \end{bmatrix} = \begin{bmatrix} 0 \\ 0 \end{bmatrix} \Rightarrow \begin{bmatrix} \mu_1 \\ \mu_2 \end{bmatrix} = \begin{bmatrix} 0.3431 \\ 0.4142 \end{bmatrix} > 0$$

We therefore conclude that PF, DF, CS, and independence of the gradients of the active constraints are satisfied and point B is a KKT point.

We will try to elaborate further the KKT conditions given by (3.58) and (3.60) (the DF conditions) for the optimization problem with inequality constraints (3.55). We will consider the nonlinear inequality constraints $g_1(\mathbf{x})$ and $g_2(\mathbf{x})$ shown in Fig. 3.4. Their gradients, when multiplied with non-negative numbers, generate a "convex cone" as shown in the same figure. In Fig. 3.4a, the case of a KKT point is shown as the negative of the gradient of the objective

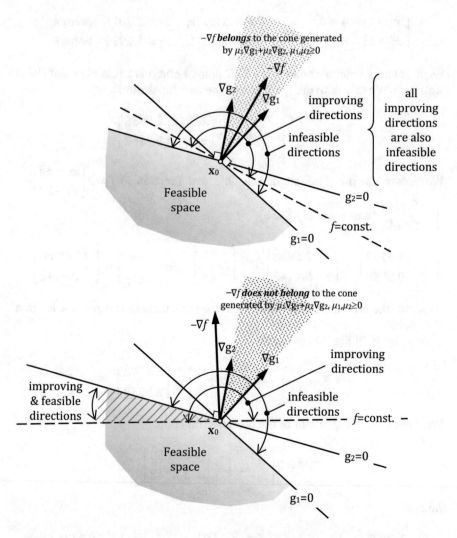

Fig. 3.4 The negative of the gradient of the objective function must belong to the "convex cone" generated by the gradients of the active constraints at any KKT point

function belongs to the "convex cone" spanned by $\nabla g_1(\mathbf{x}_0)$ and $\nabla g_2(\mathbf{x}_0)$. Note that all directions forming an angle between $\pm 90°$ with $-\nabla f(\mathbf{x}_0)$ for which:

$$f(\mathbf{x} - \mathbf{x}_0) - f(\mathbf{x}_0) \approx \nabla^T f(\mathbf{x}_0)(\mathbf{x} - \mathbf{x}_0) \leq 0 \qquad (3.62)$$

also point toward the infeasible region and are not acceptable directions of movement (even though they may result in decreasing values of the objective function). We conclude that when a feasible point $\mathbf{x}_0$ has been located for which there is no direction of movement that is also feasible and improving, then we have located a KKT point which is also a candidate solution for Problem (3.55).

However, if a feasible point $\mathbf{x}_0$ has been located for which $-\nabla f(\mathbf{x}_0)$ does not belong to the convex cone of $\nabla g_1(\mathbf{x}_0)$ and $\nabla g_2(\mathbf{x}_0)$ (as shown in Fig. 3.4b), then we can find feasible directions that are also improving directions, as shown in Fig. 3.4b. Conditions (3.58) and (3.60) cannot be satisfied simultaneously, and this point is not a KKT point or a candidate solution to Problem (3.55).

Problems with inequality constraints only are scarce in engineering as material and energy balance constraints as well as most constitutive laws are expressed in the form of equality constraints. Inequality constraints are in most cases linear and pose bounds on the variables appearing in the problem. There are also cases where nonlinear inequality constraints are also important. Several interesting recreational problems with significant technological applications can be expressed in the form of inequality constrained optimization problems. We will briefly present a couple of them before moving forward to state the optimality conditions for the general NLP problem in which both equality and inequality constraints are present.

The first one is the well-known circle packing problem. In a general setting, a circle packing is an optimized arrangement of n arbitrary sized circles inside a container, such as a rectangle or a circle, with the restriction that no two circles overlap. The quality of the packing is typically measured by the size of the container to pack a predefined number of circles. Circle packing of uniform-size circles has received a considerable amount of attention in the open literature. Its mathematical formulation is deceiving as it gives the impression that is a relatively easy problem to solve. The reality is different, and the problem is extremely difficult to solve even in the simplest cases of small number of cycles of uniform sizes.

A potential formulation of the circle packing problem is shown in Fig. 3.5. The aim is to minimize the radius of the enclosing circle R. The constraints are the non-overlapping, pairwise constraints which are equivalent to the requirement that the distance of the centers between any two circles is greater than two times their radius:

$$2r \le \sqrt{\left(x_i - x_j\right)^2 + \left(y_i - y_j\right)^2}, \quad \forall i,j; \quad i < j \tag{3.63}$$

The formulation is complete by adding the requirement that all circles are completely inside the enclosing circle with radius R:

$$\sqrt{x_i^2 + y_i^2} + r \le R, \quad \forall i \tag{3.64}$$

The complete formulation for the case of predefined number of circles with non-uniform radius r_i is given by:

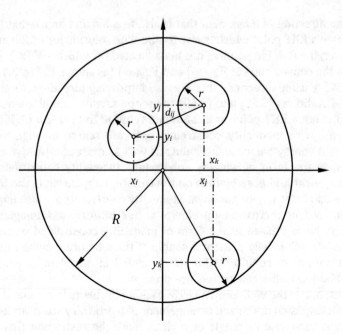

Fig. 3.5 The circle packing problem of uniform circles in a circle centered at the origin

$$\min_{R, x_i, y_i} \quad R$$

$$\text{s.t.} \tag{3.65}$$

$$r_i + r_j \le \sqrt{\left(x_i - x_j\right)^2 + \left(y_i - y_j\right)^2}, \quad \forall i, j; \quad i < j$$

$$\sqrt{x_i - y_i^2} + r_i \le R, \quad \forall i$$

Analytic result for the cases up to 19 circles of uniform size is presented by S. Kravitz (Mathematics Magazine, 40(2), 1967, pp. 65–71) which can be used to validate the results of any numerical study. A closely related problem is that of packing of spheres and the trim loss minimization problem which is shown graphically in Fig. 3.6. The trim loss problem, or the cutting stock problem in the specific case, is related to developing an optimal plan for "cutting" patterns for standard size and number of materials so as to satisfy demand and minimize waste material production or trim loss. The resulting mathematical formulation is again an inequality constrained optimization problem (the notation used is shown in Fig. 3.6):

Fig. 3.6 The cutting stock
problem

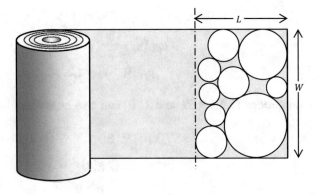

$$\min_{x_i, y_i, L} \quad L$$

$$\text{s.t.} \hspace{5cm} (3.66)$$

$$r_i + r_j \le \sqrt{\left(x_i - x_j\right)^2 + \left(y_i - y_j\right)^2}, \quad \forall i,j; \quad i < j$$

$$\left.\begin{array}{c} r_i \le x_i \le L - r_i \\ r_i \le y_i \le W - r_i \end{array}\right\} \forall i$$

$$\sum_i \pi r_i^2 \le WL$$

3.5   General Nonlinear Programming Problems

We now consider the more general form of constrained optimization problems
where both equality and inequality constraints are involved:

$$\min_{\mathbf{x}} f(\mathbf{x})$$

$$\text{s.t.} \hspace{5cm} (3.67)$$

$$\mathbf{h}(\mathbf{x}) = \mathbf{0}$$

$$\mathbf{g}(\mathbf{x}) \le \mathbf{0}$$

where $\mathbf{x}$ is a n-th dimensional vector, $f{:}R^n \to R$, $\mathbf{h}{:}R^n \to R^m$, and $\mathbf{g}{:}R^n \to R^p$ are
differentiable functions (well-behaved function in the general sense). Let $\mathbf{x}$ be a
feasible point, i.e., $\mathbf{h}(\mathbf{x}) = \mathbf{0}$ and $\mathbf{g}(\mathbf{x}) \le \mathbf{0}$, and the gradients of the equality
constraints and the active inequality constrains are linearly independent at $\mathbf{x}$. If
$\mathbf{x}$ solves Problem (3.67) locally, then there exist unique scalars $\lambda_i \in R$, i = 1,
2, ..., m and non-negative scalars μ_j, j = 1, 2, ..., p, such that:

$$\nabla f(\mathbf{x}) + \sum_{i=1}^{m} \lambda_i \nabla h_i(\mathbf{x}) + \sum_{j=1}^{p} \mu_j \nabla g_j(\mathbf{x}) = \mathbf{0} \qquad (3.68)$$

$$\mu_j g_j(\mathbf{x}) = 0, \quad \forall j = 1,2, \ \ldots, \ p \qquad (3.69)$$

$$\mu_j \geq 0, \quad \forall j = 1,2, \ \ldots, \ p \qquad (3.70)$$

Conditions (3.68, 3.69, and 3.70) can also be written in a more compact form:

$$\nabla f(\mathbf{x}) + \nabla^T \mathbf{h}(\mathbf{x})\boldsymbol{\lambda} + \nabla^T \mathbf{g}(\mathbf{x})\boldsymbol{\mu} = \mathbf{0} \qquad (3.68')$$

$$\boldsymbol{\mu}^T \mathbf{g}(\mathbf{x}) = 0 \qquad (3.69')$$

$$\boldsymbol{\mu} \geq \mathbf{0} \qquad (3.70')$$

where $\nabla \mathbf{h}(\mathbf{x})$ is the m-by-n Jacobian matrix of the equality constraints (defined in Eq (3.11)) and $\nabla \mathbf{g}(\mathbf{x})$ is the p-by-n Jacobian matrix of the inequality constraints. These conditions are referred to as the general necessary KKT conditions for Problem (3.67) and are, as you can realize, of paramount importance in optimization theory. It is important to note that a KKT point is only a candidate local solution to problem (3.67). A useful result is also the following: if point $\mathbf{x}_0$ solves Problem (3.67), then is also the solution to the following linear approximation problem:

$$\min_{\mathbf{x}} f(\mathbf{x}) + \nabla f(\mathbf{x}_0)(\mathbf{x} - \mathbf{x}_0)$$
$$\text{s.t.} \qquad\qquad (3.71)$$
$$\nabla^T \mathbf{h}(\mathbf{x}_0)(\mathbf{x} - \mathbf{x}_0) = \mathbf{0}$$
$$\mathbf{g}(\mathbf{x}_0) + \nabla^T \mathbf{g}(\mathbf{x}_0)(\mathbf{x} - \mathbf{x}_0) \leq \mathbf{0}$$

The KKT conditions are only necessary in the general case and are also sufficient when the problem has mild convexity properties. This is almost never the case in engineering problems and in order to check the sufficient conditions, we need to rely on the hessian of the Lagrangian function (which apparently needs to be p.d.):

$$\mathcal{L}(\mathbf{x}, \boldsymbol{\lambda}, \boldsymbol{\mu}) = f(\mathbf{x}) + \sum_{i=1}^{m} \lambda_i h_i(\mathbf{x}) + \sum_{j \in J(\mathbf{x})} \mu_j g_j(\mathbf{x}), \quad J(\mathbf{x}) = j : g_j(\mathbf{x}) = 0$$
$$(3.72)$$

i.e.:

$$\nabla^2 \mathcal{L}(\mathbf{x}, \boldsymbol{\lambda}, \boldsymbol{\mu}) = \nabla^2 f(\mathbf{x}) + \sum_{i=1}^{m} \lambda_i \nabla^2 h_i(\mathbf{x}) + \sum_{j \in J} \mu_j \nabla^2 g_i(\mathbf{x}) \qquad (3.73)$$

must satisfy $\mathbf{d}^T(\nabla^2\mathcal{L})\mathbf{d} > 0$. Clearly this is not a practical way to check "manually" the properties of a KKT point, but normally this is left to the software to validate and inform us on the results.

An algorithm that can be used to solve small-scale equality and inequality constrained optimization problems is based on the following steps:

STEP 0: Set $k = 1$, and assume that all inequality constraints are inactive at the optimum point or $J_A^k = \varnothing$, where J_A^k is the set of the active constraints at iteration k.

STEP 1: Solve the equality constrained optimization problem to determine the solution $(\mathbf{x}^k, \boldsymbol{\lambda}^k, \boldsymbol{\mu}^k)$:

$$\nabla f(\mathbf{x}) + \sum_i \lambda_i \nabla h_i(\mathbf{x}) + \sum_{j \in J^k} \mu_j \nabla g_j(\mathbf{x}) = \mathbf{0}$$

$$h_i(\mathbf{x}) = 0, i \in I$$

$$g_j(\mathbf{x}) = 0, j \in J_A^k$$

STEP 2: If $g_j(\mathbf{x}^k) \leq 0$, $\forall j$ and $\mu_j \geq 0, j \in J_A^k$, then a KKT point has been found, and the algorithm terminates (apparently $\mu_j = 0, j \notin J_A^k$). Otherwise, continue.

STEP 3: Remove the inequality constraint with the most negative Lagrange multiplier from the set of the active constraints, and add to J_A all violated inequality constraints. Set $k = k + 1$ and return to STEP 1.

To demonstrate the application of the algorithm, we consider again the optimal operation of the power plant studied in Example 3.3, but with the power demand increased to 550 kW (i.e., $P_D = 550$ kW). The mathematical formulation of the optimal operation of the two power plants including the constraints on their minimum and maximum capacity is the following:

$$\min_{P_1, P_2} \quad \left(1.0 + 0.025\, P_1 + 0.000020\, P_1^2\right)$$

$$+ \left(0.2 + 0.025\, P_2 + 0.000075\, P_2^2\right)$$

s.t.   (3.74)

$$h_1 = P_1 + P_2 - 550 = 0$$

$$g_1 = -P_1 + 150 \leq 0$$

$$g_2 = -P_2 + 50 \leq 0$$

$$g_3 = P_1 - 400 \leq 0$$

$$g_4 = P_2 - 200 \leq 0$$

By neglecting all inequality constraints $(J_A^1 = \varnothing)$, we can solve the equality constrained problem (as it was done in Example 3.3) to obtain $P_1 = 434.21$ kW and $P_2 = 115.79$ kW. The third inequality constraint is violated, and we update the set of active constraints $(J_A^2 = \{3\})$. We then solve the equality constrained problem:

$$\min_{P_1, P_2} \quad (1.0 + 0.025\,P_1 + 0.000020\,P_1^2)$$
$$+ (0.2 + 0.025\,P_2 + 0.000075\,P_2^2)$$
$$\text{s.t.} \qquad\qquad\qquad\qquad (3.75)$$
$$h_1 = P_1 + P_2 - 550 = 0$$
$$g_3 = P_1 - 400 = 0$$

which has the trivial solution $P_1 = 400$ kW, $P_2 = 150$ kW that satisfies all inequality constraints. To determine the corresponding Lagrange multiplier μ_3, we solve the following problem:

$$\lambda \nabla h + \mu_3 \nabla g_3 = -\nabla f$$

or:

$$\lambda \begin{bmatrix} 1 \\ 1 \end{bmatrix} + \mu_3 \begin{bmatrix} 1 \\ 0 \end{bmatrix} = - \begin{bmatrix} 0.025 + 0.000040\,P_1 \\ 0.025 + 0.000150\,P_2 \end{bmatrix}$$

It follows directly that $\lambda = -(0.025 + 0.00015\,P_2) = -0.0475, \mu_3 = 0.00605 > 0$.

3.6 Numerical Solution of Nonlinear Programming Problems

We will now present a short introduction to one of the most successful numerical methods for solving NLP problems known as successive (or sequential or recursive) quadratic programming or SQP. SQP methods employ Newton's method and take advantage of the KKT conditions for the original problem. In each iteration, the subproblem solved turns out to be the minimization of a quadratic approximation to the (Lagrangian) objective function optimized subject to the linear approximation of the constraints. To present the basic steps of this method, we will consider the equality constrained problem studied in Sect. 3.2. The conditions for optimality derived in Sect. 3.2 are the following (applied at point $\mathbf{x}^k$, where k is the iteration counter):

$$\nabla_{\mathbf{x}}L(\mathbf{x}^k, \lambda) = \nabla f(\mathbf{x}^k) + \sum_{i=1}^{m} \lambda_i \nabla h_i(\mathbf{x}^k) = \mathbf{0} \tag{3.76}$$

$$\nabla_{\lambda}L(\mathbf{x}^k, \lambda) = \mathbf{h}(\mathbf{x}^k) = \mathbf{0} \tag{3.77}$$

We have also seen that the hessian matrix is given by (see Eq. (3.13)):

$$\nabla^2 L = \begin{bmatrix} \nabla^2 f(\mathbf{x}) + \sum_{i=1}^{m} \lambda_i \nabla^2 h_i(\mathbf{x}) & \nabla^T \mathbf{h}(\mathbf{x}) \\ \nabla \mathbf{h}(\mathbf{x}) & \mathbf{0} \end{bmatrix} \tag{3.78}$$

It is reminded that the Newton's step is determined by multiplying the step by the hessian matrix and equating the result to minus the gradient of the function, i.e.:

$$\begin{bmatrix} \nabla^2 f(\mathbf{x}) + \sum_{i=1}^{m} \lambda_i \nabla^2 h_i(\mathbf{x}) & \nabla^T \mathbf{h}(\mathbf{x}) \\ \nabla \mathbf{h}(\mathbf{x}) & \mathbf{0} \end{bmatrix} \cdot \begin{bmatrix} \delta \mathbf{x}^k \\ \delta \lambda^k \end{bmatrix} = - \begin{bmatrix} \nabla f(\mathbf{x}^k) + \sum_{i=1}^{m} \lambda_i \nabla h_i(\mathbf{x}^k) \\ \mathbf{h}(\mathbf{x}^k) \end{bmatrix}$$

$$\tag{3.79}$$

We can analyze these equations to the following:

$$\left(\nabla^2 f(\mathbf{x}^k) + \sum_{i=1}^{m} \lambda_i^k \nabla^2 h_i(\mathbf{x}^k) \right) \delta \mathbf{x}^k + \nabla^T \mathbf{h}(\mathbf{x}^k) \delta \lambda^k = -\nabla f(\mathbf{x}^k) - \nabla^T \mathbf{h}(\mathbf{x}^k) \lambda^k$$

which can also be written as:

$$\left(\nabla^2 f(\mathbf{x}^k) + \sum_{i=1}^{m} \lambda_i^k \nabla^2 h_i(\mathbf{x}^k) \right) \delta \mathbf{x}^k + \nabla^T \mathbf{h}(\mathbf{x}^k) \lambda^{k+1} = -\nabla f(\mathbf{x}^k) \tag{3.80}$$

By defining the hessian of the Lagrangian with respect to $\mathbf{x}$:

$$\mathbf{H}_L(\mathbf{x}^k, \lambda^k) = \nabla^2 f(\mathbf{x}^k) + \sum_{i=1}^{m} \lambda_i^k \nabla^2 h_i(\mathbf{x}^k) \tag{3.81}$$

and substituting back into (3.79), we obtain:

$$\begin{bmatrix} \mathbf{H}_L(\mathbf{x}^k, \lambda^k) & \nabla^T \mathbf{h}(\mathbf{x}^k) \\ \nabla \mathbf{h}(\mathbf{x}^k) & \mathbf{0} \end{bmatrix} \cdot \begin{bmatrix} \delta \mathbf{x}^k \\ \lambda^{k+1} \end{bmatrix} = - \begin{bmatrix} \nabla f(\mathbf{x}^k) \\ \mathbf{h}(\mathbf{x}^k) \end{bmatrix} \tag{3.82}$$

Observe that (3.82) corresponds to applying the KKT conditions to the following quadratic programming problem:

$$\min_{\delta \mathbf{x}^k} f\left(\mathbf{x}^k\right) + \nabla^T f\left(\mathbf{x}^k\right)\delta \mathbf{x}^k + \frac{1}{2}\left(\delta \mathbf{x}^k\right)^T \mathbf{H}_{\mathcal{L}}\left(\mathbf{x}^k, \lambda^k\right)\delta \mathbf{x}^k$$

$$\text{s.t.} \tag{3.83}$$

$$\mathbf{h}\left(\mathbf{x}^k\right) + \nabla \mathbf{h}\left(\mathbf{x}^k\right)\delta \mathbf{x}^k = 0$$

A naïve SQP algorithm can be as follows (see also how Newton's method was applied to unconstrained optimization problems):

Step 0: Select an initial estimate of the solution $\mathbf{x}^0 \in R^n$ and λ^0 at iteration $k = 0$.

Step 1: Calculate $\nabla f(\mathbf{x}^k)$, $\mathbf{h}(\mathbf{x}^k)$, $\nabla \mathbf{h}(\mathbf{x}^k)$ and $\mathbf{H}_{\mathcal{L}}(\mathbf{x}^k, \lambda^k)$.

Step 2: Solve (3.82) to obtain $\mathbf{x}^{k+1} = \mathbf{x}^k + \delta \mathbf{x}^k$ and λ^{k+1}.

Step 3: If $||\delta \mathbf{x}^k|| < \varepsilon$ then STOP else set $k = k + 1$ and return to Step 1.

The algorithm of the SQP method is similar to the general "naïve" Newton's algorithm presented in Chap. 2 (Sect. 2.2). The only difference is the problem solved to determine the improving direction in Step 2. As you may expect, the algorithm is greatly improved by using a line search algorithm to satisfy several convergence criteria. In these one-dimensional search techniques, a modified Lagrangian function is employed, usually called merit function, which takes into account the need to maintain feasibility and at the same time improve the objective function. When the algorithm terminates, a KKT point has been discovered which is a candidate solution to (3.3). When both inequality and equality constraints are considered, the SQP algorithm is much more complicated, but the basic idea remains the same.

3.7   Application Examples

Example 3.5 *Optimal Operation of a Fuel Cell*

We now return to Example 3.2 (see also Example 1.5) and Formulation (3.36) that applies to the optimal operation of a fuel cell. We have concluded that Formulation (3.36) needs to be solved numerically. We will build our SQP implementation, but before doing so, we will follow the steps of the SQP algorithm. First, we define the objective function and the equality constraint, their gradient, Jacobian, and hessians and select an initial point:

```
>> clear
>> f =inline('-x(1)*x(2)') % the objective function
```

```
f = Inline function:
    f(x) = -x(1)*x(2)

>> gf=inline('[-x(2); -x(1)]') % gradient of the objective function

gf = Inline function:
     gf(x) = [-x(2); -x(1)]

>> Hf=inline('[0 -1;-1 0]') % hessian of the objective function

Hf = Inline function:
     Hf(x) = [0 -1;-1 0]

>> h =inline('x(1)-0.85+0.145*x(2)-0.0376*log(1-x(2)/3.85)')

h = Inline function:
    h(x) = x(1)-0.85+0.145*x(2)-0.0376*log(1-x(2)/3.85)

>> Jh=inline('[1 0.145+0.0376/(3.85-x(2))]') % Jacobian of h

Jh = Inline function:
     Jh(x) = [1 0.145+0.0376/(3.85-x(2))]

>> Hh=inline('[0 0;0 0.0376/(3.85-x(2))^2]') % Hessian of h

Hh = Inline function:
     Hh(x) = [0 0;0 0.0376/(3.85-x(2))^2]

>> x=[0.5;2.5]

x = 0.5000
    2.5000
```

We then evaluate all functions and determine the step toward an improved solution:

```
>> fun=feval(f,x)

fun =  -1.2500

>> gfun = feval(gf,x)

gfun =  -2.5000
        -0.5000
```

(continued)

```
>> Hfun = feval(Hf,x)

Hfun =    0   -1
         -1    0

>> ceq=feval(h,x)

ceq =    0.0519

>> Jceq=feval(Jh,x)

Jceq =    1.0000   0.1729

>> Hceq=feval(Hh,x)

Hceq =  0         0
        0   0.0206

>> lamda=x(2); % see Eq (3.38a)
>> s=-[Hfun+lamda*Hceq Jceq'; Jceq zeros(1,1)]\[gfun;ceq] %
(3.82)

s =  -0.0589
      0.0402
      2.5402

>> x=x+s(1:2)

x = 0.4411
    2.5402

>> lamda=s(3)
lamda = 2.5402
```

We now evaluate the Euclidean norm of the step taken:

```
>> norm(s(1:2))
ans =   0.0713
```

This is considered significant and we repeat the process. Two additional iterations are only needed to obtain a step with Euclidean norm of the order of 10^{-8}, and the algorithm terminates. The solution agrees with the one found in Chap. 1.

We take advantage of the experience gained by solving this example to build a general-purpose SQP solver for problems with one inequality constraint, which is given in Table 3.1. To solve the problem of the optimal operation of the fuel cell, we only need to define the initial point and the inline functions (as we have just done) and then call the mySQP m-file to obtain the following results:(the results are slightly different as the initial guess for the Lagrange multiplier is $\lambda^0 = 0$). This concludes our first case study, and before introducing the build-in SQP optimizer in MATLAB®, we will present another interesting example.

Example 3.6 *Optimal Operation of Power Systems*

We return now to the optimal design of power generation units studied in Example 3.3 and in Sect. 3.5. The complete mathematical formulation is given by (3.74). We will first introduce the NLP solver available in MATLAB and then use the solver to solve formulation (3.74). The general NLP optimizer in MATLAB® is fmincon:

xopt=**fmincon**(fun,x0,A,b,Aeq,beq,LB,UB,nonlcon)
finds a local minimum of a function of several variables under linear and nonlinear constraints. Starts at x0 and attempts to find a local minimizer xopt of the function fun subject to the linear inequality constraints A x≤b, linear equality constraints Aeq x=beq , lower bounds LB and upper bounds UB on x and nonlinear inequality constraints g and equality constraints h: [g,h]=nonlcon(x).

To demonstrate how fmincon can be used to solve Formulation (3.74), we note that the equality constraint is linear, and the same holds true for the inequality constraints. To solve the problem, we first define the objective function as an inline function:

```
>> clear all
>>
f=inline('(1+0.025*x(1)+0.000020*x(1)^2)+(0.2+0.025*x(2)+0.000075
*x(2)^2)')

f =
Inline function:
f(x) =
(1+0.025*x(1)+0.000020*x(1)^2)+(0.2+0.025*x(2)+0.000075*x(2)^2)
```

We then define the linear equality constraint $\left(\begin{bmatrix} 1 & 1 \end{bmatrix} \begin{bmatrix} P_1 \\ P_2 \end{bmatrix} = 550 \right)$:

Table 3.1 A basic implementation of SQP algorithm for equality constrained problems (1 equality constraint only)

```
function [x,fun,lambda] = mySQP(f,gf,Hf,h,Jh,Hh,x0)
% mySQP (only works for only 1 equality constraint)
% x   is an n-th dimensional vector            (n ny 1)
% f   is the objective function : R^n->R       (1 by 1)
% gf is the gradient of the objective function (n by 1)
% Hf is the hessian of the objective function  (n by n)
% h   is the equality constraint               (1 by 1)
% Jh is the Jacobian of the equality constraint (1 by n)
% Hh is the hessian of the equality constraint (n by n)
% x0 is the initial guess

n=length(x);            % number of elements of x vector
m=1;                    % only one equality constraint is allowed
k=0; lambda=0;          % initialize algorithm
tolx=1e-6;              % convergence tolerance in x-step
maxiter=100;            % maximum number of iterations allowed
fprintf('Iter        f(x)  ||Step_{x}||\n')

while ( k==0 | (norm(s(1:n))>tolx & k<maxiter) )
    fun = feval( f,x); % evaluate objective function
    gfun  = feval(gf,x); % evaluate  gradient  of  objective
function
    Hfun = feval(Hf,x); % evaluate Hessian of objective function
    ceq = feval( h,x); % evaluate equality constraint
    Jceq  =  feval(Jh,x);   %  evaluate  Jacobian  of  equality
constraint
    Hceq   =   feval(Hh,x);   %   evaluate   Hessian   of   equality
constraint
                        % calculate newton step :
    s=-[Hfun+lambda*Hceq Jceq'; Jceq zeros(m,m)]\[gfun;ceq];
    x=x+s(1:n);          % update solution
    lambda=s(n+1);       % update multiplier
    k=k+1;               % update counter
    fprintf('%4u %12.8f %12.8f\n',k,fun,norm(s(1:n)))
end
```

```
>> x=[0.5;2.5];
>> [xopt,funopt,lagramult]=mySQP(f,gf,Hf,h,Jh,Hh,x)
Iter          f(x)  ||Step_{x}||
   1   -1.25000000   0.07562817
   2   -1.12060968   0.00636423
   3   -1.12056140   0.00000747
   4   -1.12056030   0.00000000

xopt = 0.4412
       2.5399
funopt = -1.1206
lagramult = 2.5399
```

```
>> Aeq=[1 1];
>> beq=550;
```

and finally the linear inequality constraints

$$\left(\begin{bmatrix} -1 & 0 \\ 0 & -1 \\ 1 & 0 \\ 0 & 1 \end{bmatrix} \begin{bmatrix} P_1 \\ P_2 \end{bmatrix} \leq \begin{bmatrix} -150 \\ -50 \\ 400 \\ 200 \end{bmatrix} \right):$$

```
>> A=[-1 0;0 -1;1 0;0 1];
>> b=[-150; -50; 400; 200];
```

fmincon can now be used to solve the problem (with the unconstrained optimum as the initial point):

```
>> options=optimset('Display','Iter');
>> fmincon(f,[434.21;115.79],A,b,Aeq,beq,[],[],[],options)

                                   First-order      Norm of
 Iter F-count          f(x) Feasibility optimality      step
    0       3   1.972632e+01   3.421e+01  9.729e-08
    1       7   1.972632e+01   3.418e+01  3.893e-06  3.668e-02
    2      10   1.980822e+01   4.848e+00  4.404e-03  4.149e+01
    3      13   1.983755e+01   0.000e+00  5.133e-03  6.867e+00
    4      16   1.983919e+01   0.000e+00  3.085e-03  3.539e-01
    5      19   1.984873e+01   0.000e+00  1.147e-02  2.018e+00
    6      22   1.984267e+01   0.000e+00  4.966e-03  1.271e+00
    7      25   1.983755e+01   0.000e+00  5.118e-05  1.101e+00
    8      28   1.983750e+01   0.000e+00  2.032e-06  1.069e-02
    9      31   1.983750e+01   0.000e+00  2.003e-08  4.377e-04

Local minimum found that satisfies the constraints.

Optimization completed because the objective function is
non-decreasing in feasible directions, to within the value of the
optimality tolerance, and constraints are satisfied to within the
value of the constraint tolerance.

<stopping criteria details>

ans = 400.0000
      150.0000
```

An alternative way to solve the same problem is to consider the inequality constraints as bound constraints $\left(\begin{bmatrix} 150 \\ 50 \end{bmatrix} \leq \begin{bmatrix} P_1 \\ P_2 \end{bmatrix} \leq \begin{bmatrix} 400 \\ 200 \end{bmatrix} \right):$

```
>> clear all
>>
f=inline('(1+0.025*x(1)+0.000020*x(1)^2)+(0.2+0.025*x(2)
+0.000075 *x(2)^2)');
>> Aeq=[1 1];
>> beq=550;
>> lb=[150;50];
>> ub=[400;200];
>> options=optimset('Display','Iter');
>> fmincon(f,[434.21;115.79],[],[],Aeq,beq,lb,ub,[],options)

Your initial point x0 is not between bounds lb and ub; FMINCON
shifted x0 to strictly satisfy the bounds.

                                            First-order      Norm of
Iter F-count            f(x) Feasibility    optimality          step
   0        3   1.825973e+01   3.520e+01    4.709e-04
   1        6   1.834628e+01   3.313e+01    6.294e-04     1.468e+00
   2        9   1.852192e+01   2.903e+01    1.244e-03     4.095e+00
   3       12   1.983752e+01   0.000e+00    5.598e-03     2.903e+01
   4       15   1.983764e+01   0.000e+00    5.601e-03     2.532e-02
   5       18   1.984203e+01   0.000e+00    5.701e-03     9.456e-01
   6       21   1.984338e+01   1.137e-13    5.732e-03     2.874e-01
   7       24   1.984053e+01   0.000e+00    2.421e-03     6.087e-01
   8       27   1.983752e+01   0.000e+00    2.290e-04     6.508e-01
   9       30   1.983771e+01   0.000e+00    2.079e-04     4.189e-02
  10       33   1.983754e+01   0.000e+00    4.018e-05     3.642e-02
  11       36   1.983750e+01   0.000e+00    4.105e-07     8.651e-03

Local minimum found that satisfies the constraints.

ans = 399.9999
      150.0001
```

Yet another equivalent way to solve the problem (and arguably the more general) is to use an m-file to define a MATLAB function that calculates the inequality and equality constraints:

```
function [g,h]=cnstr(P)
g(1) = -P(1)+150;
g(2) = -P(2)+ 50;
g(3) =  P(1)-400;
g(4) =  P(2)-200;
h(1) =  P(1)+P(2)-550;
>> clear all
```

(continued)

```
>>
f=inline('(1+0.025*x(1)+0.000020*x(1)^2)+(0.2+0.025*x(2)
+0.000075 *x(2)^2)');
>> options=optimset('Display','Iter');
>> fmincon(f,[434.21;115.79],[],[],[],[],[],[],@cnstr,
options)
```

				First-order	Norm of
Iter	F-count	f(x)	Feasibility	optimality	step
0	3	1.972632e+01	3.421e+01	9.688e-08	
1	7	1.972632e+01	3.418e+01	3.895e-06	3.668e-02
2	10	1.980822e+01	4.848e+00	4.404e-03	4.149e+01
3	13	1.983755e+01	7.349e-08	5.133e-03	6.867e+00
4	16	1.983919e+01	5.157e-09	3.085e-03	3.539e-01
5	19	1.984873e+01	4.397e-08	1.147e-02	2.018e+00
6	22	1.984267e+01	1.336e-08	4.966e-03	1.271e+00
7	25	1.983755e+01	7.895e-09	5.118e-05	1.101e+00
8	28	1.983750e+01	1.137e-12	2.032e-06	1.068e-02
9	31	1.983750e+01	6.821e-13	2.003e-08	4.377e-04

```
Local minimum found that satisfies the constraints.

ans = 400.0000
      150.0000
```

In addition to the value of the vector xopt of optimization variables, fmincon returns the value of the objective function fopt, a value flag that describes the exit condition of fmincon, a structure output with information about the optimization process, a structure lambda with fields containing the Lagrange multipliers at the solution x, a vector g with the gradient of fun at the solution xopt, and a matrix H with the hessian of fun at the solution xopt:

```
>> [xopt,fopt,flag,output,lambda,g,H]=fmincon(f,
[434.21;115.79],[],[],[],[],[],[],@cnstr,options);
>> xopt

xopt = 400.0000
       150.0000

>> fopt

fopt = 19.8375

>> flag

flag = 1
```

(continued)

```
>> lambda.eqnonlin

ans =  -0.0475

>> lambda.ineqnonlin

ans =   0.0000
        0.0000
        0.0065
        0.0000

>> g

g =    0.0410
       0.0475

>> H

H =    0.5000   0.5000
       0.5000   0.5001
```

Note that the values of the Lagrange multipliers agree with the ones derived analytically in Sect. 3.5.

Example 3.7 *Optimal Fin Design for Heat Dissipation*

We will study the case of designing four fins located externally to a 1-m-long tube ($H = 1$ m) transferring a condensing fluid at $T_0 = 95$ °C, as shown in Fig. 3.7. Each fin can be up to $m = 1.25$ kg and is to be constructed from stainless steel with density $\rho_s = 7832$ kg/m^3 and thermal conductivity $k = 58.3$ W/(m K). The temperature of the surrounding air is $T_\infty = 15$ °C, and the heat transfer coefficient is $h = 100$ W/(m^2 K). Determine the length L and thickness t of the fins to maximize the heat dissipation. This is a classical engineering design problem in heat transfer. To solve the problem, we need to find how the heat dissipation rate depends on the geometric dimensions of the fin. From any undergraduate book on heat transfer, we can see that in order to find this dependence, we need to perform an energy balance in a differential volume and finally express the heat transfer problem through a linear second order ODE. Its solution gives the temperature profile and finally the heat dissipation rate:

Fig. 3.7 A tube equipped with four external straight steel fins with rectangular profile

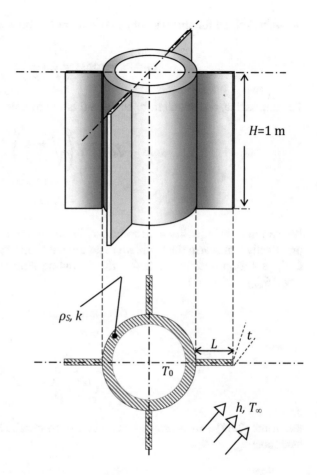

$$\dot{q}_{\text{fin}} = kHt(T_0 - T_\infty)m \tanh{(mL)}, \quad m = \sqrt{\frac{h}{k} \cdot \frac{2}{t}} \qquad (3.84)$$

which can be written as:

$$f(t, L) = \frac{\dot{q}_{\text{fin}}}{\sqrt{2k}H(T_0 - T_\infty)} = \sqrt{t} \tan h\left(\sqrt{\frac{2h}{k}}\frac{L}{\sqrt{t}}\right) \qquad (3.85)$$

As the fin needs to have a prespecified mass m, it follows that:

$$HtL\rho_s = m$$

or, when solved for the area of the profile of the fin A_p:

$$tL = A_p = \text{const} = \frac{1}{H}\left(\frac{m}{\rho_s}\right)$$

To summarize, the problem to be solved is the following:

$$\min_{t,L} \; -\sqrt{t}\,\tanh\left(\sqrt{\frac{2h}{k}}\,\frac{L}{\sqrt{t}}\right) \tag{3.86}$$

$$\text{s.t.}$$

$$tL - A_p = 0$$

We can use the equality constraint to eliminate one of the two variables and practically transform the problem to an unconstrained problem. By eliminating L, the solution obtained satisfies the following equation (Aziz, A., *Appl Mech Rev*, 45(5), 1992, p. 155):

$$6\xi = \sinh(2\xi) \tag{3.87}$$

where:

$$\xi = A_p\sqrt{\frac{2h}{k}}\cdot\frac{1}{t^{3/2}} \tag{3.88}$$

Equation (3.87) can be solved numerically to obtain $\xi = 1.4192$, and (3.88) becomes:

$$t = 0.9977\sqrt[3]{A_p^2\frac{h}{k}} \tag{3.89}$$

and:

$$L = 1.0023\sqrt[3]{\frac{k}{h}A_p} \tag{3.90}$$

If we treat the problem in its initial form, we need to define the Lagrangian function:

$$\mathcal{L}(t, L, \lambda) = -\sqrt{t}\,\tan h\left(\sqrt{\frac{2h}{k}}\,\frac{L}{\sqrt{t}}\right) + \lambda(tL - A_p) \tag{3.91}$$

The partial derivatives with respect to t, L, and λ are:

$$-\frac{1}{2\sqrt{t}} \tanh\left(\sqrt{\frac{2h}{k}}\frac{L}{\sqrt{t}}\right) + \operatorname{sech}^2\left(\sqrt{\frac{2h}{k}}\frac{L}{\sqrt{t}}\right)\left(\sqrt{\frac{2h}{k}}\frac{L}{2t}\right) + \lambda L = 0 \quad (3.92a)$$

$$-\operatorname{sech}^2\left(\sqrt{\frac{2h}{k}}\frac{L}{\sqrt{t}}\right)\sqrt{\frac{2h}{k}} + \lambda t = 0 \quad (3.92b)$$

$$tL - A_p = 0 \quad (3.92c)$$

We may use the last equality constraint to eliminate L (and we also use the definition of ξ to simplify the final expression):

$$-\tanh(\xi) + \xi \cdot \operatorname{sech}^2(\xi) + 2\lambda\frac{A_p}{\sqrt{2}} = 0 \quad (3.93a)$$

$$2\xi \cdot \operatorname{sech}^2(\xi) - 2\lambda\frac{A_p}{\sqrt{2}} = 0 \quad (3.93b)$$

We add the two equations to obtain:

$$3\xi \operatorname{sech}^2(\xi) = \tanh(\xi) \quad (3.94)$$

The last one can be transformed to (3.87) using the identities $\operatorname{sech}(\xi) = 1/\cosh(\xi)$ and $2\sinh(\xi)\cosh(\xi) = \sinh(2\xi)$. The result is the same irrespective of the strategy followed, as expected.

To use fmincon, we need to build two m-files, one for the objective function and one for the nonlinear constraints:

```
function f=ObjFIN(x)
f = -sqrt(x(1))*tanh(sqrt(2*100/58.3)*x(2)/sqrt(x(1)));
```

```
function [g,h]=ConstrFIN(x)
g = [];
h = x(1)*x(2)-(1.25/7832)/1;
```

We then use fmincon to solve the problem:

```
>> fmincon(@ObjFIN, [0.01;0.01],[],[],[],[],[0;0],[],
@ConstrFIN)

Local minimum found that satisfies the constraints.

ans =   0.0035
        0.0454
```

The solution obtained ($t = 0.0035$ m and $L = 0.0454$ m) agrees well with the analytic solution proposed by Aziz (see Eqs. (3.89) and (3.90)):

$$t = 0.9977 \sqrt[3]{\left(\frac{m/\rho_s}{H}\right)^2 \frac{h}{k}} = 0.9977 \sqrt[3]{\left(\frac{1.25/7832}{1}\right)^2 \frac{100}{58.3}} = 0.0035$$

$$L = 1.0023 \sqrt[3]{A_p \frac{k}{h}} = 1.0023 \sqrt[3]{\left(\frac{1.25/7832}{1}\right) \frac{58.3}{100}} = 0.0454$$

After having applied the SQP optimizer available in MATLAB once, we need to repeat the solution process, but this time, we will follow a more detailed approach. This approach facilitates better understanding, easier debugging, and easier generalization or modification of the problem at hand. The m-file that calculates the objective function is modified to incorporate extensive comments. It is a good practice to add comments that describe the code as comments allow others to understand your code and you can refresh your memory when you return to it later:

```
function f=ObjFIN(x)

% A. EXPLAIN THE MEANING OF EACH ELEMENT OF x
% x(1) is the thickness of the fin in m
% x(2) is the length of the fin in m

% B. ASSIGN VALUES TO ALL MODEL PARAMETERS
ht= 100; % heat transfer coefficient in W/(m2 K)
k = 58.3; % thermal conductivity
m = 1.25; % mass of the fin in kg
rhos=7832;% density of the steel in kg/m3
H = 1;   % height of the fin in m
Ap= (m/rhos)/H; % profile area in m2

% C. DEFINE OBJECTIVE FUNCTION
f = -sqrt(x(1))*tanh(sqrt(2*ht/k)*x(2)/sqrt(x(1)));
```

Note that in the proposed structure, we divide the m-file in three parts. The first part involves only comments that inform the user about the physical variables that correspond to the elements of the x vector. In the second part, all constants appearing in the model are defined, and numerical values are assigned to them. It is important, apart from the numerical values, to indicate carefully the units used in each constant. In the third part of the m-file, the objective function is defined.

The same idea is followed for the m-file that calculates the equality and inequality constraints:

```
function [g,h]=ConstrFIN(x)
% A. EXPLAIN THE MEANING OF EACH ELEMENT OF x
% x(1) is the thickness of the fin in m
% x(2) is the length of the fin in m

% B. ASSIGN VALUES TO ALL MODEL PARAMETERS
ht= 100; % heat transfer coefficient in W/(m2 K)
k = 58.3; % thermal conductivity
m = 1.25; % mass of the fin in kg
rhos=7832;% density of the steel in kg/m3
H = 1;   % height of the fin in m
Ap= (m/rhos)/H;

% C. DEFINE NONLINEAR INEQUALITY CONSTRAINTS
g = [];

% D. DEFINE NONLINEAR EQUALITY CONSTRAINTS
h = x(1)*x(2)-Ap;
```

Note that parts A and B are the same as in the m-file that defines the objective function. The inequality constraints are defined in part C and the nonlinear equality constraints in part D. We need now to build an m-file with all the commands necessary to use the two m-files presented above in fmincon to solve the problem:

```
clear
% A. EXPLAIN THE MEANING OF EACH ELEMENT OF x
% x(1) is the thickness of the fin in m
% x(2) is the length of the fin in m

% B. ASSIGN VALUES TO ALL MODEL PARAMETERS
ht= 100; % heat transfer coefficient in W/(m2 K)
k = 58.3; % thermal conductivity
m = 1.25; % mass of the fin in kg
rhos=7832;% density of the steel in kg/m3
H = 1;   % height of the fin in m
Ap= (m/rhos)/H;

x0=[ 0.01;
  0.01];
A =[];b =[];
Aeq=[];beq=[];
LB=[0;0]; UB=[1;1];
options=optimoptions
('fmincon','Algorithm','sqp','Display','iter');
```

(continued)

```
[xopt,fopt,flag,outinfo,lambda,grad,hessian]=...
  fmincon(@ObjFIN,x0,A,b,Aeq,beq,LB,UB,@ConstrFIN,options)

eig(hessian)
lambda.eqnonlin
xopt_analytic=[0.9977*(Ap^2*ht/k)^(1/3);1.0023*(Ap*k/ht)^
(1/3)]
```

The response obtained is the following:

```
>> runOptFIN

Iter Func-count       Fval Feasibility StepLength   Normof First-order
                                                     step  optimality
   0         3 -1.831276e-02  5.960e-05  1.000e+00  0.000e+00  1.790e+00
   1         7 -3.365186e-02  9.609e-05  7.000e-01  1.318e-02  7.804e-01
   2        12 -3.823316e-02  8.732e-05  4.900e-01  2.611e-02  3.624e+00
   3        15 -5.230511e-02  3.762e-06  1.000e+00  2.744e-03  4.533e+00
   4        18 -5.272058e-02  2.100e-08  1.000e+00  6.969e-04  1.028e-01
   5        24 -5.272283e-02  5.817e-09  3.430e-01  5.121e-04  2.554e-02
   6        27 -5.272348e-02  1.147e-10  1.000e+00  3.784e-05  1.873e-04
   7        30 -5.272350e-02  1.798e-14  1.000e+00  5.000e-07  3.851e-06
Local minimum found that satisfies the constraints.
Optimization completed because the objective function is
non-decreasing in feasible directions, to within the default value
of the optimality tolerance, and constraints are satisfied to within
the default value of the constraint tolerance.
xopt =  0.0035
        0.0454
fopt = -0.0527
flag =    1
...
grad = -5.0015
       -0.3869
hessian =
  691.3546 -103.0276
 -103.0276   15.5001
ans =
     0.1435
   706.7112
ans =
  110.1147
xopt_analytic =
   0.0035
   0.0454
```

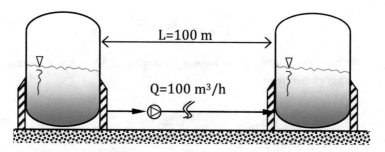

Fig. 3.8 Description of the liquid pumping system

Note that the solution obtained agrees well with the analytic solution and that the hessian has positive eigenvalues. The Lagrange multiplier of the nonlinear equality constraint is equal to 110.1147.

Example 3.8 *Optimal Design of a Liquid Pumping System*
In Fig. 3.8, the basic elements of a liquid transfer system are shown. Liquid with density $\rho = 1000$ kg/m^3 and viscosity $\mu = 0.001$ kg/(m s) is to be pumped between two storage tanks that are located $L = 100$ m apart, with flow rate $Q = 100$ m^3/h. The unit will operate for $t_y = 8322$ h/y, and the cost of electricity is $c_{el} = 0.1$ \$/kWh. A smooth pipe is to be used with diameter between 0.03 and 0.30 m with an installed cost that depends on the diameter of the pipe used:

$$C_{PIPE}[\$] = 1500 \, d^{1.5} L, L,d \text{ in } m \tag{3.95}$$

The equipment cost involves also the cost of the pump used which is given by:

$$C_{PUMP}[\$] = 750 \, w^{0.3}, w \text{ in } kW \tag{3.96}$$

w is the electrical power needed to drive the pump (in kW). To calculate the electrical power, we take into consideration the friction losses through the straight pipe:

$$\Delta p = f\left(\frac{L}{d}\right) \cdot \frac{1}{2}\rho u^2, \Delta p \text{ in Pa} \tag{3.97}$$

where u is the velocity of the liquid in the pipe and f is the Moody friction factor calculated (for smooth pipes) as a function of the Reynolds number by the equation (Re > 5000):

$$f = \frac{0.316}{\text{Re}^{0.25}}, \quad \text{Re} = \frac{ud\rho}{\mu} \tag{3.98}$$

Finally, if the pressure drop is known, the power to drive a pump with efficiency η_e is given by:

$$w = \frac{Q \cdot \Delta p}{\eta_e} \tag{3.99}$$

The aim is to select the pipe diameter so as to minimize the following approximate total annual cost (TAC) of the piping system:

$$\text{TAC} = \frac{1}{3}(C_{\text{PIPE}} + C_{\text{PUMP}}) + wt_y c_{\text{el}} \tag{3.100}$$

If we put everything together, we obtain the following optimization problem:

$$\min_{d,u,\text{Re},f,\Delta p,w} \frac{1}{3}\left(1500\ d^{1.5}L + 750w^{0.3}\right) + wt_y c_{el}$$

$$\text{s.t.} \tag{3.101}$$

$$\frac{\pi d^2}{4}u - Q = 0$$

$$\text{Re} - \frac{ud\rho}{\mu} = 0$$

$$f - \frac{0.316}{\text{Re}^{0.25}} = 0$$

$$\Delta p - f\frac{L}{d}\cdot\frac{1}{2}\rho u^2 = 0$$

$$w = \frac{Q \cdot \Delta p}{\eta_e}$$

The model is complete when we also consider the bound constraints:

$$0.03 \leq d \leq 0.3, \quad 0 \leq u, \quad 5000 \leq \text{Re}, \quad 10^{-6} \leq f \leq 10^{-1}, \quad 0 \leq \Delta p, \quad 0 \leq w$$

The problem at hand has six optimization variables and five equality constraints, and there is only one degree of freedom. The meaning is that if we assign value to one of the variables, then all remaining variables can be determined by solving the equations. To see that this is indeed the case for the problem under investigation, assume that we assign a value to the pipe diameter d, and then we can determine:

- The fluid velocity $u = Q/(\pi d^2/4)$
- The Reynolds number from its definition $Re = ud\rho/\mu$
- The friction factor from $f = 0.316/Re^{0.25}$
- The pressure drop from $\Delta p = f(L/d)0.5\,\rho u^2$
- The electrical power consumption $w = \Delta pQ/\eta_e$
- The TAC from Eq. (3.100)

The practical usefulness of this property is that we may simply use many values for the pipe diameter within the permissible range and compare the TAC values to locate the minimum. The result of these calculations is presented in Fig. 3.9, from which it is clear that there is a global minimum at approximately $d = 0.13$ m.

Another practical use is that we can use the same equations to derive tight upper and lower bound for all variables. To demonstrate that, let us assume that the pipe diameter obtains its lowest value $d = 0.03$ m. Then we may determine the fluid velocity $u = Q/(\pi d^2/4) = (100/3600)/(\pi 0.03^2/4) = 39.3$ m/s. This is the maximum velocity that can be obtained as it corresponds to minimum diameter. If we had used the maximum diameter

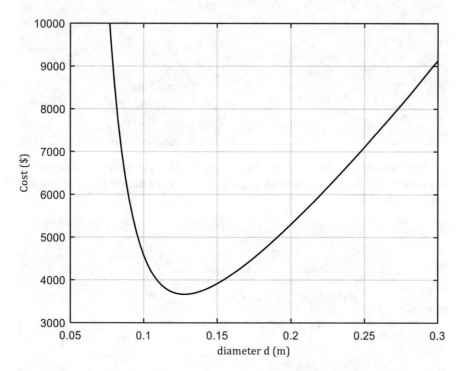

Fig. 3.9 Cost as a function of the pipe diameter for the fluid pumping case study

$d = 0.3$ m, then the velocity obtained is 0.393 m/s, and this is clearly the minimum velocity that can be obtained. As Reynolds number is inversely proportional to the diameter, the minimum diameter corresponds to the maximum Reynolds number, etc.

The following m-file is a possible implementation for the calculation of the objective function:

```
function f=pumping_obj(x)

% A. EXPLAIN THE MEANING OF EACH ELEMENT OF x
% x(1) is the diameter         d  in m
% x(2) is the velocity         u  in m/s
% x(3) is the Reynolds number Re    [-]
% x(4) is the friction factor f     [-]
% x(5) is the pressure drop    Dp in kPa
% x(6) is the power            w  in W

% B. ASSIGN VALUES TO ALL MODEL PARAMETERS
L    = 100;        % length of pipe in m
Q    = 100/3600;   % volumetric flowrate in m3/s
rho  = 1000;       % density in kg/m3
miu  = 0.001;      % viscosity in kg/(m s)
hta_e= 0.8;        % pump efficiency fractional
ty   = 8322;       % operating hours per year
c_el = 0.1;        % electricity cost in $/kWh

% C. DEFINE OBJECTIVE FUNCTION
f=(1/3)* ( 1500*x(1)^1.5*L + 750*(x(6)/1000)^0.3 ) +...
(x(6)/1000)*ty*c_el;
```

The following m-file is a possible implementation for the calculation of the inequality and equality constraints (parts A and B are omitted as they are the same as above):

```
function [g,h]=pumping_cnstr(x)

% A. EXPLAIN THE MEANING OF EACH ELEMENT OF x
...
% B. ASSIGN VALUES TO ALL MODEL PARAMETERS
...
% C. DEFINE NONLINEAR INEQUALITY CONSTRAINTS
g = [];

% D. DEFINE NONLINEAR EQUALITY CONSTRAINTS
h(1) = x(2)*(pi*x(1)^2/4) - Q;
h(2) = miu*x(3) - x(2)*x(1)*rho;
h(3) = x(4)*x(3)^(1/4) - 0.316;
h(4) = x(5)*x(1) - x(4)*L*0.5*rho*x(2)^2;
```

It is interesting to note the way that we have written the equality constraints trying always to avoid division by an optimization variable that may cause numerical problems. Also note that as Eq. (3.99) is a linear equality constraint, it does not appear in the m-file. The main program is as follows:

```
clear
% A. EXPLAIN THE MEANING OF EACH ELEMENT OF x
...
% B. ASSIGN VALUES TO ALL MODEL PARAMETERS
...

% C. ASSIGN INITIAL VALUES AND BOUNDS TO ALL MODEL PARAMETERS
LB(1)=0.03;
x0(1)=0.25;
UB(1)=0.30;

LB(2)=Q/(pi*UB(1)^2/4);
x0(2)=Q/(pi*x0(1)^2/4);
UB(2)=Q/(pi*LB(1)^2/4);

LB(3)=LB(1)*LB(2)*rho/miu;
x0(3)=x0(1)*x0(2)*rho/miu;
UB(3)=UB(1)*UB(2)*rho/miu;

LB(4)=0.316/UB(3)^(1/4);
x0(4)=0.316/x0(3)^(1/4);
UB(4)=0.316/LB(3)^(1/4);

LB(5)=0;
x0(5)=x0(4)*(L/x0(1))*0.5*rho*x0(2)^2 ;
UB(5)=50000;

LB(6)=LB(5)*Q/hta_e;
x0(6)=x0(5)*Q/hta_e;
UB(6)=UB(5)*Q/hta_e;

% D. LINEAR INEQUALITY CONSTRAINTS
A=[]; b=[];

% E. LINEAR EQUALITY CONSTRAINTS
Aeq=[0 0 0 0 -Q hta_e]; beq=0; % Eq (3.99)

options = optimoptions('fmincon','Display','iter');
fmincon(@pumping_obj,x0,A,b,Aeq,beq,LB,UB,@pumping_cnstr,
options);
```

```
>> format short g
>> run_pumping
                                          First-order      Norm of
   Iter F-count            f(x) Feasibility optimality        step
      0       7  7.105599e+03   3.553e-15   6.806e-02
      1      14  7.105594e+03   4.615e-10   6.806e-02   6.811e-02
      2      21  7.105571e+03   1.165e-08   6.806e-02   3.405e-01
      3      28  7.105455e+03   2.925e-07   6.806e-02   1.703e+00
   ...
     26     217  3.663773e+03   5.461e-02   7.676e+01   4.436e+02
     27     224  3.663796e+03   2.285e-04   2.189e+00   2.869e+01
     28     231  3.663797e+03   2.582e-09   9.870e-03   9.661e-02

Local minimum found that satisfies the constraints.

ans = 0.12773
         2.1678
     2.769e+05
      0.013776
         25342
        879.92
```

Example 3.9 *Equilibrium Composition of Ethane Decomposition*

In Fig. 3.10, a fixed bed reactor used to decompose ethane in the presence of steam at 1000 K and 1 bar is shown. If thermodynamic equilibrium is achieved, then the composition of the product stream can be estimated based on classical equilibrium thermodynamics as has been discussed in Example 3.4. The resulting mathematical formulation is as follows:

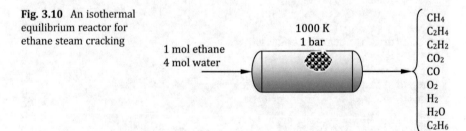

Fig. 3.10 An isothermal equilibrium reactor for ethane steam cracking

$$\min_{n_i \geq 0} \quad G(n_i) = \sum_{i=1}^{NC} n_i \frac{\Delta G_{f,i}^0}{RT} + \ln \; n_i - \ln \sum_{i=1}^{NC} n_i$$

$$\text{s.t.} \hspace{6cm} (3.102)$$

$$\sum_{i=1}^{NC} n_i \; a_{i,k} = a_k^0, \qquad\qquad k = 1, 2, \ldots, NE$$

where n_i are the mole of chemical species i at the product stream (there are NC chemical species present), $\Delta G_{f,i}^0$ is the standard state Gibbs free energy of formation of chemical species i, T is the absolute temperature of the reaction, and R is the ideal gas constant. a_{ik} denote the number of atoms of element k in chemical species i, and NE are the elements present in the system. The atoms of element k present in the feed are a_k^0.

An interesting application of the chemical equilibrium is that of hydrogen production from hydrocarbons. In a particular example, a stream that contains 4 mol of water per mol of ethane (C_2H_6) is fed to the fixed bed reactor shown in Fig. 3.10. The components that are detected (or believed that can be detected) in the product stream are shown in Fig. 3.10. Our aim is to solve the mathematical Formulation (3.102) to estimate the composition of the product stream at equilibrium.

To perform the calculation, we need to determine $\Delta G_{f,i}^0/RT$ for all components. This is a long calculation to be performed by hand, but the details are explained in most undergraduate books on chemical thermodynamics. The results of the calculation for $T = 1000$ K are shown in Table 3.2 together with all other parameters appearing in Formulation (3.102).

The first step in solving the problem in MATLAB is to define the objective function which is achieved through the following m-file:

```
function obj=ChemEquil(n,G0)
obj=n'*(G0+log(n))-sum(n)*log(sum(n));
```

where G0 denotes $\Delta G_{f,i}^0/RT$ (the condensed form at which we have written the objective functions needs the attention of the reader). We then use the following script to solve the problem (as all equality constraints are linear, there is no need to define an m-file for the nonlinear equality/inequality constraints):

Table 3.2 Parameters appearing in Formulation (3.102) for the cracking of ethane

$a_{i,k}$	$i = CH_4$	$i = C_2H_4$	$i = C_2H_2$	$i = CO_2$	$i = CO$	$i = O_2$	$i = H_2$	$i = H_2O$	$i = C_2H_6$	a_k^0
$k = C$	1	2	2	1	1	0	0	0	2	2
$k = H$	4	4	2	0	0	0	2	2	6	14
$k = O$	0	0	0	2	1	2	0	1	0	4
n_i^0								4	1	
$\frac{\Delta G_i^0}{RT}$	2.4703	14.2933	20.4388	-47.5942	-24.0675	0	0	-23.1795	13.1664	

```
clear
% A. EXPLAIN THE MEANING OF EACH ELEMENT OF n
% n(i),i=1,2,..,9 are the mol of component i
% i= 1:CH4 2:C2H4 3:C2H2 4:CO2 5:CO 6:O2 7:H2 8:H2O 9:C2H6

% B. ASSIGN VALUES TO ALL MODEL PARAMETERS
% G0/RT
G0=[ 2.4703423
    14.2933065
    20.4388260
   -47.5942455
   -24.0674782
     0.0000000
     0.0000000
   -23.1794789
    13.1664487];

% C. MODEL LINEAR INEQUALITY AND EQUALITY AND BOUND CONSTRAINTS
A = [] ; b =[];
%    CH4 C2H4 C2H2 CO2 CO  O2  H2  H2O C2H6
Aeq=[1  2    2    1   1   0   0   0   2 ; % C
     4  4    2    0   0   0   2   2   6 ; % H
     0  0    0    2   1   2   0   1   0]; % O
beq=[ 2; % C
     14; % H
      4]; % O
% D. ASSIGN INITIAL VALUES AND BOUNDS TO ALL MODEL PARAMETERS
LB=eps*ones(9,1);
for i=1:9
    UB(i,1)=min( beq./Aeq(:,i) );
end

% initial guess
n0=[ eps  eps  eps  2  eps  eps  7  eps  eps]';

options=optimoptions('fmincon','Algorithm','sqp','Display',
'iter');
[x,f]=fmincon(@(n)
ChemEquil(n,G0),n0,A,b,Aeq,beq,LB,UB,[],options)
```

The response obtained in MATLAB is the following:

```
>> runChemEq
Iter Func- Fval            Feasibility Step Length   Norm of First-order
     count                                              step  optimality
   0  10 -9.995585e+01     3.553e-15   1.000e+00   0.000e+00  4.910e+01
   1  20 -1.005245e+02     8.882e-15   1.000e+00   6.114e+00  1.825e+01
   2  30 -1.014197e+02     4.441e-16   1.000e+00   1.491e+00  1.223e+01
   3  50 -1.014219e+02     1.776e-15   2.825e-02   4.140e-02  1.445e+01
 ...
  24 300 -1.042672e+02     1.776e-15   1.000e+00   3.222e-06  2.034e-01
  25 310 -1.042672e+02     1.776e-15   1.000e+00   1.618e-06  6.394e-02

Local minimum possible. Constraints satisfied.

x =  0.062545005772458
     0.000000093992436
     0.000000000000000
     0.552064310254511
     1.385390078518543
     0.000000000000000
     5.364427873293356
     1.510481300972434
     0.000000208734807

f = -1.042671973068029e+02
```

The total moles in the product stream are 8.8749 mol, and the composition is 60% H_2, 17% water, 15.6% carbon monoxide, 6.2% carbon dioxide, and 0.7% methane. The moles in the product stream are also shown in Fig. 3.11.

Example 3.10 *Optimization of a Heat Exchanger Network*

In Fig. 3.12, a heat exchanger network is shown. A hot stream with a product of flow rate and heat capacity of $FC_{p,H} = 10$ kW/°C, feed temperature of $T_{H,in} = 200$ °C, and final temperature $T_{H,out} = 50$ °C exchanges heat with a cold stream with a product of flow rate and heat capacity of $FC_{p,C} = 30$ kW/°C, feed temperature of $T_{C,in} = 100$ °C, and final temperature $T_{C,out} = 150$ °C. High-pressure steam (hps) with constant temperature $T_{hps} = 250$ °C and cooling water (cw) with inlet temperature of $T_{cw,in} = 15$ °C and exit temperature $T_{cw,out} = 25$ °C are also available.

Fig. 3.11 An isothermal equilibrium reactor for ethane steam cracking, summary of problem solution (the moles of components not shown are less than 0.0001 mol)

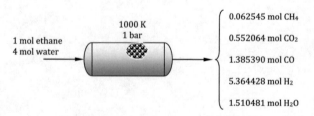

1 mol ethane
4 mol water

1000 K
1 bar

0.062545 mol CH₄

0.552064 mol CO₂

1.385390 mol CO

5.364428 mol H₂

1.510481 mol H₂O

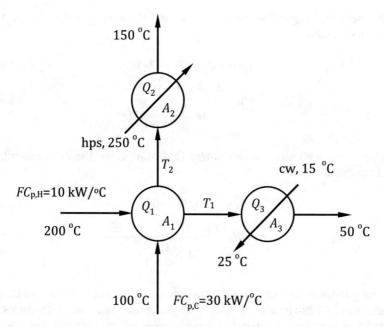

Fig. 3.12 The heat exchanger network case study

The intermediate temperatures T_1 and T_2 (see Fig. 3.12) determine the heat load Q_1 (in kW) of the process-to-process heat exchanger. If these temperatures are different than the desired final temperatures of the hot and cold streams, then low-pressure steam (heat load Q_2) and/or cooling water (heat load Q_3) can be used to achieve the targets.

To develop the process model, we start with the process-to-process heat exchanger for which we can develop two energy balances, one for the hot stream:

$$Q_1 = FC_{p,H}(200 - T_1) \tag{3.103}$$

and one for the cold stream:

$$Q_1 = FC_{p,C}(T_2 - 100) \tag{3.104}$$

We also have an equation that expresses the heat load of the heat exchanger as a function of the available heat transfer area A_1, the overall heat transfer coefficient U_1, and the logarithmic mean temperature difference:

$$Q_1 = A_1 U_1 \frac{(200 - T_2) - (T_1 - 100)}{\ln \dfrac{(200 - T_2)}{(T_1 - 100)}} \tag{3.105}$$

For the heat exchanger that heats up the cold stream to its destination temperature, we have:

$$Q_2 = FC_{p,C}(150 - T_2) \tag{3.106}$$

$$Q_2 = A_2 U_2 \frac{(250 - T_2) - (250 - 150)}{\ln \dfrac{(250 - T_2)}{(250 - 150)}} \tag{3.107}$$

In a similar way, for the heat exchanger that cools down the hot stream to its target temperature:

$$Q_3 = FC_{p,H}(T_1 - 50) \tag{3.108}$$

$$Q_3 = A_3 U_3 \frac{(T_1 - 25) - (50 - 15)}{\ln \dfrac{(T_1 - 25)}{(50 - 15)}} \tag{3.109}$$

As T_1 may become smaller than the incoming temperature of the cold stream, we also need to introduce an inequality constraint that also introduces the minimum temperature approach ΔT_{min} between any two streams observed in the network:

$$T_1 \geq 100 + \Delta T_{min} \tag{3.110}$$

The objective function to minimize is an approximate total annual operating cost of the network that consists of the annualized installed equipment cost and the cost of utilities:

$$\min_{\substack{T_1, T_2 \\ Q_1, Q_2, Q_3 \\ A_1, A_2, A_3}} f\left(\frac{\$}{y}\right) = 2800\left(A_1^{0.65} + A_2^{0.65} + A_3^{0.65}\right) + \left(c_{lps}Q_2 + c_{CW}Q_3\right) \cdot t_y \tag{3.111}$$

where $t_y = 3 \cdot 10^7$ s/y is the operating time, $c_{lps} = 10 \cdot 10^{-6}$ \$/kJ is the cost of the low-pressure steam, and $c_{cw} = 1.6 \cdot 10^{-6}$ \$/kJ is the unit cost of the cooling water. The overall heat transfer coefficients are $U_1 = U_2 = U_3 = 1$ kW/(m^2 °C).

By selecting the minimum temperature approach ΔT_{min}, we can calculate the temperature T_1: $T_1 = 100 + \Delta T_{min}$ and then solve the mathematical equations in the order that have been presented to calculate all parameters appearing in the model. By selecting values for $\Delta T_{min} \in [0.1, 10]$, the model is solved, the cost is calculated, and the results are shown in Fig. 3.13. The minimum cost solution corresponds to $T_1 = 101.826$ °C, and the corresponding cost is 237,866.6 \$/y.

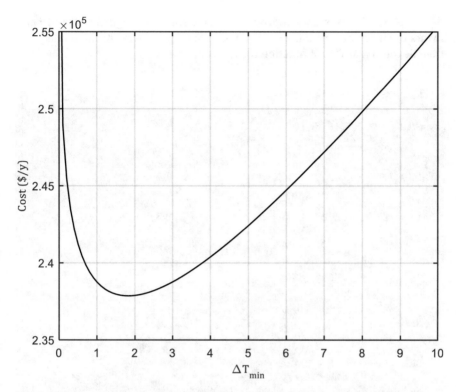

Fig. 3.13 The approximate total annual cost as a function of the minimum temperature approach for the heat exchanger network shown in Fig. 3.12

We first create the MATLAB function that calculates the cost:

```
function f=HeatExcNetConstr(x)
% global to local
variables................................
T1 = x(1); % intermediate temperature of the hot stream deg C
T2 = x(2); % intermediate temperature of the cold stream deg C
Q1 = x(3); % heat load in the process-to-process heat exch. kW
Q2 = x(4); % heat load in the process-to-lps heat exch. in kW
Q3 = x(5); % heat load in the process-to-cw heat exch. in kW
A1 = x(6); % area of the process-to-process heat exc. in sq m
A2 = x(7); % area of the process-to-lps heat exc. in sq m
A3 = x(8); % area of the process-to-cw heat exc. in sq m
% cost function in $/y
f=2800*(A1^0.65+A2^0.65+A3^0.65)+30*10*Q2+30*1.6*Q3; f=f/1e5;
```

It could have been easier to define an inline function to achieve the same objective, but the definition of a dedicated MATLAB function offers greater flexibility to the user and improved transparency (through the use of comments to clarify the meaning of each variable or parameter of the model). We then develop a MATLAB function for the constraints:

```
function [g,h]=HeatExcNetConstr(x)
% global to local variables
. . . . . . . . . . . . . . . . . . . . . . . . . . . . . . . . . . . .
T1 = x(1); % intermediate temperature-hot stream in deg C
T2 = x(2); % intermediate temperature-cold stream in deg C
Q1 = x(3); % heat load process-to-process heat exch. in kW
Q2 = x(4); % heat load process-to-lpd   heat exch. in kW
Q3 = x(5); % heat load process-to-cw    heat exch. in kW
A1 = x(6); % area of the process-to-process heat exc. in m2
A2 = x(7); % area of the process-to-lps heat exc. in sq m
A3 = x(8); % area of the process-to-cw  heat exc. in sq m
% constants
. . . . . . . . . . . . . . . . . . . . . . . . . . . . . . . . . . . . . . . . . . . . .
THIN=200;   % inlet temperature of the hot stream in deg C
THOUT=50;   % target temperature of the hot stream in deg C
FCpH=10;    % "FCP" of the hot stream in kW/deg C
TCIN=100;   % inlet temperature of the cold stream in deg C
TCOUT=150;  % target temperature of the cold stream in deg C
FCpC=30;    % "FCP" of the cold stream in kW/deg C
TLPS=250;   % temperature of the lps in deg C
TCWIN=15;   % inlet temperature of the cooling water in deg C
TCWOUT=15;  % outlet temperature of the cooling water in deg C
U1=1;       % overall heat transfer coef. in kW/(m^2 deg C)
U2=1;       % overall heat transfer coef. in kW/(m^2 deg C)
U3=1;       % overall heat transfer coef. in kW/(m^2 deg C)
% inequality constraints
. . . . . . . . . . . . . . . . . . . . . . . . . . . . . . . . . . . . .
g=[];
% equality constraints
. . . . . . . . . . . . . . . . . . . . . . . . . . . . . . . . . . . . .
h(1) = Q1 - FCpH * (THIN - T1);
h(2) = Q1 - FCpC * (T2  - TCIN);
h(3) = Q1*(log(THIN-T2)-log(T1-TCIN))-...
       A1*U1*((THIN-T2)-(T1-TCIN));

h(4) = Q2 - FCpC * (TCOUT - T2);
h(5) = Q2*(log(TLPS-T2)-log(TLPS-TCOUT))-...
       A2*U2*((TLPS-T2)-(TLPS-TCOUT));

h(6) = Q3 - FCpH * (T1 - THOUT);
h(7) = Q3*(log(T1-TCWOUT)-log(THOUT-TCWIN))-...
       A3*U3*((T1-TCWOUT)-(THOUT-TCWIN));
```

Finally, fmincon is used to obtain the solution:

```
clear all
% initial values
..................................................
T1 = 110; % intermediate temperature stream in deg C
T2 = 130; % intermediate temperature stream in deg C
Q1 = 900; % heat load process-to-process heat exch. in kW
Q2 = 600; % heat load process-to-lpd   heat exch. in kW
Q3 = 600; % heat load process-to-cw    heat exch. in kW
A1 = 10;  % area of the process-to-process heat exc. in sq m
A2 = 10;  % area of the process-to-lps heat exc. in sq m
A3 = 10;  % area of the process-to-cw heat exc. in sq m

x0=[T1 T2 Q1 Q2 Q3 A1 A2 A3]';

lb=[100 100 0  0  0  0  0  0]';
ub=[200 150 1500 1500 1500 1000 1000 1000]';

optn=optimset('Display','Iter')
fmincon(@HeatExcNetCost,x0,[],[],[],[],lb,ub,
@HeatExcNetConstr,optn)
```

```
>> runHeatExcNet
                                       First-order     Norm of
 Iter F-count        f(x)  Feasibility  optimality       step
    0       9  2.463214e+00   1.151e+03   7.362e-04
    1      18  2.551154e+00   1.387e+00   2.833e-03  1.963e+01
    2      27  2.551365e+00   7.627e-07   2.836e-03  2.441e-02
    3      36  2.551363e+00   1.925e-08   2.836e-03  1.139e-03
...
   20     191  2.373218e+00   1.648e-04   2.190e-05  2.029e-02
   21     200  2.373218e+00   1.462e-04   2.310e-07  1.909e-02

Local minimum found that satisfies the constraints.

Optimization completed because the objective function is
non-decreasing in feasible directions, to within the value of the
optimality tolerance, and constraints are satisfied to within the
value of the constraint tolerance.

ans = 101.8252  ← T1  [°C]
      132.7249  ← T2  [°C]
      981.7475  ← Q1  [kW]
      518.2525  ← Q2  [kW]
      518.2525  ← Q3  [kW]
       54.1061  ← A1  [m²]
        4.7806  ← A2  [m²]
        9.0855  ← A3  [m²]
```

Learning Summary

The more general form of constrained optimization problems where both equality and inequality constraints are involved is known as the NLP problem:

$$\min_{\mathbf{x}} f(\mathbf{x})$$

$$\text{s.t.}$$

$$\mathbf{h}(\mathbf{x}) = \mathbf{0}$$

$$\mathbf{g}(\mathbf{x}) \le \mathbf{0}$$

If $\mathbf{x}$ is a feasible point, i.e., $\mathbf{h}(\mathbf{x}) = \mathbf{0}$ and $\mathbf{g}(\mathbf{x}) \le \mathbf{0}$, the gradients of the equality constraints and the active inequality constrains are linearly independent at $\mathbf{x}$ and $\mathbf{x}$ solves the NLP problem locally, then there exist unique scalars $\lambda_i \in R$, $i = 1, 2, \ldots, m$ and non-negative scalars $\mu_j, j = 1, 2, \ldots, p$, such that:

$$\nabla f(\mathbf{x}) + \sum_{i=1}^{m} \lambda_i \nabla h_i(\mathbf{x}) + \sum_{j=1}^{p} \mu_j \nabla g_j(\mathbf{x}) = \mathbf{0}$$

$$\mu_j g_j(\mathbf{x}) = 0, \quad \forall j = 1, 2, \ldots, p$$

$$\mu_j \ge 0, \quad \forall j = 1, 2, \ldots, p$$

These are necessary conditions and are known as KKT conditions.

When only equality constraints are involved, then the NLP problem can be solved using the SQP algorithm that is based on solving the following linear system of equations at iteration k:

$$\begin{bmatrix} \mathbf{H}_L(\mathbf{x}^k, \lambda^k) & \nabla^T \mathbf{h}(\mathbf{x}^k) \\ \nabla \mathbf{h}(\mathbf{x}^k) & \mathbf{0} \end{bmatrix} \cdot \begin{bmatrix} \delta \mathbf{x}^k \\ \lambda^{k+1} \end{bmatrix} = - \begin{bmatrix} \nabla f(\mathbf{x}^k) \\ \mathbf{h}(\mathbf{x}^k) \end{bmatrix}$$

where the hessian of the Lagrangian with respect to $\mathbf{x}$ is the following:

$$\mathbf{H}_L(\mathbf{x}^k, \lambda^k) = \nabla^2 f(\mathbf{x}^k) + \sum_{i=1}^{m} \lambda_i^k \nabla^2 h_i(\mathbf{x}^k)$$

A naïve SQP algorithm can be implemented in the following steps:

Step 0: Select an initial estimate of the solution $\mathbf{x}^0 \in R^n$ and λ^0 at iteration $k = 0$.
Step 1: Calculate $\nabla f(\mathbf{x}^k)$, $\mathbf{h}(\mathbf{x}^k)$, $\nabla \mathbf{h}(\mathbf{x}^k)$ and $\mathbf{H}_L(\mathbf{x}^k, \lambda^k)$.
Step 2: Solve (3.82) to obtain $\mathbf{x}^{k+1} = \mathbf{x}^k + \delta \mathbf{x}^k$ and λ^{k+1}.
Step 3: If $||\delta \mathbf{x}^k|| < \varepsilon$, then STOP else set $k = k + 1$, and return to Step 1.

When both inequality and equality constraints are considered, the SQP algorithm is much more complicated, but the basic idea remains the same.

Terms and Concepts

You must be able to discuss the concept of:

Lagrangian function
Lagrange multipliers
KKT conditions when applied to general NLP problems
SQP algorithms for solving NLP problems

Problems

3.1 Study the paper by Kravitz, S., *Packing Cylinders into Cylindrical Containers,* Mathematics Magazine, 40(2), p. 65–71, 1967. Select 2, 4, 7, 8, 16, 19, and 32 circles of unit radius, and solve Formulation (3.65). Compare your results with the results reported in the paper, and commend on how the difficulty scales with the number of circles considered.

3.2 Consider now the case of unit circle packing in a square problem. Develop the formulation based on the formulation used in Problem 3.1. Solve the problem for $N = 3, 4, \ldots, 8$ circles.

3.3 P. Biswas and D. Kunzru (Int. J. of Hydrogen Energy, 32, pp. 969–980, 2007) study the steam reforming (SR) of ethanol for hydrogen production over $Ni/Ce_{1-x}Zr_xO_2$ catalyst prepared by co-precipitation and incipient wetness impregnation technique. In Table 4 of the paper, the results of the experimental study are summarized. Use Formulation (3.102) to investigate whether they achieve equilibrium in their experiments or not. Prepare a short report.

3.4 A stream of benzene (molecular mass: 78.113 and liquid density 11.2145 $kmol/m^3$) and toluene (molecular mass: 92.14 and liquid density 9.3553 $kmol/m^3$), 60% in benzene, has a molar flow rate of $F = 100$ kmol/h. Benzene and toluene form an ideal mixture, and their vapor-liquid equilibrium (VLE) can be described by the Raoult's law. The saturation pressure of benzene (B) and toluene (T) is given by the following Antoine equations:

$$P_B^{\text{sat}} = \exp\left(9.22143 - \frac{2755.642}{T - 53.989}\right)$$

$$P_T^{\text{sat}} = \exp\left(9.38490 - \frac{3090.783}{T - 53.963}\right)$$

where T is the absolute temperature in K and the saturation pressure is given in bar. You have been asked to design a flash separation unit to obtain a vapor stream with at least 70% benzene and recovery of benzene of at least 50%.

The objective is to determine the operating pressure P (1 bar $\leq P \leq 10$ bar) and operating temperature (360 K $\leq T \leq 450$ K) so as to minimize the installed cost of the vessel C (in \$), which is given by:

$$C = 12500\, D\, H^{0.8}$$

where D is the diameter of the vessel and H its height (both in m). You need to take the following constraints into consideration:

1. The velocity of the vapor stream inside the vessel u_V (in m/s) must satisfy the following constraint:

$$u_V = 0.0305 \sqrt{\frac{\rho_L - \rho_V}{\rho_V}}$$

   where ρ_V is the density of the vapor and ρ_L is the density of the liquid (both in kg/m^3).
2. The ratio of the height of the vessel over its diameter must be between 3 and 5, i.e., $3 \leq H/D \leq 5$.
3. The height of the vessel that is not covered by liquid must be at least 1.67 m, i.e., $H_V > 1.67$ m.
4. The liquid holdup is determined by the requirement that the liquid has a residence time of 10 min (based on the liquid product stream).
Assume that the vessel has flat ends.

3.5 In Fig. P3.5, a heat exchanger network design problem is shown involving the determination eight mass flow rates $(F_1, F_2, \ldots, F_8)$, four temperatures (T_3, T_4, T_5, T_6), and the areas of the two heat exchangers $(A_1$ and $A_2)$. The aim is to determine these parameters so as to minimize the cost of the two heat exchangers given by:

$$C = 5000\left(A_1^{0.6} + A_2^{0.6}\right)$$

The feed stream has a heat capacity of 1 kJ/(kg °C) and the temperature approach any two streams that exchange heat cannot be less than

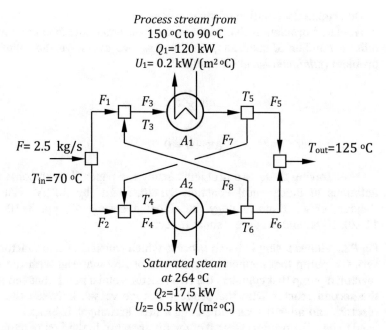

Process stream from
150 °C to 90 °C
Q_1=120 kW
U_1= 0.2 kW/(m² °C)

F_1 F_3 T_5 F_5

T_3

A_1 F_7

F= 2.5 kg/s T_{out}=125 °C

T_{in}=70 °C

A_2 F_8

T_4

F_2 F_4 T_6 F_6

Saturated steam
at 264 °C
Q_2=17.5 kW
U_2= 0.5 kW/(m² °C)

Fig. P3.5 A heat exchanger network design problem

20 °C. It is reminded that the area of the heat exchangers is related to the heat load Q (in kW), overall heat transfer coefficient U (in kW/(m² °C), and mean logarithmic temperature difference ΔT_{lm} through the following equation:

$$Q = AU\Delta T_{lm} = AU \frac{(T_{h,in} - T_{c,out}) - (T_{h,out} - T_{c,in})}{\ln \frac{(T_{h,in} - T_{c,out})}{(T_{h,out} - T_{c,in})}}$$

where the index h denotes hot stream, the index c cold stream, the index in at input state, and out at output state.

3.6 (Left and right inverse of a non-square matrix). Consider matrix $\mathbf{A}$ with m rows and n columns and $m > n$. In that case, the system of linear equations $\mathbf{Ax} = \mathbf{b}$ does not have a unique solution as there are more equations than unknowns (inconsistent). We can, however, obtain a least squares solution of the system by solving the following minimum norm problem:

$$\min_x \mathbf{e}^T \mathbf{e}$$

s.t.

$$\mathbf{e} = \mathbf{b} - \mathbf{Ax}$$

Determine the solution of the problem.

A related problem is that when $m < n$ (less equations than unknowns, infinite number of solutions). In this case we can solve the following problem (*minimum norm solution*):

$$\min_{\mathbf{x}} \frac{1}{2}\mathbf{x}^T\mathbf{x}$$

s.t.

$$\mathbf{Ax} - \mathbf{b} = 0$$

After solving these two problems, study the paper "Linear Constraint Relations in Biochemical Reaction Systems: I&II" by R.T.J.M. van der Heijden, et al., Biotechnology and Bioengineering, 43, pp. 3–10 and 11–20, 1994, and prepare a short report.

3.7 Fig P3.7 shows a simple batch process which consists of two reactors in series, a pump that connects the storage of raw material with the first reactor, a pump that connects the two reactors, and a pump that connects the second reactor with the product storage vessel. Between the two reactors and after the second pump, a heat exchanger is located that is used to heat the product stream of the first reactor. In the first reactor, the raw material R is transformed to an intermediate product I, which is then transformed to the product P in the second reactor. There is a need to produce M_y kg/y of the product mix P. If N_b is the number of batches delivered in 1 year and the time between the delivery of two consecutive batches is τ, then the amount of product mix delivered at each batch M must satisfy the following constraints:

$$M \cdot N_b = M_y$$

$$\tau \cdot N_b \le t_y \text{ or } \tau \cdot M_y \le t_y M$$

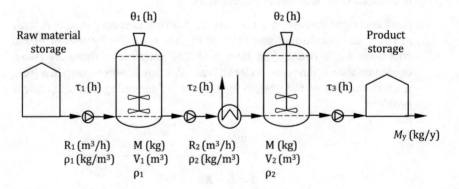

Fig. P3.7 A simple batch plant

where t_y is the operating hours in 1 year (8300 h/year). To complete the formulation express mathematically, the following constraints:

1. The volume of the material in each reactor (M/ρ) must be less than the 80% of the overall reactor volume.
2. The time needed to load or upload the material in or from each reactor (τ_1, τ_2, τ_3) must be enough so as to deliver the mass processed in one batch M, while the volumetric flowrate through pump i is R_i (in m^3/h).
3. The processing time of each reactor is the sum of the loading time (τ_1 for the first reactor and τ_2 for the second reactor), the processing time (θ_1 for the first reactor and θ_2 for the second reactor), and the uploading time (τ_2 for the first reactor and τ_3 for the second reactor).
4. The cycle time τ (the time between the completion of two consecutive batches) is greater or equal to the processing time of each reactor.

The installed equipment cost consists of the cost of the two reactors and the cost of the three pumps:

$$C = a_1 V_1^{n_1} + a_2 V_2^{n_2} + \beta_1 R_1^{m_1} + \beta_2 R_2^{m_2} + \beta_3 R_3^{m_3}$$

where a_i, n_i, β_i, and m_i are constants.

It is given that : $t_y = 8300$ h/y, $M_y = 10{,}000{,}000$ kg/y.
$\rho_1 = 800$ kg/m^3, $\rho_2 = 1000$ kg/m^3.
$\theta_1 = 5$ h, $\theta_2 = 6$ h.
$a_1 = 50000$ \$, $n_1 = 0.6$, $a_2 = 60000$ \$, $n_2 = 0.4$.
$\beta_1 = \beta_2 = \beta_3 = 1000$ \$, $m_1 = m_2 = m_3 = 0.6$.

Determine the cycle time τ, the two reactor volumes V_1 and V_2, and the tree transfer rates R_1, R_2, and R_3 so as to minimize the cost of installed equipment.

3.8 The flue gas from a power plant is available at 1 bar and 180 °C and has a volumetric flow rate of $F_{fg} = 10$ m^3/s (molar density $\rho = 0.02655$ kmol/m^3). Its composition is approximately 10% CO_2, 72% N_2, 3% O_2, and 15% H_2O and has an approximately constant molar heat capacity $Cp_{fg} = 31$ kJ/(kmol °C). The flue gas is fed to a carbon capture production unit. At the first step, the flue gas needs to be cooled down to 60 °C. Three options are available for the cooling medium (U is the estimated overall heat transfer coefficient):

(i) Cooling water ($T_{CW,\ in} = 25$ °C, $T_{CW,\ out} = 35$ °C, cost 0.35 \$/GJ, $U = 0.1$ kW/(m^2 °C))
(ii) Low-temperature refrigerant at -20 °C, cost 8 \$/GJ, $U = 0.2$ kW/(m^2 °C)
(iii) Very-low-temperature refrigerant at -50 °C, cost 13 \$/GJ, $U = 0.3$ kW/(m^2 °C)

Fig. P3.9 A single-effect
evaporator system

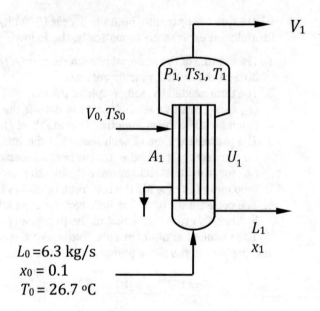

$L_0 = 6.3$ kg/s
$x_0 = 0.1$
$T_0 = 26.7\ ^\circ C$

The cost of installed shell and tube heat exchangers in $ is
$C_{he} = 5000A^{0.9}$, where A is the heat transfer area (in m^2) and the capital
charge factor is CCF $= 0.2$. Design the plant that has the minimum total
annual cost (the plant operates for $t_y = 8300$ h/y).

3.9 A single-evaporator system is shown in Fig. P3.9. The following data are
provided:

Property	Equation
Boiling point rise (°C)	BPR $= (1.8 + 6.2x)x$
Heat capacity of solids (kJ/(kg °C))	cps $= 1.84$
Overall heat transfer coef. (kJ/(h m^2 °C))	$U(x) = \exp\left(8.14 - 0.078 \ln(x) + 0.26[\ln(x)]^2\right)$
Specific enthalpy of sat. liquid water (kJ/kg)	$h_w(T) = 9.14 + 3.8648T + 3.3337 \cdot 10^{-3}T^2 - 9.8 \cdot 10^{-6}T^3$
Specific enthalpy of sat. water vapor (kJ/kg)	$H_w(T) = 2502 + 1.8125\,T + 2.585 \cdot 10^{-4}T^2 - 9.8 \cdot 10^{-6}T^3$
Specific enthalpy of solution (kJ/kg)	$h(x, T) = (1 - x)h_w(T) + xcp_sT$

The condition of the feed is given in Fig. P3.9. Our aim is to increase the
solid content of the liquid product stream to at least 50% solids. Design
the system to minimize the following approximate total annual cost
(TAC):

$$TAC = CCF \cdot F_{BM} \cdot C_{evap} + V_0 t_y c_s$$

$$\log_{10} C_{eval} = 3.9119 + 0.8627 \cdot \log_{10}(A_1) - 0.0088 \cdot [\log_{10}(A_1)]^2$$

where C_{evap} is the purchasing cost of the evaporator in $, $CCF = 1/3$ (1/y) is the capital charge factor, $F_{BM} = 2.25$ is the bare module factor, A_1 is the heat transfer area of the evaporator in m², V_0 is the steam (saturated steam at $Ts_0 = 160$ °C) consumption in the first effect in kg/h, t_y is the on-stream time (8322 h/y), and $c_s = 0.03$ $/kg is the unit steam cost. Other variables of interest are:

T_1: temperature in the evaporator in °C
P_1: pressure in the evaporator in bar
V_1: vapor stream coming out from the evaporator in kg/s
L_1: liquid stream coming out from the evaporator in kg/s
w_1: mass fraction of solids in the evaporator
     The pressure, the temperature, and the boiling point rise in the evaporator are related through the following equations:

$$\ln P_1 = 11.622 - \frac{3798.3}{Ts_1 + 227.03}$$

$$T_1 = Ts_1 + BPR(x_1)$$

where Ts_1 is the saturation temperature of pure water in °C. The evaporator can operate in a pressure between $P = 0.1$ bar and $P = 1$ bar.

3.10 Ramanathan και Gaudy (Biotechnology and Bioengineering, XI, pp. 207–237, 1969) used the continuous experimental bioreactor of Fig. P3.10 and collected the following experimental data at steady state by varying the dilution rate:

i	D_i (h^{-1})	X_i (mg/L)	S_i (mg/L)
1	0.042	1589	221
2	0.056	2010	87
3	0.083	1993	112
4	0.167	1917	120
5	0.333	1731	113
6	0.500	1787	224
7	0.667	676	1569

where D is the dilution rate ($D = F/V$, F is the volumetric flow rate of the feed and V the volume of the reactor), X is the biomass concentration, and S is the substrate concentration (substrate at the feed $S_0 = 3000$ mg/L).

Fig. P3.10 A continuous
bioreactor

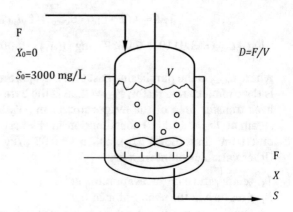

F

$X_0=0$

$S_0=3000$ mg/L

$D=F/V$

V

F

X

S

Assume that the Monod equation:

$$\mu = \mu_{max} \cdot \frac{S}{K_s + S}$$

is a valid approximation for the system under study (μ is the specific
growth rate and μ_{max} and K_s are the parameters of the Monod equation).
The yield coefficient $Y_{x/s}$ ($= r_x/r_s$) is used to relate the concentration of
the substrate and the biomass at steady state. Prove that a mathematical
model for the process at steady state consists of the following algebraic
equations:

$$D = \frac{\mu_{max} S}{K_s + S} \quad \text{or} \quad S = \frac{DK_s}{\mu_{max} - D}$$

$$X = Y_{x/s} \cdot (S_0 - S)$$

Define the error matrix:

$$E\left(\mu_{max}, K_s, Y_{\frac{x}{s}}\right) = \begin{bmatrix} e_{1,1} & e_{1,2} \\ e_{2,1} & e_{2,2} \\ \vdots & \vdots \\ e_{k,1} & e_{k,2} \\ \vdots & \vdots \\ e_{7,1} & e_{7,2} \end{bmatrix} = \begin{bmatrix} S_1 - \dfrac{D_1 K_s}{\mu_{max} - D_1} & X_1 - Y_{x/s} \cdot (S_0 - S_1) \\ S_2 - \dfrac{D_2 K_s}{\mu_{max} - D_2} & X_2 - Y_{x/s} \cdot (S_0 - S_2) \\ \vdots & \vdots \\ S_k - \dfrac{D_k K_s}{\mu_{max} - D_k} & X_k - Y_{x/s} \cdot (S_0 - S_k) \\ \vdots & \vdots \\ S_7 - \dfrac{D_7 K_s}{\mu_{max} - D_7} & X_7 - Y_{x/s} \cdot (S_0 - S_7) \end{bmatrix}$$

Determine the unknown parameters (μ_{max}, K_s, $Y_{x/s}$) of the model by
minimizing $J = \det |E^T E|$, known as the determinant criterion.

Chapter 4
Linear Programming

4.1 Introduction to Linear Programming

In the previous chapter, we discussed optimization problems with many optimization variables in the presence of nonlinear equality and inequality constraints:

$$\min_{\mathbf{x}} f(\mathbf{x})$$
$$\text{s.t.}$$
$$\mathbf{h}(\mathbf{x}) = 0$$
$$\mathbf{g}(\mathbf{x}) \leq 0$$
$$\mathbf{x}_L \leq \mathbf{x} \leq \mathbf{x}_U$$
(4.1)

where $\mathbf{x}$ is a n-th dimensional vector, $\mathbf{h}$ is vector valued function ($\mathbf{h}: \mathrm{R}^n \rightarrow \mathrm{R}^m$), and $\mathbf{g}$ is also a vector valued function ($\mathbf{g}: \mathrm{R}^n \rightarrow \mathrm{R}^p$) (as always f is a real valued function $f: \mathrm{R}^n \rightarrow \mathrm{R}$). In this chapter, we will focus on an important special case where all constraints and the objective function are linear leading to a linear programming (LP) problem:

$$\min_{\mathbf{x}} \ \mathbf{c}^T \mathbf{x}$$
$$\text{s.t.}$$
$$\mathbf{A}\mathbf{x} = \mathbf{b}$$
$$\mathbf{C}\mathbf{x} \leq \mathbf{d}$$
$$\mathbf{x}_L \leq \mathbf{x} \leq \mathbf{x}_U$$
(4.2)

where $\mathbf{A}$ is a m-by-n constant matrix, $\mathbf{C}$ is a p-by-n constant matrix, $\mathbf{c}$ is a n dimensional vector, $\mathbf{b}$ is a m dimensional constant vector, and $\mathbf{d}$ is a

© The Author(s), under exclusive license to Springer Nature Switzerland AG 2022

I. K. Kookos, *Practical Chemical Process Optimization*, Springer Optimization and Its Applications 197, https://doi.org/10.1007/978-3-031-11298-0_4

p dimensional constant vector. Despite the limitations inherent in building linear models for real-world processes, the practical usefulness and applications of the LP formulation have been enormous in the last 70 years to deserve a special chapter. We will first present some representative examples of the application of LP models to chemical processes and then discuss the most well-known method for solving LP problems known as the simplex method. Again, we need to remind the reader that our approach is not to dive into the underlying mathematics but to focus on the results that are most important from the practical point of view.

4.2 Examples of LP Formulations from the Chemical Industry

We will present two representative examples of the use of LP models in the chemical and related industries. The first one is from the petroleum refineries and the second one from the cement and clinker production. We will first consider a classical example of blending in petroleum refineries. Petroleum refineries process crude oil (a mixture of hydrocarbons that also contains impurities such as sulfur, metals, and nitrogen compounds) and produce a multitude of commercial products (such as gaseous and liquid fuels and raw materials for the petrochemicals industry). As the proportions of different molecules are different in each crude oil (there are practically as many different crude oil qualities as there are different oil fields), each crude has its own specific physical and chemical properties and product yields. A light crude oil contains a higher proportion of the smaller (light) molecules when compared to a heavy crude which contains a high proportion of the very large molecules. Crude classifications are, to some extent, arbitrary, and the most well-known are based on gravity (light, medium, and heavy crudes) and sulfur content (sweet/low-sulfur, middle-, and sour/high-sulfur crudes).

A modern refinery (see Fig. 4.1) consists of fractionating operations, chemical operations (to change the chemical structure of intermediates), cracking operations (to "break down" complex molecules into "simple" molecules), and blending operations to produce final products and contaminant removal processes. Crude processing begins at the atmospheric crude oil distillation unit (CDU) or primary distillation unit which separates crude oil into different fractions based on ranges of their boiling points. The CDU is fed with desalted crude that has been heated in a train of heat exchangers and in a furnace (to a temperature of around 360 °C and a pressure of around 2 bar). The CDU is often easy to identify as it is the largest column in a refinery. The top product passes through a further series of smaller columns, which separate it into light products, propane, and butane. Side products such as light naphtha (used as petrochemical feedstock or as a gasoline component), heavy naphtha (which is

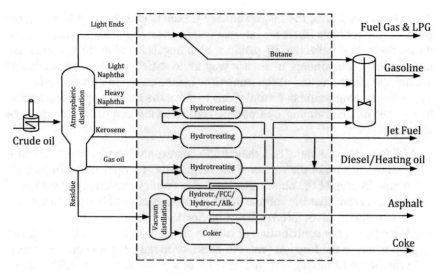

Fig. 4.1 Structure of a petroleum refinery

usually upgraded to be used as gasoline component), kerosene, and gas oil cuts are also drawn off the CDU. The residual heavy liquid or residue is the bottoms product and is fed to a column operating under vacuum (vacuum distillation unit or VDU) to lower the boiling points and to improve recovery of light components. These products (vacuum distillates) are mixed with the products of the CDU or are fed to downstream processes for further processing. The main products are:

- Liquefied petroleum gas or LPG which is a group of light C_2–C_4 hydrocarbon gases.
- Automotive and aviation gasoline that have boiling points up approximately to 250 °C.
- Kerosene is a light petroleum distillate used for heating and lighting.
- Diesel fuels used in compression ignition engines.
- Fuel oils that are mainly used in space heating and are available in several grades starting from No. 1 fuel oil (similar to kerosene) to No. 2 fuel oil (similar to diesel fuel). Heavier grades of No. 3 and 4 can also be available.
- Lubricants
- Asphalt used in the construction industry.
- Petroleum coke used as fuel in several industries.

Almost all refinery products are obtained through some sort of blending of intermediate products. When computers and modern instrumentation for online monitoring and control were not available, all blending operations were batch operations. Nowadays, even small refineries use online computer-based blending because the equipment is relatively inexpensive and cost savings are significant as the volumes of the products are extremely high.

Apart from lubricants, the major refinery products produced by blending are gasoline and jet fuels (fuels for internal combustion engines), heating oils, and diesel fuels. The objective of product blending is to allocate the available intermediate components in such a way as to satisfy product demands and quality specifications and at the same time maximize overall profit. To be able to use linear programming formulations to solve the blending problems posed in the petroleum refining operations, the following requirements must be satisfied:

- *Additivity*, i.e., the value of a magnitude corresponding to a whole object is equal to the sum of the values of the magnitudes corresponding to its parts for any division of the object into parts. For example, additivity of volume of gasoline means that the volume of the gasoline produced is equal to the sum of the volumes of all intermediates blended.
- *Linearity*, i.e., the contribution of each specific intermediate varies in direct proportion to the fraction (volume, mass, or molar) of the component used in the blend (doubling, e.g., the fraction of a particular intermediate in the blend doubles its contribution to a particular property of the final product).

Let's now consider an oversimplified version of crude oil selection. Let us assume that a particular refinery uses two crude oil suppliers, $CrOil_1$ and $CrOil_2$, to produce LPG, gasoline (GSL), diesel (DSL), and asphalt (ASP). Let us also assume that the data shown in Table 4.1 are available to assist engineers to decide which mix of $CrOil_1$ and $Croil_2$ to select.

Note that the two alternative crude oils have different yields relative to the four products selected. There is also an important difference in the relative prices of the potential products and a less significant difference in the cost of crude oils. To model the problem at hand, we denote the amount of crude oil $CrOil_1$ by m_{CrOil1} and the amount of crude oil $CrOil_2$ by m_{CrOil2} (in kt). The LPG produced m_{LPG} is given by the following linear equality constraint:

$$m_{LPG} = 0.3m_{CrOil1} + 0.2m_{CrOil2} \tag{4.3}$$

Similar equalities hold for the other products:

Table 4.1 Data for the oversimplified crude oil mix selection problem

Product	Crude oil yields Y_{ji} (t product/t crude)		Relative selling price	Maximum (kt)
	$CrOil_1$	$CrOil_2$		
LPG	0.3	0.2	9	1.2
GSL	0.2	0.3	8	1.2
DSL	0.2	0.2	7	0.9
ASP	0.3	0.3	1	1.8
Relative cost	3.25	1.9		

$$m_{GSL} = 0.2m_{CrOil1} + 0.3m_{CrOil2} \qquad (4.4)$$

$$m_{DSL} = 0.2m_{CrOil1} + 0.2m_{CrOil2} \qquad (4.5)$$

$$m_{ASP} = 0.3m_{CrOil1} + 0.3m_{CrOil2} \qquad (4.6)$$

If we take into account the restriction in the maximum amount that can be produced (due to limited storage capacity of market share), we may write:

$$0.3m_{CrOil1} + 0.2m_{CrOil2} \leq 1.2 \qquad (4.7)$$

$$0.2m_{CrOil1} + 0.3m_{CrOil2} \leq 1.2 \qquad (4.8)$$

$$0.2m_{CrOil1} + 0.2m_{CrOil2} \leq 0.9 \qquad (4.9)$$

$$0.3m_{CrOil1} + 0.3m_{CrOil2} \leq 1.8 \qquad (4.10)$$

or:

$$1.5m_{CrOil1} + m_{CrOil2} \leq 6 \qquad (4.11)$$

$$m_{CrOil1} + 1.5m_{CrOil2} \leq 6 \qquad (4.12)$$

$$m_{CrOil1} + m_{CrOil2} \leq 4.5 \qquad (4.13)$$

$$m_{CrOil1} + m_{CrOil2} \leq 6 \qquad (4.14)$$

Equations (4.11), (4.12), (4.13), and (4.14) can also be written in matrix form:

$$\begin{bmatrix} 1.5 & 1 \\ 1 & 1.5 \\ 1 & 1 \end{bmatrix} \begin{bmatrix} m_{CrOil1} \\ m_{CrOil2} \end{bmatrix} \leq \begin{bmatrix} 6 \\ 6 \\ 4.5 \end{bmatrix} \qquad (4.15)$$

where we have omitted (4.14) as it is automatically satisfied if (4.13) is satisfied. The aim is to assign values to the two optimization variables m_{CrOil1} and m_{CrOil2} to maximize the expected profit defined as the revenues minus the cost of purchasing the crude oils:

$$f = 9m_{LPG} + 8m_{GSL} + 7m_{DSL} + m_{ASP} - 3.25m_{CrOil1} - 1.9m_{CrOil2} \qquad (4.16)$$

We can use Eqs. (4.3, 4.4, 4.5, and 4.6) to eliminate the amounts of the specific products and express the objective as a function of the two optimization variables only. The final formulation is as follows:

$$\max_{m_{\text{CrOil1}},m_{\text{CrOil2}}} f = 2.75m_{\text{CrOil1}} + 4m_{\text{CrOil2}}$$

$$\text{s.t.} \tag{4.17}$$

$$1.5m_{\text{CrOil1}} + m_{\text{CrOil2}} \leq 6$$

$$m_{\text{CrOil1}} + 1.5m_{\text{CrOil2}} \leq 6$$

$$m_{\text{CrOil1}} + m_{\text{CrOil2}} \leq 4.5$$

$$0 \leq m_{\text{CrOil1}}, m_{\text{CrOil2}}$$

This is a linear programming problem, and in this chapter, we will present the methods available for its solution (including their MATLAB® implementation).

We will now show that a seemingly unrelated problem has an almost identical mathematical formulation. We will briefly present the case of the Portland cement clinker (PCC) production. PCC is produced from two natural raw materials—calcium carbonate (limestone) and aluminum silicate (clay, shale, etc.). These two natural raw materials complement each other to give rise to the compounds or phases present in the clinker in the required quantities. Several corrective materials such as bauxite, laterite, iron ore or blue dust, sand, or sandstone, etc. are used to compensate the specific chemical shortfalls in the composition of the raw mix. A clinker comprises four major compounds or phases: 50–55% tricalcium silicate (C3S) or "alite," 25–30% dicalcium silicate (C2S) or "belite," 9–11% tricalcium aluminate (C3A), and 12–15% tetracalcium aluminoferrite (C4AF) or "brownmillerite." In terms of the four major oxides present in the clinker, the phase percentage of the clinker that was presented corresponds approximately to 62–65% CaO, 19–21% SiO_2, 4–6% Al_2O_3, and 3–5% Fe_2O_3. Generally, three to five component mixes are prepared, in which the proportion of limestone varies from about 80 to 95%, depending on its quality. The requirements of the four major oxides are usually expressed by the following three ratios:

Alumina ratio or modulus (AR):

$$AR = \frac{m^c_{Al_2O_3}}{m^c_{Fe_2O_3}} \tag{4.18}$$

Silica ratio or modulus (SR):

$$SR = \frac{m^c_{SiO_2}}{m^c_{Al_2O_3} + m^c_{Fe_2O_3}} \tag{4.19}$$

Lime saturation factor (LSF):

$$LSF = \frac{m^c_{CaO}}{2.8m^c_{SiO_2} + 1.1m^c_{Al_2O_3} + 0.65m^c_{Fe_2O_3}} \tag{4.20}$$

where m_i^C denotes the mass of oxide i in the clinker. If the mass fraction of the oxide in the raw material r is $\Omega_{i,r}$, then the mass of the oxide i in the raw materials mix is:

$$m_i^C = \sum_{r=1}^{NR} \Omega_{i,r} m_r \qquad (4.21)$$

where m_r is the mass of the raw material r used in the raw materials mix. Finally, a basis for the calculations needs to be defined and is usually 1000 kg or 1 t of clinker:

$$\sum_{i=1}^{NO} m_i^C = 1000 \qquad (4.22)$$

where NO denotes the number of oxides present in the mix. Other elements such as alkalis, metals, and SO_3 as well as ash due to fuel burned to supply the necessary energy are present in the raw material, but we ignore them to simplify the presentation. We will consider a specific case study to facilitate clarity. The details are presented in Table 4.2. Four different raw materials are considered with different composition and cost characteristics. Loss on ignition or LOI refers to the % of volatile material that ends up to the gaseous stream and does not become part of the solid product (clinker).

The raw material mix usually is selected to satisfy specific amounts of each oxide in the mixture and minimize the cost of raw materials used. The percentage of the oxides is expressed in an indirect way through the alumina ratio AR, silica ratio SR, and lime saturation factor LSF by setting upper (denoted by subscript U) and lower (denoted by subscript L) bounds. If we select the AR, for example, then $AR_L \leq AR \leq AR_U$, or:

$$AR_L \leq \frac{m_{Al_2O_3}^C}{m_{Fe_2O_3}^C} \leq AR_U \qquad (4.23)$$

or:

Table 4.2 Raw materials composition and relative cost for the clinker production process

Oxides	$\Omega_{i,r}$	Limestone ($r = L$)	Clay ($r = C$)	Bauxite ($r = B$)	Pyrites ($r = P$)
Oxides (i)	SiO_2	0.10	0.55	0.01	0.05
	Al_2O_3	0.02	0.10	0.40	0.03
	Fe_2O_3	0.01	0.05	0.35	0.85
	CaO	0.50	0.13	0.15	0.02
	Loss on ignition	0.37	0.15	0.07	0.02
	Relative Cost	4	1	40	25

$$m_{Al_2O_3}^c - AR_U m_{Fe_2O_3}^c \leq 0 \tag{4.24a}$$

$$AR_L m_{Fe_2O_3}^c - m_{Al_2O_3}^c \leq 0 \tag{4.24b}$$

By taking Eq. (4.21) into consideration, we obtain:

$$\sum_{r \in \{L, C, B, P\}} \Omega_{Al_2O_3,r} m_r - AR_U \sum_{r \in \{L, C, B, P\}} \Omega_{Fe_2O_3,r} m_r \leq 0 \tag{4.25a}$$

$$-\sum_{r \in \{L, C, B, P\}} \Omega_{Al_2O_3,r} m_r + AR_L \sum_{r \in \{L, C, B, P\}} \Omega_{Fe_2O_3,r} m_r \leq 0 \tag{4.25b}$$

Finally:

$$\sum_{r \in \{L, C, B, P\}} \left(\Omega_{Al_2O_3,r} - AR_U \Omega_{Fe_2O_3,r} \right) m_r \leq 0 \tag{4.26a}$$

$$\sum_{r \in \{L, C, B, P\}} \left(AR_L \Omega_{Fe_2O_3,r} - \Omega_{Al_2O_3,r} \right) m_r \leq 0 \tag{4.26b}$$

Similar linear inequality constraints can be developed for the SR and LSF. These linear inequalities together with the linear equality (4.22) define the feasible space of the problem. The objective function (cost minimization) is the following:

$$\min_{m_L, m_C, m_B, m_P} \sum_{r \in \{L, C, B, P\}}^{NR} c_r m_r = c_L m_L + c_C m_C + c_B m_B + c_P m_P \tag{4.27}$$

where c_r is the relative unit cost of raw material r.

4.3 Graphical Solution of Linear Programming Problems

In this section, we will present the graphical solution of the crude oil selection problem in order to get some insight in the nature of the problem. We consider again the LP formulation for the crude oil selection given by Formulation (4.17). We will start by finding what the feasible space of the LP problem is. As we have only two variables, it is relatively easy to discover what the feasible space is. We draw the inequalities as equalities and then exclude the half space that does not satisfy the inequality. If we consider the general inequality constraint $a_1 x_1 + a_2 x_2 \leq b$, then we can draw the linear equation $a_1 x_1 + a_2 x_2 = b$ as it passes through the points $(0, b/a_2)$ and $(b/a_1, 0)$. This line separates R^2 into two half spaces. We then consider the origin (point $(0, 0)$), and if it satisfies the inequality, then the half space that includes the origin belongs to the feasible space. We repeat for all inequalities to find the feasible space of the overall problem.

We begin by considering the first inequality:

$$1.5m_{CrOil1} + m_{CrOil2} \leq 6 \qquad (4.28)$$

The corresponding equality passes through the points (4, 0) and (0, 6). The origin satisfies the constraint as $0 < 6$, and we obtain Fig. 4.2a where we have also taken into consideration the non-negativity constraints $0 \leq m_{CrOil1}$ and $0 \leq m_{CrOil2}$. We then add the second linear inequality constraint:

$$m_{CrOil1} + 1.5m_{CrOil2} \leq 6 \qquad (4.29)$$

The corresponding line passes through the points (0, 4) and (6, 0). The combined result is shown in Fig. 4.2b. We then add the final inequality constraint:

$$m_{CrOil1} + m_{CrOil2} \leq 4.5 \qquad (4.30)$$

to obtain Fig. 4.3. In Fig. 4.3, in addition to the feasible space, the objective function is also given. Points that have equal values of the objective function are shown as dashed straight lines. As the value of the function becomes smaller, the line of the objective function moves away from the origin. It is important to note that all lines of equal values of the objective function are parallel to each other. As (in this book) we express all optimization problems as minimization problems, the objective function shown is the negative of the objective function given in Formulation (4.17).

It is important to note that the optimal point is located when the line of the objective function is moved in the improving direction until it only has a single point in the feasible region. As we will see shortly, this is a general characteristic of LP problems as the solution is always located in a "corner" or vertex point of the feasible region. In this particular example, the optimal point is the intersection of the following two constraints:

$$m_{CrOil1} + 1.5m_{CrOil2} = 6$$

$$m_{CrOil1} + m_{CrOil2} = 4.5$$

If we subtract the second from the first, we obtain $m_{CrOil2} = 3$ kt and then $m_{CrOil1} = 1.5$ kt. It is also important to note that all "corner" points in the convex polygon formed in the R^2 space shown in Fig. 4.3 correspond to the simultaneous solution of the two constraints that are active (satisfied as equalities) at this point. We may therefore construct a quick method to solve algebraically small problems. We may locate all corner points by examining all possible combinations of two constraints and then solve the resulting linear systems, evaluate the solution, and then determine the objective function. This sounds as an extremely efficient method, but you only need to recall the

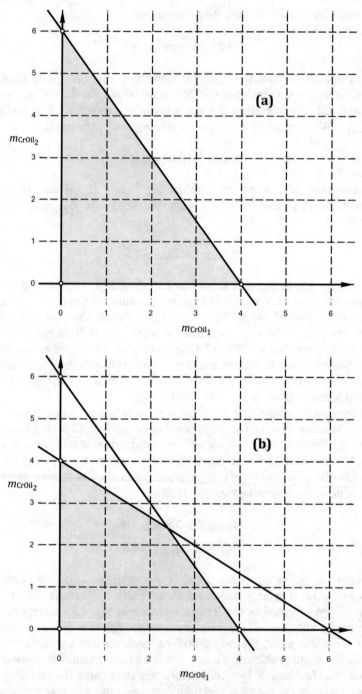

Fig. 4.2 Step toward sketching the feasible space of the crude oil selection problem

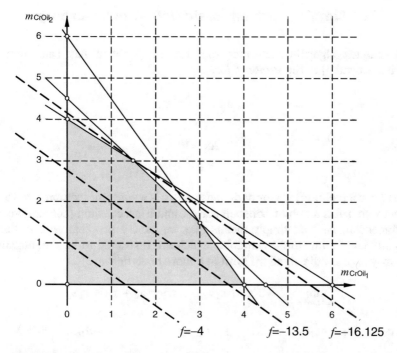

Fig. 4.3 Graphical solution of the crude mix optimization problem

number of alternative ways in which you may select n elements among the m members of a set to realize that the number of corner points can be extremely high. This number is given by:

$$\binom{m}{n} = \frac{m!}{n!(m-n)!} \tag{4.31}$$

(the binomial formula for counting combinations). If, for example, we have a problem with 10 variables and 30 inequality constraints (a relatively small case study), then we need to investigate more than 30 million corner points! This is clearly not a feasible strategy, and a cleverer approach is needed in order to be able to attack real-world problems with thousands of variables and thousands or even millions of constraints. This is simplex method developed by G. B. Dantzig.

4.4   The Simplex Method: Basic Definitions and Steps

The simplex algorithm assumes that the LP problem has been defined (or transformed) in the *standard form*:

$$\min_{x} \ f = c^T x$$
$$\text{s.t.} \tag{4.32}$$
$$Ax = b$$
$$x \geq 0$$

where x is a (column) vector of dimension n, c is a constant (column) vector of dimension n, A is a m-by-n constant matrix, and b is a constant (column) vector of dimension m. This simply means that we have a problem with m linear equality constraints and n variables which are restricted to be non-negative. We may express the standard form in a more analytic way:

$$\min_{x_1, x_2, \ldots, x_n} \ f = c_1 x_1 + c_2 x_2 + \ldots + c_k x_k + \ldots + c_n x_n$$

$$\text{s.t.}$$

$$
\begin{array}{ccccccccccc}
a_{11}x_1 & + & a_{12}x_2 & + & \ldots & + & a_{1k}x_k & + & \ldots & + & a_{1n}x_n & = & b_1 \\
a_{21}x_1 & + & a_{22}x_2 & + & \ldots & + & a_{2k}x_k & + & & + & a_{2n}x_n & = & b_2 \\
\vdots & & \vdots & & \vdots & & \vdots & & & & \vdots & & \vdots \\
a_{r1}x_1 & + & a_{r2}x_2 & + & \ldots & + & a_{rk}x_k & + & \ldots & + & a_{rn}x_n & = & b_r \\
\vdots & & \vdots & & & & \vdots & & & & \vdots & & \vdots \\
a_{m1}x_1 & + & a_{m2}x_2 & + & \ldots & + & a_{mk}x_k & + & \ldots & + & a_{mn}x_n & = & b_m
\end{array}
$$

$$x_1 \geq 0, \quad x_2 \geq 0, \ \ldots, \ x_k \geq 0, \ \ldots, \ x_n \geq 0$$

$$\tag{4.33}$$

We will demonstrate the steps of the simplex algorithm through the crude oil selection problem that we have just solved graphically. We can recall that the initial formulation as given by Eq. (4.17) is not in the standard form as (a) the problem is a maximization problem and (b) the constraints are inequality constraints. Point a is easy to resolve as maximizing f is equivalent to minimizing $-f$. To transform the inequality problem into an equality problem, we introduce one non-negative variable s_i for each constraint i that can be thought of as the distance from being an active inequality (these are called slack variables). By performing these two modifications, we obtain the following formulation:

$$\min_{m_{CrOil1},\ m_{CrOil2}} f = -2.75 m_{CrOil1} - 4 m_{CrOil2} + 0 s_1 + 0 s_2 + 0 s_3$$

$$\text{s.t.} \tag{4.34}$$

$$1.5 m_{CrOil1} + m_{CrOil2} + s_1 = 6.0$$

$$m_{CrOil1} + 1.5 m_{CrOil2} + s_2 = 6.0$$

$$m_{CrOil1} + m_{CrOil2} + s_3 = 4.5$$

$$0 \le m_{CrOil1},\ m_{CrOil2}\ s_1,\ s_2,\ s_3$$

We define the following column vector to facilitate clarity and generality of our presentation:

$$\mathbf{x} = \begin{bmatrix} x_1 \\ x_2 \\ x_3 \\ x_4 \\ x_5 \end{bmatrix} = \begin{bmatrix} m_{CrOil1} \\ m_{CrOil2} \\ s_1 \\ s_2 \\ s_3 \end{bmatrix} \tag{4.35}$$

Then (4.34) can be written as:

$$
\begin{array}{rrrrrrrrrrr}
- & 2.75x_1 & - & 4x_2 & + & 0x_3 & + & 0x_4 & + & 0x_5 & = & f \\
+ & 1.5x_1 & + & x_2 & + & x_3 & & & & & = & 6 \\
+ & x_1 & + & 1.5x_2 & & & + & x_4 & & & = & 6 \\
+ & x_1 & + & x_2 & & & & & + & x_5 & = & 4.5
\end{array}
\tag{4.36}
$$

This general form is typical of general LP problems formulated in the LP standard form. It is interesting and important to note that this representation has a special and very important structure:

$$\begin{bmatrix} \mathbf{c}^T & \mathbf{0} \\ \mathbf{A} & \mathbf{I} \end{bmatrix} \cdot \begin{bmatrix} \mathbf{x}_N \\ \mathbf{x}_B \end{bmatrix} = \begin{bmatrix} f \\ \mathbf{b} \end{bmatrix} \tag{4.37}$$

What is particularly desirable about this structure is that if $\mathbf{b} \ge \mathbf{0}$, we can immediately obtain an *initial feasible* solution, i.e., a solution that satisfies all problem constraints. This is of great practical importance. To demonstrate that, you only need to set $\mathbf{x}_N = \mathbf{0}$, and as $\mathbf{A}\mathbf{x}_N + \mathbf{I}\mathbf{x}_B = \mathbf{b}$, then $\mathbf{x}_B = \mathbf{b}$, and $f = 0$ is an initial feasible solution. The elements of the $\mathbf{x}_B$ vector are called *basic variables*, and the elements of the $\mathbf{x}_N$ vector are called the *nonbasic variables*, while $[\mathbf{x}_B{}^T,\ \mathbf{x}_N{}^T]$ is called a basic solution. $\mathbf{x}_B$ has always as many elements as equality constraints in the LP standard form, i.e., m, while $\mathbf{x}_N$ has n-m elements (number of variables minus the number of inequality constraints).

For the crude oil mix selection problem, we observe that the *initial basic feasible solution* is:

$$x = \begin{bmatrix} x_N \\ \cdots \\ x_B \end{bmatrix} = \begin{bmatrix} x_1 \\ x_2 \\ \cdots \\ x_3 \\ x_4 \\ x_5 \end{bmatrix} = \begin{bmatrix} m_{CrOil1} \\ m_{CrOil2} \\ \cdots \\ s_1 \\ s_2 \\ s_3 \end{bmatrix} = \begin{bmatrix} 0 \\ 0 \\ \cdots \\ 6 \\ 6 \\ 4.5 \end{bmatrix} = \begin{bmatrix} 0 \\ \cdots \\ b \end{bmatrix} \qquad (4.38)$$

and $f = 0$ (no crude oil is purchased, no products are produced and no revenues are generated). This point corresponds to the point $(0, 0)$ in Fig. 4.3. This is the origin and is a vertex point in the convex feasible region. This is not a coincidence. Every feasible basic solution corresponds to a vertex of the feasible space, and for every vertex, there is a basic feasible solution. As $(0, 0)$ is not the optimal point, we need to devise a method to reach the optimal point through a number of finite steps. In the simplex algorithm, we move from a vertex of the feasible space to a neighboring vertex if the objective function is improving by doing so. Neighboring vertex points have basic feasible solutions that share the same $m - 1$ basic variables and have one different basic variable. In every step of the simplex algorithm, we exchange one basic variable with one nonbasic variable, i.e., one basic variable becomes nonbasic but is replaced by a nonbasic variable (which becomes basic). So, there are three questions to be answered before developing an efficient LP solution algorithm:

- How do we identify an optimal solution (i.e., when the algorithm terminates)?
- In the case of a basic feasible solution that is not optimal, how do we select which nonbasic variable will become basic
- How do we select which basic variable will become nonbasic?

To answer the first question, we return to the crude oil mix selection and to Formulation (4.36). We note that at the initial basic solution, the nonbasic variables are zero, i.e., $x_1 = 0$ and $x_2 = 0$. As both x_1 and x_2 are non-negative variables, we can only increase them. As their coefficients are both negative, the objective function will improve (decrease) when we increase any of the nonbasic variables. We can immediately see that increasing the nonbasic variables will not result in improving the objective function when their coefficients in the objective function are all positive. This is the simple criterion for identifying the optimal solution: *all coefficients in the objective function that correspond to the nonbasic variables have to be positive at the optimum solution.*

If there are coefficients that are negative, then the vertex that corresponds to the current basic feasible solution is not the optimal solution. We need to move to a neighboring vertex. To achieve that, we can only (or allowed to) make a switch between one basic and one nonbasic variable. To decide which nonbasic variable will become basic, we examine the coefficients of the nonbasic variables in the objective function and locate the most negative coefficient (assume that this is coefficient c_p that corresponds to x_p). We select the most negative coefficient because for the same change in the variable, we achieve the maximum decrease in the objective function. For the crude oil mix selection problem, we note from Formulation (4.36) that x_1 has a coefficient of -2.75 and x_2 has a coefficient of -4. We therefore select x_2 to become basic variable.

As we increase x_2 while keeping $x_1 = 0$, the linear equalities in formulation (4.36) become:

$$
\begin{array}{rcll}
x_2 + x_3 & = & 6 & \\
1.5x_2 + x_4 & = & 6 & \quad (4.39) \\
x_2 + x_5 & = & 4.5 &
\end{array}
$$

As we increase x_2, the basic variables x_3, x_4, and x_5 must be decreased to ensure satisfaction of the equality constraints. We want to increase x_2 as far as possible, and we can decrease all basic variables, but they have to remain non-negative. At the extreme case, they become zero, and this happens when:

- $x_3 = 0 \Rightarrow x_2 = 6/1 = 6$.
- $x_4 = 0 \Rightarrow x_2 = 6/1.5 = 4$.
- $x_5 = 0 \Rightarrow x_2 = 4.5/1 = 4.5$.

The minimum among these ratios is the most restrictive condition, and the corresponding basic variable (x_4 in our case) will become nonbasic.

We have concluded that x_2 will become basic variable and x_4 will become nonbasic. We exchange their positions in (4.36):

$$
\begin{array}{rcll}
- 2.75x_1 + 0x_4 + 0x_3 - 4x_2 + 0x_5 & = & f & \\
+ 1.5x_1 + x_3 + x_2 & = & 6 & \\
+ x_1 + x_4 + 1.5x_2 & = & 6 & \quad (4.40) \\
+ x_1 + x_2 + x_5 & = & 4.5 &
\end{array}
$$

We are not finished yet as we have destroyed the structure of our table (needs to have the structure implied by Eq. (4.37)). To bring the table back into the standard format, we eliminate x_2 from all rows but the second one. This can be achieved by first dividing row 3 by the coefficient of x_2 in that row ($R3 \mapsto R3/1.5$):

$$
\begin{array}{rcll}
- 2.75x_1 + 0x_4 + 0x_3 - 4x_2 + 0x_5 & = & f & \\
+ 1.5x_1 + x_3 + x_2 & = & 6 & \\
+ \frac{2}{3}x_1 + \frac{2}{3}x_4 + x_2 & = & 4 & \quad (4.41) \\
+ x_1 + x_2 + x_5 & = & 4.5 &
\end{array}
$$

and then subtract row 3 from rows 2 (R2 $\mapsto$ R2 − R3) and 4 (R4 $\mapsto$ R4 − R3) (eliminate x_2):

$$
\begin{aligned}
-\quad 2.75x_1 \quad + \quad 0x_4 \quad + \quad 0x_3 \quad - \quad 4x_2 \quad + \quad 0x_5 \quad &= \quad f \\
+\quad \frac{5}{4}x_1 \quad - \quad \frac{2}{3}x_4 \quad + \quad x_3 \qquad\qquad\qquad\qquad &= \quad 2 \\
+\quad \frac{2}{3}x_1 \quad + \quad \frac{2}{3}x_4 \qquad\qquad\qquad + \quad x_2 \qquad\quad &= \quad 4 \\
+\quad \frac{1}{3}x_1 \quad - \quad \frac{2}{3}x_4 \qquad\qquad\qquad\qquad + \quad x_5 \quad &= \quad 0.5
\end{aligned}
\tag{4.42}
$$

We finally eliminate x_2 from the first row (R1 $\mapsto$ R1 + 4R3) (note that all coefficients of the basic variables are zero in the line corresponding to the objective function in (4.37)):

$$
\begin{aligned}
-\quad \frac{0.25}{3}x_1 \quad + \quad \frac{8}{3}x_4 \quad + \quad 0x_3 \quad + \quad 0x_2 \quad + \quad 0x_5 \quad &= \quad f + 16 \\
+\quad \frac{5}{4}x_1 \quad - \quad \frac{2}{3}x_4 \quad + \quad x_3 \qquad\qquad\qquad\qquad &= \quad 2 \\
+\quad \frac{2}{3}x_1 \quad + \quad \frac{2}{3}x_4 \qquad\qquad\qquad + \quad x_2 \qquad\quad &= \quad 4 \\
+\quad \frac{1}{3}x_1 \quad - \quad \frac{2}{3}x_4 \qquad\qquad\qquad\qquad + \quad x_5 \quad &= \quad 0.5
\end{aligned}
\tag{4.43}
$$

Note that we have restated the problem in the standard form, and the solution follows immediately by simply setting the nonbasic variables equal to zero:

$$
x = \begin{bmatrix} x_N \\ \cdots \\ x_B \end{bmatrix} = \begin{bmatrix} x_1 \\ x_4 \\ \cdots \\ x_3 \\ x_2 \\ x_5 \end{bmatrix} = \begin{bmatrix} m_{\text{CrOil1}} \\ s_2 \\ \cdots \\ s_1 \\ m_{\text{CrOil2}} \\ s_3 \end{bmatrix} = \begin{bmatrix} 0 \\ 0 \\ \cdots \\ 2 \\ 4 \\ 0.5 \end{bmatrix} = \begin{bmatrix} 0 \\ \cdots \\ b \end{bmatrix}
\tag{4.44}
$$

This new basis corresponds to the point (4, 0) in Fig. 4.3 and is again, as expected, a vertex solution. The objective function has decreased to $f = -16$. The slack variables $s_1 = 2$ and $s_3 = 0.5$ show how far away we are from "hitting" the first and third constraints. Is this solution optimal? The answer is "no" as the coefficient of x_1 in the first line of Formulation (4.43) is negative. We have therefore completed a step of the simplex algorithm, but we are not finished yet!

We have identified x_1 as the nonbasic variable to become basic by examining the condition:

$$p = \operatorname*{argmin}_{i}\{c_i\} \tag{4.45}$$

i.e., p is the index i (argument) that corresponds to the most negative c_i (apparently, if $c_p \geq 0$, we are at the optimum solution, and the algorithm would have terminated at the previous step). To locate which basic variable will become nonbasic, we perform the following calculation:

$$q = \arg\min_{i:a_{ip} > 0} \frac{b_i}{a_{ip}} \tag{4.46}$$

i.e., we calculate all ratios b_i/a_{ip} and locate the most restrictive (smaller positive ratio). For our case study:

$$
\begin{array}{rcl}
-\dfrac{0.25}{3}x_1 + \dfrac{8}{3}x_4 + 0x_3 + 0x_2 + 0x_5 &=& f+16 \\[2mm]
+\dfrac{5}{4}x_1 - \dfrac{2}{3}x_4 + x_3 &=& 2 \\[2mm]
+\dfrac{2}{3}x_1 + \dfrac{2}{3}x_4 + x_2 &=& 4 \\[2mm]
+\dfrac{1}{3}x_1 - \dfrac{2}{3}x_4 + x_5 &=& 0.5
\end{array}
\tag{4.47}
$$

- $x_3 = 0 \Rightarrow x_1 = 2/(5/4) = 8/5 = 1.6.$
- $x_2 = 0 \Rightarrow x_2 = 4/(2/3) = 6.$
- $x_5 = 0 \Rightarrow x_2 = 0.5/(1/3) = 1.5.$

It follows that x_5 will become the new nonbasic variable. It is left as an exercise to the reader to show that one additional iteration is necessary, the optimum solution shown in Fig. 4.3. This completes our presentation of the simplex algorithm which was based on the assumption that the initial formulation was given in the LP standard form given by (4.32) or (4.33). In some cases, however, this is not the case, and some preliminary steps are necessary to transform (if possible) out formulation into the standard form. There are numerous books (such as Dantzig and Thapa, *Linear Programming*, Springer, 1997) that present all details in a comprehensive way for the interested reader. Having said that, we need to inform the reader that all modern LP solvers accept formulations that are not in the standard form and perform all adjustments (if necessary) automatically. Our formulation has indicated all details of the simplex algorithm that are important from the user's point of view.

Before closing the presentation of the simplex algorithm, we need to emphasize that not all LP formulations need to have a solution or a unique solution. If, for example, we have the constraints $x < 1$ and $x > 5$ in our formulation, then our problem has no feasible solution as there is no x that is less than 1 and greater than 5 at the same time. A modern LP solver will terminate with an appropriate warning message to inform us on this unexpected feature of our formulation. A LP formulation can also have many solutions when the objective function happens to be parallel to a constraint that is active at the optimum point. All points on the constraint also correspond to the same value of the objective function. If a LP problem has a feasible solution, it will also have a basic feasible solution and, at least, one optimum point (which is always going to be a vertex point). Finally, although the standard LP formulation involves only equality constraints and non-negative variables, any modern LP solver can handle problems of the general LP form that involves both equality and inequality constraints and lower and upper bound on $\mathbf{x}$:

$$\min_{\mathbf{x}} \ f = \mathbf{c}^T \mathbf{x}$$

$$\text{s.t.} \hspace{4cm} (4.48)$$

$$\mathbf{A}\mathbf{x} \leq \mathbf{b}$$

$$\mathbf{A}_{eq}\mathbf{x} = \mathbf{b}_{eq}$$

$$LB \leq \mathbf{x} \leq UB$$

4.5   Solving LP Problems in MATLAB®

The LP solver available in MATLAB® solves the general problem given by (4.48), and its general structure is the following:

[xopt,fopt,flag,output,lambda] =**linprog**(c,A,b,Aeq,beq,LB,UB)
attempts to solve the linear programming problem given by:
min c'*x subject to: A*x <= b,Aeq*x=beq&LB<x< UB
Set A=[] and B=[] if no inequalities exist. Use empty matrices for LB and UB if no bounds exist. xopt is the optimal solution and fopt the optimum value of the objective function. When flag=1 the global solution has been found. Lambda is s structure with the Lagrange multipliers.

To demonstrate the use of linprog, we consider the crude oil mix selection problem as it is summarized by Formulation (4.17). As linprog solves minimization problems, we use the negative of the objective function:

```
>> clear
>> c=[-2.75;-4];
>> A=[1.5 1; 1 1.5;1 1]; b=[6;6;4.5];
```

(continued)

```
>> LB=[0;0]; UB=[6;6];
>> [xopt,fopt,flag,output,lamda]=linprog(c,A,b,[],[],LB,UB)

Optimal solution found.

xopt = 1.5000
       3.0000

fopt = -16.1250

flag =  1

output =
  struct with fields:
           iterations: 2
        constrviolation: 0
              message: 'Optimal solution found.'
            algorithm: 'dual-simplex'
        firstorderopt: 2.2204e-15

lamda =
  struct with fields:
        lower: [2×1 double]
        upper: [2×1 double]
        eqlin: []
      ineqlin: [3×1 double]

>> lamda.ineqlin

ans =    0
      2.5000
      0.2500
```

The solution obtained in MATLAB agrees with the solution obtained graphically in Fig. 4.3 and the solution obtained by the "manual" application of the simplex algorithm. Note that the Lagrange multipliers for the second and third constraints that are active are different than zero. To demonstrate how they can be calculated, we may apply Eq. (3.6):

$$\nabla f + \sum_{i=1}^{p} \mu_i \nabla g_i = \nabla(c_1 x_1 + c_1 x_1) + \mu_2 \nabla(a_{21} x_1 + a_{22} x_2)$$
$$+ \mu_3 \nabla(a_{31} x_1 + a_{32} x_2) = \mathbf{0}$$

or:

$$\begin{bmatrix} c_1 \\ c_2 \end{bmatrix} + \mu_2 \begin{bmatrix} a_{21} \\ a_{22} \end{bmatrix} + \mu_3 \begin{bmatrix} a_{31} \\ a_{32} \end{bmatrix} = \mathbf{0} \qquad (4.49)$$

This can also be written as:

$$\begin{bmatrix} a_{21} & a_{31} \\ a_{22} & a_{32} \end{bmatrix} \begin{bmatrix} \mu_2 \\ \mu_3 \end{bmatrix} = - \begin{bmatrix} C_1 \\ C_2 \end{bmatrix} \tag{4.50}$$

We substitute the numerical values to obtain:

$$\begin{bmatrix} 1 & 1 \\ 1.5 & 1 \end{bmatrix} \begin{bmatrix} \mu_2 \\ \mu_3 \end{bmatrix} = \begin{bmatrix} 2.75 \\ 4 \end{bmatrix} \tag{4.51}$$

which has the solution $[\mu_1 \ \mu_2] = [2.5 \ 0.25]$. As the Lagrange multipliers are positive, this is a KKT point, and given the fact that the problem is convex, it follows that it is the global optimal solution.

We now return to the clinker production problem for which we have to finalize the mathematical formulation. We start by using the data in Table 4.2 to write Eq. (4.27) in the following form:

$$\min_{m_L, m_C, m_B, m_P} 4m_L + m_C + 40m_B + 25m_P \tag{4.52}$$

The mass of the oxides in the clinker is obtained from Eq. (4.21) by substituting the numbers (mass fractions $\Omega_{i,r}$) given in Table 4.2:

$$\begin{bmatrix} m^c_{SiO_2} \\ m^c_{Al_2O_3} \\ m^c_{Fe_2O_3} \\ m^c_{CaO} \end{bmatrix} = \begin{bmatrix} 0.10 & 0.55 & 0.01 & 0.05 \\ 0.02 & 0.10 & 0.40 & 0.03 \\ 0.01 & 0.05 & 0.35 & 0.85 \\ 0.50 & 0.13 & 0.15 & 0.02 \end{bmatrix} \begin{bmatrix} m_L \\ m_C \\ m_B \\ m_P \end{bmatrix}$$

or:

$$\begin{bmatrix} 0.10 & 0.55 & 0.01 & 0.05 & -1 & 0 & 0 & 0 \\ 0.02 & 0.10 & 0.40 & 0.03 & 0 & -1 & 0 & 0 \\ 0.01 & 0.05 & 0.35 & 0.85 & 0 & 0 & -1 & 0 \\ 0.50 & 0.13 & 0.15 & 0.02 & 0 & 0 & 0 & -1 \end{bmatrix} \begin{bmatrix} m_L \\ m_C \\ m_B \\ m_P \\ m^c_{SiO_2} \\ m^c_{Al_2O_3} \\ m^c_{Fe_2O_3} \\ m^c_{CaO} \end{bmatrix} = \begin{bmatrix} 0 \\ 0 \\ 0 \\ 0 \end{bmatrix}$$

$$\tag{4.53}$$

We also need to add constraint (4.22) which expresses the fact that the mass of the produced clinker (oxides) is 1000 kg:

$$[0\ 0\ 0\ 0\ 1\ 1\ 1\ 1] \begin{bmatrix} m_L \\ m_C \\ m_B \\ m_P \\ m_{SiO_2}^c \\ m_{Al_2O_3}^c \\ m_{Fe_2O_3}^c \\ m_{CaO}^c \end{bmatrix} = 1000 \tag{4.54}$$

We then express the upper and lower bound constraints on AR, SR, and LSF (see Eqs. (4.18), (4.19), and (4.20)):

$$-m_{Al_2O_3}^c + AR_L m_{Fe_2O_3}^c \le 0 \tag{4.55a}$$

$$m_{Al_2O_3}^c - AR_U m_{Fe_2O_3}^c \le 0 \tag{4.55b}$$

$$-m_{SiO_2}^c + SR_L m_{Al_2O_3}^c + SR_L m_{Fe_2O_3}^c \le 0 \tag{4.56a}$$

$$m_{SiO_2}^c - SR_U m_{Al_2O_3}^c - SR_U m_{Fe_2O_3}^c \le 0 \tag{4.56b}$$

$$2.8LSF_L m_{SiO_2}^c + 1.1LSF_L m_{Al_2O_3}^c + 0.65LSF_L m_{Fe_2O_3}^c - m_{CaO}^c \le 0 \tag{4.57a}$$

$$-2.8LSF_U m_{SiO_2}^c - 1.1LSF_U m_{Al_2O_3}^c - 0.65LSF_U m_{Fe_2O_3}^c + m_{CaO}^c \le 0 \tag{4.57b}$$

where the subscript L denotes lower bound and the subscript U upper bound (see also Eqs. (4.23) and (4.24) on how Eqs. (4.55a) and (4.55b) can be derived, and the approach is exactly the same for Eqs. (4.56) and (4.57)). The inequality constraints can also be written in the following matrix form:

$$\begin{bmatrix} 0 & 0 & 0 & 0 & 0 & -1 & AR_L & 0 \\ 0 & 0 & 0 & 0 & 0 & +1 & -AR_U & 0 \\ 0 & 0 & 0 & 0 & -1 & SR_L & SR_L & 0 \\ 0 & 0 & 0 & 0 & +1 & -SR_U & -SR_U & 0 \\ 0 & 0 & 0 & 0 & 2.8LSF_L & 1.1LSF_L & 0.65LSF_L & -1 \\ 0 & 0 & 0 & 0 & -2.8LSF_U & -1.1LSF_U & -0.65LSF_U & +1 \end{bmatrix} \begin{bmatrix} m_L \\ m_C \\ m_B \\ m_P \\ m_{SiO_2}^c \\ m_{Al_2O_3}^c \\ m_{Fe_2O_3}^c \\ m_{CaO}^c \end{bmatrix} \le \begin{bmatrix} 0 \\ 0 \\ 0 \\ 0 \\ 0 \\ 0 \end{bmatrix}$$

$$\tag{4.58}$$

We have therefore completed the formulation, and we can solve the problem in MATLAB (which we organize in a script file due to a large volume of information necessary to use linprog for this case study):

```
c=[4 1 40 25 0 0 0 0];
Omega=[0.10 0.55 0.01 0.05;
       0.02 0.10 0.40 0.03;
       0.01 0.05 0.35 0.85;
       0.50 0.13 0.15 0.02];
Aeq=[Omega        -eye(4);
   zeros(1,4) ones(1,4)]; % see eq (4.5.5) & (4.5.6)
beq=[zeros(1,4) 1]';
ARL = 1.0;  ARU = 2.7;
SRL = 1.5;  SRU = 2.5;
LSFL= 0.85; LSFU= 1.02;
A=[ 0            -1         ARL          0;
    0             1        -ARU          0;
   -1           SRL         SRL          0
    1          -SRU        -SRU          0;
   2.8*LSFL  1.1*LSFL   0.65*LSFL  -1;
  -2.8*LSFU -1.1*LSFU  -0.65*LSFU   1];
A=[ zeros(6,4) A]; % see eq (4.5.9)
b=zeros(6,1);
LB=zeros(8,1);  UB=[];
[x,f,flag,output,lambda]=linprog(c,A,b,Aeq,beq,LB,[])
```

```
>> clinker
Optimal solution found.

x = 1.2551
    0.2150
    0.0040
    0.0287
    0.2452
    0.0490
    0.0490
    0.6567
f = 6.1103
flag = 1
```

It follows that at the optimal solution, 1,255.1 kg of limestone (83.5%), 215 kg of clay (14.3%), 4 kg of bauxite (0.27%), and 28.7 kg of pyrite (1.91%) are mixed to produce 1000 kg of clinker. The mixture contains 245.2 kg of SiO_2, 49 kg of Al_2O_3, 49 kg of Fe_2O_3, and 656.7 kg of CaO (1000 kg in total). The AR, SR, and LSF obtain the values:

$$AR = \frac{m^c_{Al_2O_3}}{m^c_{Fe_2O_3}} = \frac{49}{49} = 1 = AR_L$$

$$SR = \frac{m^c_{SiO_2}}{m^c_{Al_2O_3} + m^c_{Fe_2O_3}} = \frac{245.2}{98} = 2.5 = SR_U$$

$$LSF = \frac{m^c_{CaO}}{2.8 m^c_{SiO_2} + 1.1 m^c_{Al_2O_3} + 0.65 m^c_{Fe_2O_3}}$$

$$= \frac{656.7}{2.8 \cdot 245.2 + 1.1 \cdot 48 + 0.65 \cdot 48} = 0.85 = LSF_L$$

As the corresponding linear inequalities are active, the corresponding Lagrange multipliers must be positive:

```
>> lambda.ineqlin

ans = 23.2086 <--- ARL  < AR  is active
           0 <--- AR   < ARU
           0 <--- SRL  < SR
     18.2160 <--- SR   < SRU is active
      1.5723 <--- LSFL < LSF is active
           0 <--- LSF  < LSFU
```

4.6   Classical LP Formulations

There are some classical LP formulations of great importance as they usually form the basis for building formulations for realistic case studies. Most of them have been the subject of extensive research, and, for some of them, specialized algorithms have been developed that outperform general LP solution algorithms. We will present just a sample of these problems with the aim to improve the modeling skills of the reader.

Example 4.1 *The Transportation Problem and Related Problems*

In the transportation problem, we consider m origins (or sources) and n available destinations (or sinks). Each origin i can supply S_i units of a commodity, and each destination j demands D_j units of the same commodity. The general structure of the transportation problem is shown in Fig. 4.4, where c_{ij} is the unit transportation cost corresponding to the link (i, j). The variables of the problem are denoted by x_{ij} and correspond to the amount of commodity shipped from origin i to destination j. If we assume that the total demand equals the total supply (if this is not the case, a dummy origin or a dummy destination can be assumed to satisfy the requirement), then the transportation problem has the following LP formulation:

Fig. 4.4 Representation of the transportation problem

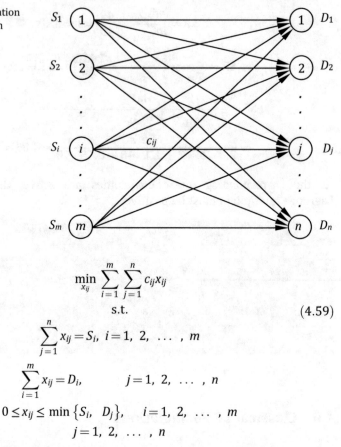

$$\min_{x_{ij}} \sum_{i=1}^{m} \sum_{j=1}^{n} c_{ij} x_{ij}$$

s.t.

$$\sum_{j=1}^{n} x_{ij} = S_i, \quad i = 1, \ 2, \ \dots, \ m$$

$$\sum_{i=1}^{m} x_{ij} = D_i, \qquad j = 1, \ 2, \ \dots, \ n$$

$$0 \le x_{ij} \le \min\{S_i, \ D_j\}, \quad i = 1, \ 2, \ \dots, \ m$$
$$j = 1, \ 2, \ \dots, \ n$$

(4.59)

The first m equality constraints state that the total amount shipped from origin i must always be equal to the available supply from this origin S_i. The remaining n equality constraints ensure that the demand D_j of each destination j is satisfied. If we take a problem with three origins and three destinations, then we may write the equations in the following matrix form (only the nonzero elements are shown to indicate the special structure of the $\mathbf{A}_{eq}$ matrix):

$$
\begin{bmatrix}
1 & 1 & 1 & & & & & & \\
& & & 1 & 1 & 1 & & & \\
& & & & & & 1 & 1 & 1 \\
1 & & & 1 & & & 1 & & \\
& 1 & & & 1 & & & 1 & \\
& & 1 & & & 1 & & & 1
\end{bmatrix}
\begin{bmatrix}
x_{11} \\ x_{12} \\ x_{13} \\ x_{21} \\ x_{22} \\ x_{23} \\ x_{31} \\ x_{32} \\ x_{33}
\end{bmatrix}
=
\begin{bmatrix}
s_1 \\ s_2 \\ s_3 \\ d_1 \\ d_2 \\ d_3
\end{bmatrix}
\qquad (4.60)
$$

This structure can be generated quite efficiently in MATLAB if the number of origins and the destinations are known together with the supply at each source and the demand at each destination. The following m-file uses this information together with the unit cost for each arc arranged in a m by n matrix, generates the matrix of the equality constraints, and then uses `linprog` to solve the problem:

```
function [xopt,fopt,flag]=transportationLP(C,S,D)
% build and solve a transportation model
% C is the m-by-n matrix with the unit costs
% S is the m dimensional column vector with the supplies
% D is the n dimensional column vector with the demands
if (sum(S)-sum(D)) % if not balanced give a warning
   fprintf('Supply is not equal to demand'); pause
   xopt=[];fopt=[];flag=-inf;   return;
end
m=length(S) ; % number of origins
n=length(D) ; % number of destinations

c=reshape(C',m*n,1);

Aeq=[ kron(eye(m),ones(1,n));
      kron(ones(1,m),eye(n))];

beq=[S;D]';

[xopt,fopt,flag]=linprog(c,[],[],Aeq,beq,zeros(m*n,1));
```

The following script can be used to solve the transportation problem shown in Fig. 4.5:

Fig. 4.5 Representative
example of a
transportation problem

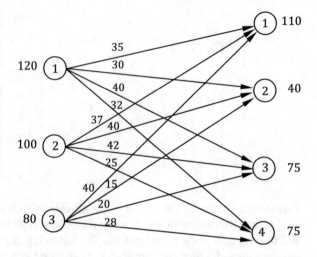

```
clear
C=[ 35 30 40 32;
    37 40 42 25;
    40 15 20 28];
S=[120 100 80]';
D=[110 40 75 75]';
[xopt,fopt,flag]=transportationLP(C,S,D)
```

```
>> run_tranportationLP_1
Optimization terminated.

xopt = 85.0000 <-- x11
       35.0000 <-- x12
        0.0000
        0.0000
       25.0000 <-- x21
        0.0000
        0.0000
       75.0000 <-- x24
        0.0000
        5.0000 <-- x32
       75.0000 <-- x33
        0.0000

fopt = 8.4000e+03
flag = 1
```

In the transportation problem, we consider m origins and n destinations
where each origin i can supply S_i units of a commodity, and each destination

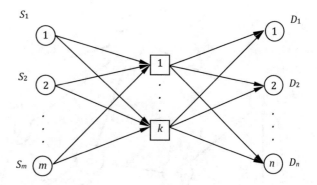

Fig. 4.6 Representation of the transshipment problem

j demands D_j units of the same commodity. A closely related problem is the assignment of a set of objects (such as workers) to a set of assignments or jobs. This is the well-known *assignment problem*, a special case of the transportation problem with $n = m$ and $S_i = D_j = 1$ (equal number of objects and assignments and unit supply or demand). A specialized algorithm for solving the assignment problem is the Hungarian algorithm, but it can also be solved as an LP problem.

The most interesting problem related to the transportation problem is that of the transshipment model. In the transportation model, the assumption is that each point is either an origin or a destination (objects are available at the origins and are required in the destinations). There are many practical applications with intermediate points, which are not points where the objects are available or required. These intermediate points are points where the objects can be transshipped before they are transported to their final destinations (see Fig. 4.6). A classical problem is that of municipal solid waste collection systems where the waste is first collected and then transshipped to local collection points. At these intermediate points, some initial sort of processing takes place, before finally the waste is transported to its final destination. Transshipment problems as well as transportation and assignments models can also be seen as special cases of a more general class of problems called *network models*.

Example 4.2 *Network Models*

Networks consist of *nodes* (denoted by circles) and *arcs* (denoted by lines connecting the nodes). When an arc (i, j) connects nodes i and j, then some form of material, energy, information, or general object can "flow" between the nodes. If the arc is a *directed arc*, then its direction is denoted by putting an arrow at the end of the arc into the terminal node. No such restriction applies to an *undirected arc*, and "flow" can be in both directions (a single line is used). Lower and/or upper bounds (minimum and maximum capacity bounds) can exist to the amount of "flow" in each arc.

A representative network with 10 nodes and 15 undirected arcs is shown in Fig. 4.7. This can be thought of as roads connecting ten cities or pipelines

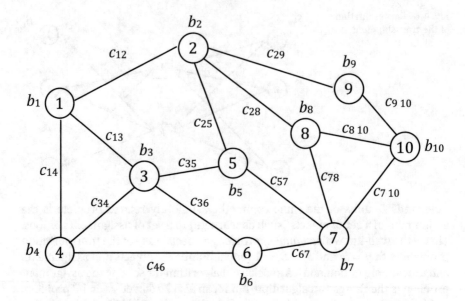

Fig. 4.7 Representation of the transshipment problem

connecting storage facilities with production and consumption facilities. To represent these cases, we associate the parameter c_{ij} to the arc (i, j) connecting node i and node j to denote some kind of unit cost (or shipping cost or weight) to objects moving between nodes i and j. We also associate a number b_i to each node denoting either the available supply of an object ($b_i > 0$) or the required demand ($b_i < 0$) for the object, i.e., $b_i > 0$ for sources and $b_i < 0$ for destinations or sinks. When $b_i = 0$, then node i is an intermediate (or transshipment) node. We also associate with each arc (i, j) the variable x_{ij} to denote the amount of "flow" on the arc. We finally assume that $\sum_{i=1}^{n} b_i = 0$, i.e., the total supply equals the total demand (if this is not the case, we can add a dummy node to satisfy the constraint).

We can now present the general LP formulation for a network problem which is the following:

$$\min_{x_{ij}} \sum_{i=1}^{m} \sum_{j=1}^{n} c_{ij} x_{ij}$$

$$\text{s.t.} \tag{4.61}$$

$$\sum_{j=1}^{n} x_{ij} - \sum_{j=1}^{n} x_{ji} = b_i, \qquad i = 1, 2, \ldots, n$$

$$0 \leq x_{ij} \leq UB_{ij}, \qquad i, j = 1, 2, \ldots, n$$

The equality constraint is a balance constraint and expresses the fact that objects need to be conserved in nodes (cannot be destroyed or generated from nothing). The first summation is the total amount leaving node i, while the second summation is the total amount that is fed to node i. Their difference must be equal to b_i, i.e., the supply or demand at the node. This constraint is also known as the flow conservation or Kirchhoff equation.

Two well-known cases of the general network model are the models of the maximum flow and the model of the shortest route. The maximum flow formulation between nodes 1 and n is the following:

$$\max_{x_{ij}} f$$

$$\text{s.t.} \tag{4.62}$$

$$\sum_{j=1}^{n} x_{ij} - \sum_{j=1}^{n} x_{ji} = \begin{cases} +f, & i=1 \\ 0, & 1<i<n \\ -f, & i=n \end{cases}$$

$$0 \le x_{ij} \le UB_{ij}, \quad i, j = 1, 2, \ldots, n$$

Observe that there is only one source (node 1) and only one destination (node n). The formulation seeks to calculate the maximum flow available at node 1 that can be transported through the network to the destination.

The shortest route formulation is the following:

$$\min_{x_{ij}} \sum_{i=1}^{n} \sum_{j=1}^{n} \text{dist}_{ij} x_{ij}$$

$$\text{s.t.} \tag{4.63}$$

$$\sum_{j=1}^{n} x_{ij} - \sum_{j=1}^{n} x_{ji} = \begin{cases} +1, & i=1 \\ 0, & 1<i<n \\ -1, & i=n \end{cases}$$

$$0 \le x_{ij} \le 1, \quad i, j = 1, 2, \ldots, n$$

Note that we have replaced the unit cost c_{ij} by dist_{ij} to stress the fact that we are minimizing the total distance covered. Even though we do not impose any integrality constraint on x_{ij}, the solution is always $x_{ij} = 1$ or $x_{ij} = 0$ (binary). This is an amazing characteristic of the particular formulation and is due to the fact that the $\mathbf{A}_{eq}$ matrix is unimodular. In Fig. 4.8, a hypothetical network is shown with the distances between the nodes indicated. The distances are also given in the following matrix:

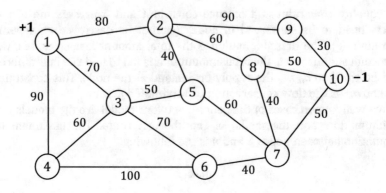

Fig. 4.8 Example case study for the shortest route problem

$$\textbf{\textit{dist}} = \begin{bmatrix}
0 & 80 & 70 & 90 & M & M & M & M & M & M \\
80 & 0 & M & M & 40 & M & M & 60 & 90 & M \\
70 & M & 0 & 60 & 50 & 70 & M & M & M & M \\
90 & M & 60 & 0 & M & 100 & M & M & M & M \\
M & 40 & 50 & M & 0 & M & 60 & M & M & M \\
M & M & 70 & 100 & M & 0 & 40 & M & M & M \\
M & M & M & M & 60 & 40 & 0 & 40 & M & 50 \\
M & 60 & M & M & M & M & 40 & 0 & M & 50 \\
M & 90 & M & M & M & M & M & M & 0 & 30 \\
M & M & M & M & M & M & 50 & 50 & 30 & 0
\end{bmatrix} \qquad (4.64)$$

The M symbol is used to denote a nonexistent connection between the corresponding nodes. For the LP implementation, a sufficiently large number can be used for M that will make the selection of the corresponding arc particularly unattractive (remember that we are solving a minimization problem). This technique is known as the Big-M method. A potential implementation of the shortest route model is given in the m-file that follows:

```
function [xopt,fopt,flag]=ShortestRouteLP(dist)

% Solving the Shortest Route problem
% Using LP formulation
% dist is the n-by-n matrix with the distances between nodes

[n,dummy]=size(dist) ; % number of nodes
```

(continued)

```
c=reshape(dist',n*n,1)

Aeq= kron(eye(n),ones(1,n)) + kron(ones(1,n),-eye(n));

beq=zeros(n,1); beq(1)=1; beq(n)=-1;

[xopt,fopt,flag]=...
        linprog(c,[],[],Aeq,beq,zeros(n*n,1),ones(n*n,1));

return
```

The solution to the example shown in Fig. 4.8 can be obtained as follows:

```
clear
M=200;
C=[ M 80 70  90  M   M   M  M   M   M;
    80  M  M   M  40   M  M 60  90   M;
    70  M  M  60 50  70  M  M   M   M;
    90  M 60   M  M 100  M  M   M   M;
     M 40 50   M  M   M 60  M   M   M;
     M  M 70 100  M   M 40  M   M   M;
     M  M  M   M 60  40  M 40   M  50;
     M 60  M   M  M   M 40  M   M  50;
     M 90  M   M  M   M  M  M   M  30;
     M  M  M   M  M   M 50 50  30   M];
[xopt,fopt,flag]=ShortestRouteLP(C)
```

The solution obtained (too long to be included) involves moving from node 1 to node 2, then to node 8, and finally to node 10 with a minimum distance of 190.

Example 4.3 *Cutting Stock (or Trim Loss) Models*

Cutting stock problems (or trim loss problems) deal with the problem of producing objects of prespecified size(s) and shape(s) from given initial objects so as to minimize the loss (unused material) or some more general and detailed cost-related objective function. The problem can involve 1D, 2D, or even 3D objects. As an example, we consider the case of 1D objects shown in Fig. 4.9. We assume that rods of three different lengths are available: 10 cm, 7 cm, and 4 cm. We want to produce rods with sizes 5 cm (150 pieces) and 3 cm (100 pieces), and we want to minimize the loss while satisfying the demand. Each cut we perform has a (labor and machine setting related) cost associated with it. One way to formulate the problem is to enumerate all cutting patterns that are compatible with our requirements. The six different cutting patterns that we

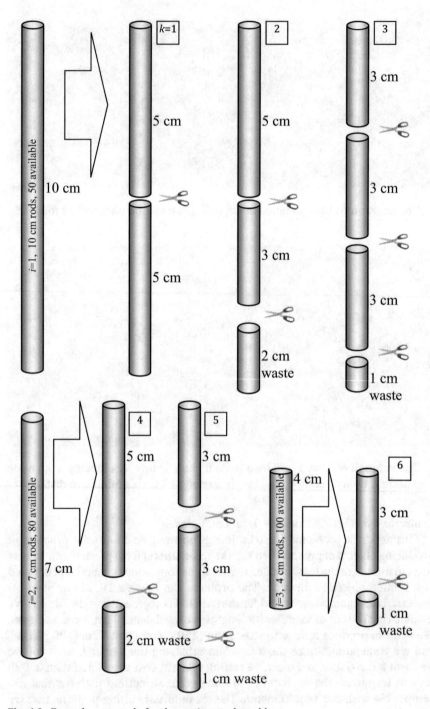

Fig. 4.9 Example case study for the cutting stock problem

can use to produce the requested rods from the ones available are shown in Fig. 4.9 (numbers in squares denote the number of the specific cutting pattern).

To develop a skeleton formulation of the problem at hand, let's use the index i to denote the available rods ($i = 1, 2, 3$) and S_i to denote the supply available of rod i. We will also be using index j to denote the type of the rod in demand ($j = 1$ for the 5 cm rods and $j = 2$ for the 3 cm rods) and D_j the demand of rod type j. We will be using the index k to denote the six different cutting patterns shown in Fig. 4.9 and Y_{jk} to denote the number of rods of type j produced from cutting pattern k. In our example:

$$
\begin{array}{c c c c c c c}
Y_{jk} & 1 & 2 & 3 & 4 & 5 & 6 \\
1 & \left[\begin{array}{c} 2 \\ 0 \end{array}\right. & \begin{array}{c} 1 \\ 1 \end{array} & \begin{array}{c} 0 \\ 3 \end{array} & \begin{array}{c} 1 \\ 0 \end{array} & \begin{array}{c} 0 \\ 2 \end{array} & \left.\begin{array}{c} 0 \\ 1 \end{array}\right]
\end{array}
\tag{4.65}
$$

Finally, the variable x_k denotes the number of times cutting pattern k is used in our solution and the constant matrix $\mathbf{A}$ with elements $a_{ik} = 1$ when cutting pattern k is produced from rod i, and $a_{ik} = 0$ otherwise. In our case:

$$
\begin{array}{c c c c c c c}
a_{ik} & 1 & 2 & 3 & 4 & 5 & 6 \\
1 & \left[\begin{array}{c} 1 \end{array}\right. & 1 & 1 & & & \\
2 & & & & 1 & 1 & \\
3 & & & & & & \left.1\right]
\end{array}
\tag{4.66}
$$

The constraints of the formulation are the following:

$$
\sum_{k=1}^{6} a_{ik} x_k \leq S_i, \quad i = 1,2,3
\tag{4.67}
$$

$$
\sum_{k=1}^{6} Y_{jk} x_k = D_j, \quad j = 1,2
\tag{4.68}
$$

The first constraint ensures that we will not be using more rods than the ones available and the second one that demand is satisfied. A possible objective function is the cost of rods used to satisfy the demands and the cost of cutting the rods. We assume a basis of the unit cost which is the cost of the rods of length 4 cm (assumed $c_3 = 1$ monetary unit). Then, we set arbitrarily the relative cost of the 7 cm rods to $c_2 = 1.5$ and the relative cost of the 10 cm rods to $c_1 = 2$. Each cutting operation is assigned a cost of $c_{cut} = 0.1$ monetary units relative to the unit cost of the 4 cm rods. The objective function is:

$$f = 2\,(x_1 + x_2 + x_3) + 1.5\,(x_4 + x_5) + x_6$$
$$+ 0.1(x_1 + 2x_2 + 3x_3 + x_4 + 2x_5 + x_6)$$

or:

$$f = \sum_{i=1}^{3} c_i \sum_{k=1}^{6} a_{ik} x_k + c_{\text{cut}} \sum_{k=1}^{6} \kappa_k x_k \qquad (4.69)$$

where κ_k is the number of cuts necessary to implement cutting pattern k. Finally, we assume that the demand is 150 for the 5 cm rods and 100 for the 3 cm rods. The problem can be solved in MATLAB as follows:

```
clear
A=[1 1 1 0 0 0;
   0 0 0 1 1 0;
   0 0 0 0 0 1];
S=[50; 80; 100];
Aeq=[2 1 0 1 0 0;
     0 1 3 0 2 1];
D=[150; 100];
c=[2.1 2.2 2.3 1.6 1.7 1.1];
[xopt,fopt,flag]=linprog(c,A,S,Aeq,D,zeros(6,1))
```

```
>> CuttingStock

Optimal solution found.

xopt = 50.0000
            0
            0
       50.0000
       30.0000
       40.0000

fopt = 280

flag = 1
```

The optimal solution involves cutting 50 10 cm rods according to pattern 1 to produce 100 5 cm rods and cutting 50 7 cm rods into 5 cm rods to satisfy fully the demand for 150 5 cm rods. Thirty 7 cm rods are cut into sixty 3 cm

Table 4.3 Data for the employee scheduling case study

Slot	00:00– 04:00	04:00– 08:00	08:00– 12:00	12:00– 16:00	16:00– 20:00	20:00– 24:00
Slot ID (i)	1	2	3	4	5	6
Min. number of employees (d_i)	4	6	18	14	10	6

rods, and 40 4 cm rods are cut into 40 3 cm rods to satisfy the demand for 100 3 cm rods. The overall cost is 280 monetary units relative to the unit cost of 4 cm rods. Before closing this example, it is important to emphasize that the variables appearing in the mathematical formulation for the one-dimensional cutting stock problem are integer variables (as only integer patterns can be cut). The solution obtained by linear programming may involve non-integer solution in which case rounding to the nearest integer that satisfies the demand can be an acceptable solution.

Example 4.4 *Employee Scheduling Models*

In this application example, we will consider a simplified scheduling problem that assigns employees to working shifts. We will consider the more basic form of the problem where time slots have been defined and minimum requirements for workers have been identified for each time slot. An example is shown in Table 4.3 where 4-hour time slots are considered. We assume that workers start their shift at the beginning of a time slot, and they work for two consecutive time slots (8-hour shifts). It then follows that at each time slot, the workers available are the ones that started their shift at the beginning of the current time slot or the previous time slot. In this simplified example, we will assume that all workers are identical in terms of skills required and unit cost. To present the mathematical formulation, we define the index $i = 1, 2, .., n$ to denote the time slots and d_i to denote the minimum number of employees in time slot i. We will also be using the variable x_i to denote the workers starting their shift at the beginning of time slot i. The objective function is to minimize the number of total employees necessary to satisfy the manning requirements for the n time slots:

$$\min_{x_i} \sum_{i=1}^{n} x_i = \mathbf{1}^T \mathbf{x} \tag{4.70}$$

where $\mathbf{1} = [\, 1\ 1\ 1 \ldots\ 1\,]^T$. The only constraint, apart from the non-negativity constraint on the x_i, expresses the fact that the workers that started their shift at the beginning of the current time slot or the previous time slot must satisfy the demand. We write them analytically for each time slot:

$$
\begin{array}{lllll}
\text{time slot 1:} & x_1 & & & +x_6 & \geq 4 \\
\text{time slot 2:} & x_1 & +x_2 & & & \geq 6 \\
\text{time slot 3:} & & x_2 & +x_3 & & \geq 18 \\
\text{time slot 4:} & & & x_3 & +x_4 & \geq 14 \\
\text{time slot 5:} & & & & x_4 & +x_5 & \geq 10 \\
\text{time slot 6:} & & & & & x_5 & +x_6 & \geq 6
\end{array}
\tag{4.71}
$$

In a matrix notation, we may write:

$$
\begin{bmatrix}
-1 & 0 & 0 & \cdots & 0 & -1 \\
-1 & -1 & 0 & \cdots & 0 & 0 \\
0 & -1 & -1 & \ddots & 0 & 0 \\
0 & 0 & -1 & \ddots & 0 & 0 \\
\vdots & \vdots & \ddots & \ddots & -1 & 0 \\
0 & 0 & \cdots & 0 & -1 & -1
\end{bmatrix}
\begin{bmatrix}
x_1 \\ x_2 \\ x_3 \\ \vdots \\ x_n
\end{bmatrix}
\leq -
\begin{bmatrix}
d_1 \\ d_2 \\ d_3 \\ \vdots \\ d_n
\end{bmatrix}
\tag{4.72}
$$

The solution to the problem stated can be obtained in MATLAB as follows:

```
clear
n=6;
c=ones(n,1);
D=-[4; 6; 18; 14; 10; 6];
A=diag(-ones(1,n))+diag(-ones(1,n-1),-1); A(1,n)=-1;
[xopt,fopt,flag]=linprog(c,A,D,[],[],zeros(n,1))
```

```
>> employeeScheduling
Optimal solution found.

xopt = 4
       2
      16
       4
       6
       0


fopt = 32
flag =   1
```

It follows that 32 workers need to be available; 4 of them start their shift at 00:00, 2 at 04:00, 16 at 08:00, 4 at 12:00, 6 at 16:00, and none at 20:00. The solution obtained is also shown in Fig. 4.10.

Example 4.5 *Production Scheduling Models*
In this application example, we will consider another simplified production scheduling problem. A common situation in many personal care, cosmetics, and pharmaceutical products is that their demand varies significantly during a calendar year as many of them are seasonal products (sun care products have, e.g., higher demand during spring and summer, or several flu vaccines have higher demand during autumn and winter). Let us assume that a pharmaceutical company has estimated the demand for its new flu vaccine for autumn and winter (September to February) given in Table 4.4. The company wishes to determine what is the yearly production schedule that minimizes

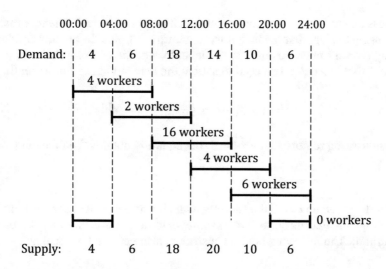

Fig. 4.10 Solution of the employee scheduling example

Table 4.4 Data for the vaccine production scheduling case study

Month	Sep	Oct	Nov	Dec	Jan	Feb
t	$t = 1$	2	3	4	5	6
Demand (unit) d_t	10,000	40,000	30,000	20,000	40,000	10,000
Production Cost in \$/unit pc_t	10	9	16	20	10	5
Storage cost in \$/unit sc_t	1	2	3	4	2	9 (Feb to Aug)

overall production cost. Apart from the varying demand, the company also faces varying production and storage costs that need to be taken into consideration.

These costs are also given in Table 4.4. The maximum production capacity is 50,000 vaccines per month. Storage capacity is also limited and cannot also exceed 100,000 vaccines.

The most important variable to determine is the monthly production p_t, $t = 1, 2, ..., n$. The next most important variable is the inventory I_t, i.e., the amount of the product kept in storage at the end of month t. Based on these two variables, we can propose the following objective function which accounts for the production and storage cost:

$$\min_{x_t, I_t} \sum_{t=1}^{6} (pc_t p_t + sc_t I_t) \qquad (4.73)$$

There is a material balance constraint which is applied at each calendar month, expressing the fact that the inventory at month $t - 1$ plus the amount produced during month t must be equal to the inventory at the end of month t plus the amount that has been delivered to satisfy the demand at the same month:

$$I_{t-1} + p_t = I_t + d_t, \quad t = 1,2,\cdots,6 \qquad (4.74)$$

It is interesting to note that for $t = 1$, the equation obtains the following form:

$$I_n + p_t = I_t + d_t, \quad t = 1 \qquad (4.75)$$

i.e., $t - 1 = n$, when $t = 1$ (the vaccines available on September are the ones left in storage in February, as we assume that no flu vaccine is sold in the meantime). The following bound constraints also need to be satisfied:

$$\left. \begin{array}{l} 0 \le p_t \le 50{,}000 \\ 0 \le I_t \le 100{,}000 \end{array} \right\}, \quad t = 1,2,\cdots,6 \qquad (4.76)$$

We define the vector of optimization variables **x** which has 12 elements; the first six are the productions for the 6 months, and the last six are the inventories:

$$x = [p_1 \quad p_2 \quad \cdots \quad p_6 \quad I_1 \quad I_2 \quad \cdots \quad I_6]^T \qquad (4.77)$$

The structure of the $\mathbf{A}_{eq}$ matrix is the following:

$$A_{eq} = \begin{bmatrix} 1 & 0 & 0 & \cdots & 0 & -1 & 0 & 0 & \cdots & 1 \\ 0 & 1 & 0 & \ddots & \vdots & 1 & -1 & 0 & \cdots & 0 \\ 0 & 0 & 1 & \ddots & 0 & 0 & 1 & -1 & \ddots & 0 \\ \vdots & \ddots & \ddots & \ddots & 0 & \vdots & \ddots & \ddots & \ddots & 0 \\ 0 & \cdots & 0 & 0 & 1 & 0 & \cdots & 0 & 1 & -1 \end{bmatrix} \qquad (4.78)$$

The solution to the problem stated can be obtained in MATLAB as follows:

```
clear
n=6; % number of scheduling periods
c=[10 9 16 20 10 5 1 2 3 4 2 9];
D=[10; 40; 30; 20; 40; 10]*1000;
Aeq=[eye(n) diag(-ones(1,n))+diag(ones(1,n-1),-1)];
Aeq(1,2*n)=1;
LB=zeros(2*n,1);
UB=[50*ones(n,1); 100*ones(n,1)]*1000;
[xopt,fopt,flag]=linprog(c,[],[],Aeq,D,LB,UB)
```

```
>> VaccineProductionScheduling

Optimal solution found.

xopt = 50000
       50000
           0
           0
       40000
       10000
       40000
       50000
       20000
           0
           0
           0


fopt = 1600000

flag = 1
```

The plant operates at maximum vaccine production capacity during September and October. The 2-month production satisfies the demand for the 4 months (September to December). Therefore, there is no vaccine production during November and December, and the demand is satisfied from the inventories. Finally, during January and February, the production matches the demand, and there is no inventory that is kept in storage at the end of February. The minimum cost is M\$1.6.

Example 4.6 *Facility Location Models*

In this application example, we will consider a simplified version of the facility location problem. The general description of the case study is shown in Fig. 4.11. Existing facilities (shown as circles) are located at positions A (0,250), B (300,500) and C (400,0) in the two-dimensional space (the units of the distances can be considered arbitrary, but you might think of them as distances in m if it helps visualize the problem). Two new facilities (shown as squares) are to be located in the two-dimensional space. The new facilities will exchange material with the existing facilities and among themselves with transportation cost that depends on their rectilinear distances. The rectilinear distance between two points is simply the sum of the difference in x and y coordinates between two points, i.e., if point i is (x_i, y_i) and point j is (x_j, y_j), then their rectilinear distance rd_{ij} is simply:

$$rd_{ij} = |x_i - x_j| + |y_i - y_j| \tag{4.79}$$

The transportation cost between the facilities is summarized in Table 4.5. Zeros in this table indicate a pair of facilities that do not exchange material (and therefore, the solution must not depend on the corresponding rectilinear distance). The cost is given in monetary units per year and unit distance covered (\$/(m y) are used just to facilitate presentation).

Fig. 4.11 Graphical representation of the facility location problem

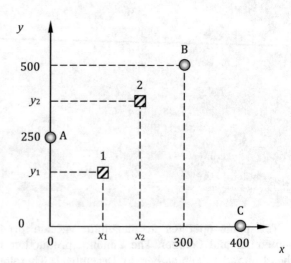

Table 4.5 Unit transportation costs between facilities in Fig. 4.11

c_{ij} $(10^2$ \$/(m y))	1	2	A	B	C
1		–	5	1	0	2
2			–	1.5	2	0

We will be using the coordinates of the two new facilities in the two-dimensional space (x_1, y_1) and (x_2, y_2) as the optimization variables. The objective function is the minimization of the transportation cost between the facilities:

$$\min_{x_1,y_1,x_2,y_2} c_{12}rd_{12} + c_{1A}rd_{1A} + c_{1C}rd_{1C} + c_{2A}rd_{2A} + c_{2B}rd_{2B} \tag{4.80}$$

The rectilinear distances are defined in Eq. (4.79), but they are nonlinear as they involve absolute values. There is fortunately an efficient technique to linearize the absolute value terms. Let's consider the rd_{12}:

$$rd_{12} = |x_1 - x_2| + |y_1 - y_2| \tag{4.81}$$

We introduce four positive variables:

$$x_1 - x_2 = X_{12}^+ - X_{12}^-, \quad X_{12}^+, X_{12}^- \geq 0 \tag{4.82a}$$

$$y_1 - y_2 = Y_{12}^+ - Y_{12}^-, \quad Y_{12}^+, Y_{12}^- \geq 0 \tag{4.82b}$$

Note that if $x_1 > x_2$, then:

$$x_1 - x_2 = X_{12}^+, \quad X_{12}^- = 0 \tag{4.83a}$$

while if $x_1 < x_2$, then:

$$x_1 - x_2 = -X_{12}^-, \quad X_{12}^+ = 0 \tag{4.83b}$$

and therefore:

$$|x_1 - x_2| = X_{12}^+ + X_{12}^- \tag{4.84}$$

We therefore conclude that the objective function can be written as:

$$\min_{x_1,y_1,x_2,y_2} c_{12}\left(X_{12}^+ + X_{12}^- + Y_{12}^+ + Y_{12}^-\right) + c_{1A}\left(X_{1A}^+ + X_{1A}^- + Y_{1A}^+ + Y_{1A}^-\right)$$

$$+ c_{1C}\left(X_{1C}^+ + X_{1C}^- + Y_{1C}^+ + Y_{1C}^-\right) + c_{2A}\left(X_{2A}^+ + X_{2A}^- + Y_{2A}^+ + Y_{2A}^-\right)$$

$$+ c_{2B}\left(X_{2B}^+ + X_{2B}^- + Y_{2B}^+ + Y_{2B}^-\right) \tag{4.85}$$

Note that we have achieved the indented "linearization" of the absolute values, but we have paid a price: the number of the variables has increased dramatically!

The mathematical formulation involves the following equality constraints:

$$\left.\begin{aligned} x_1 - x_2 &= X_{12}^+ - X_{12}^- \\ y_1 - y_2 &= Y_{12}^+ - Y_{12}^- \end{aligned}\right\} 1-2 \tag{4.86a}$$

$$\left.\begin{aligned} x_1 - x_A &= X_{1A}^+ - X_{1A}^- \\ y_1 - y_A &= Y_{1A}^+ - Y_{1A}^- \end{aligned}\right\} 1-A \tag{4.86b}$$

$$\left.\begin{aligned} x_1 - x_C &= X_{1C}^+ - X_{1C}^- \\ y_1 - y_C &= Y_{1C}^+ - Y_{1C}^- \end{aligned}\right\} 1-C \tag{4.86c}$$

$$\left.\begin{aligned} x_2 - x_A &= X_{2A}^+ - X_{2A}^- \\ y_2 - y_A &= Y_{2A}^+ - Y_{2A}^- \end{aligned}\right\} 2-A \tag{4.86d}$$

$$\left.\begin{aligned} x_2 - x_B &= X_{2B}^+ - X_{2B}^- \\ y_2 - y_B &= Y_{2B}^+ - Y_{2B}^- \end{aligned}\right\} 2-B \tag{4.86e}$$

$$x_1, x_2, X_{12}^+, X_{12}^-, X_{1A}^+, X_{1A}^-, X_{2A}^+, X_{2A}^-, X_{2B}^+, X_{2B}^-, X_{1C}^+, X_{1C}^- \geq 0 \tag{4.87a}$$

$$y_1, y_2, Y_{12}^+, Y_{12}^-, Y_{1A}^+, Y_{1A}^-, Y_{2A}^+, Y_{2A}^-, Y_{2B}^+, Y_{2B}^-, Y_{1C}^+, Y_{1C}^- \geq 0 \tag{4.87b}$$

Solution of the case study is left as an exercise to the reader.

Example 4.7 *Statistical Estimation Based on LP Formulation*

In Chap. 2, we have discussed the methodology of least squares for estimating the parameters of nonlinear models. There is a class of models in which the parameters appear linearly in the models and are known as models that are linear in the parameters. These can be in general nonlinear models, but the parameter happens to appear linearly. A classic example to clarify the concept is the Antoine equation for calculating the vapor pressure of pure components, which in its simplest form is:

$$\ln P^s = A + B\frac{1}{T} \tag{4.88}$$

This is clearly a nonlinear model between the vapor pressure P^s and the absolute temperature T. However, the two parameters of the model A and B appear linearly. If we perform experiments in which we vary the temperature and measure the corresponding saturation pressure, then we may define the variables $Y = \ln P^s$ and $X = 1/T$. In this case, our model becomes linear: $Y = A + BX$. Linear least squares can be used to determine the model parameters, but other options are available as well.

Let's assume that a general linear in the parameters model is available involving many explanatory variables X_i, $i = 1, 2, ..., n$. The general form of such a model is:

$$Y = p_0 + p_1 X_1 + p_2 X_2 + \cdots + p_n X_n \tag{4.89}$$

This model can be generalized to involve mixed terms such as $X_1 X_2$ and still be linear in the parameters, but we only keep the first-order terms to facilitate clarity of presentation. If we have k experimental pairs (X_j, Y_j) available, then we can define the corresponding errors or residuals r_j:

$$r_j = Y_j - (p_0 + p_1 X_{1j} + p_2 X_{2j} + \cdots + p_n X_{nj}), \quad j = 1, 2, \cdots, k \tag{4.90}$$

Instead of minimizing the sum of the squared residuals, we can minimize the maximum residual:

$$\min_{p_0, \ p_1, \ \cdots, \ p_n, \mu} \mu$$

$$\text{s.t.} \tag{4.91}$$

$$-\mu \leq Y_j - (p_0 + p_1 X_{1j} + p_2 X_{2j} + \ \ldots \ + p_n X_{nj}) \leq \mu,$$

$$j = 1, \ 2, \ \ldots, \ k \ \ \mu \geq 0$$

or we can minimize the sum of the absolute residuals:

$$\min_{p_0, \ p_1, \ \cdots, \ p_n, \mu} \sum_{j=1}^{k} \left(\mu_j^+ + \mu_j^- \right)$$

$$\text{s.t.} \tag{4.92}$$

$$Y_j - (p_0 + p_1 X_{1j} + p_2 X_{2j} + \ \ldots \ + p_n X_{nj}) \leq \mu_j^+ - \mu_j^-,$$

$$j = 1, \ 2, \ \ldots, \ k \ \ \ \mu_j^+ - \mu_j^- \geq 0, \quad j = 1, \ 2, \ \ldots, k$$

To clarify the LP formulations just presented, we consider the experimental data shown in Table 4.6 (the reader is referred to the original publication for the description of this interesting case study and its practical implications). We will first apply Formulation (4.91) to determine the model parameters p_0, p_1, p_2, p_3, and p_4. We start by writing down the full formulation:

$$\min_{p_0, \ p_1, \ p_2, \ p_3, \ p_4, \ \mu} \mu$$

$$\text{s.t.} \tag{4.93}$$

$$\left. \begin{array}{l} -\mu - p_0 - p_1 X_{1j} - p_2 X_{2j} - p_3 X_{3j} - p_4 X_{4j} \leq -Y_j \\[6pt] -\mu + p_0 + p_1 X_{1j} + p_2 X_{2j} + p_3 X_{3j} + p_4 X_{4j} \leq +Y_j \end{array} \right\}, \ j = 1, \ 2, \ \ldots, \ k$$

$$\mu \geq 0$$

Table 4.6 Data on the dependence of heat evolved during cement hardening as a function of composition

j	$X_{1,j}$	$X_{2,j}$	$X_{3,j}$	$X_{4,j}$	Y_j
1	4	21	26	47	114.0
2	6	7	60	26	78.5
3	6	7	33	52	95.9
4	8	11	20	56	104.3
5	8	11	47	31	87.6
6	8	10	12	68	109.4
7	9	11	12	66	113.3
8	9	11	22	55	109.2
9	15	1	52	29	74.3
10	17	3	6	71	102.7
11	18	2	22	54	93.1
12	22	1	44	31	72.5
13	23	1	34	40	83.8

Data from Woods, Steinour and Starke, Ind. & Eng. Chem., 24(11), 1932, p. 1207
X_1, % $4CaO \cdot Al_2O_3 \cdot Fe_2O_3$; X_2, % $3CaO \cdot Al_2O_3$; X_3, % $2CaO \cdot SiO_2$; X_4, % $2CaO \cdot SiO_2$; Y heat evolved during hardening for 180 days (in kcal/kg)

This formulation can be implemented in MATLAB as follows:

```
clear
data=[ 4  21 26  47   114.0
       6  7  60  26   78.5
       6  7  33  52   95.9
       8  11 20  56   104.3
       8  11 47  31   87.6
       8  10 12  68   109.4
       9  11 12  66   113.3
       9  11 22  55   109.2
       15 1  52  29   74.3
       17 3  6   71   102.7
       18 2  22  54   93.1
       22 1  44  31   72.5
       23 1  34  40   83.8];
X=data(:,1:4); Y=data(:,5);
[k,n]=size(X); n=n+1;
A=[-ones(k,1) -ones(k,1) -X; -ones(k,1) +ones(k,1) +X]
b=[-Y; +Y]
c=zeros(n+1,1); c(1)=1;
LB=-inf*ones(n+1,1); LB(1)=1;
[xopt,fopt,flag]=linprog(c,A,b,[],[],LB,[])
```

(continued)

```
>> StatEstLP
A = -1   -1    -4   -21   -26   -47
    -1   -1    -6    -7   -60   -26
    -1   -1    -6    -7   -33   -52
    -1   -1    -8   -11   -20   -56
    -1   -1    -8   -11   -47   -31
    -1   -1    -8   -10   -12   -68
    -1   -1    -9   -11   -12   -66
    -1   -1    -9   -11   -22   -55
    -1   -1   -15    -1   -52   -29
    -1   -1   -17    -3    -6   -71
    -1   -1   -18    -2   -22   -54
    -1   -1   -22    -1   -44   -31
    -1   -1   -23    -1   -34   -40
    -1    1     4    21    26    47
    -1    1     6     7    60    26
    -1    1     6     7    33    52
    -1    1     8    11    20    56
    -1    1     8    11    47    31
    -1    1     8    10    12    68
    -1    1     9    11    12    66
    -1    1     9    11    22    55
    -1    1    15     1    52    29
    -1    1    17     3     6    71
    -1    1    18     2    22    54
    -1    1    22     1    44    31
    -1    1    23     1    34    40
b =
 -114.0000
  -78.5000
  -95.9000
 -104.3000
  -87.6000
 -109.4000
 -113.3000
 -109.2000
  -74.3000
 -102.7000
  -93.1000
  -72.5000
  -83.8000
  114.0000
   78.5000
   95.9000
  104.3000
   87.6000
  109.4000
  113.3000
  109.2000
   74.3000
```

(continued)

```
   102.7000
    93.1000
    72.5000
    83.8000
Optimal solution found.
xopt =  2.9967
        80.2464
        -0.1442
         1.3998
        -0.3057
         0.3379

fopt = 2.9967
flag = 1
```

The optimal approximating model according to the minimum worst case absolute deviation is the following:

$$Y = 80.2464 - 0.1442X_1 + 1.3998X_2 - 0.3057X_3 + 0.3379X_4 \qquad (4.94)$$

The solution of formulation (4.92) is left as an exercise to the reader.

4.7   Interior Point Methods for Solving LP Problems

The reader must have been convinced about the practical importance of LP models and the wide range of real-world applications that fall in this category of constrained optimization problems. Some might have also realized that MATLAB does not use the simplex algorithm introduced in this chapter, but it uses an "interior point" algorithm. The simplex algorithm has been optimized in the last 70 years, and extensive research work has been performed to fine-tune all its implementation details. It is, in overall, a very successful algorithm, but there are some classes of problems for which simplex algorithm is particularly inefficient. This fact resulted in extended research work in the 1970s which led to the introduction of the *interior point methods*. The name is due to the fact that, in contrast to the simplex method in which we are moving across the boundary of the feasible space (from one vertex or extreme point to the next), interior point methods approach the solution from the interior of the feasible space.

To present briefly the basic idea of interior point algorithms, we start from the LP problem in the standard form:

$$\min_{\mathbf{x}} f = \mathbf{c}^T \mathbf{x}$$

$$\text{s.t.} \tag{4.95}$$

$$\mathbf{Ax} = \mathbf{b}$$

$$-\mathbf{x} \leq \mathbf{0}$$

We define its Lagrangian function:

$$\mathcal{L}(\mathbf{x}, \boldsymbol{\lambda}) = \mathbf{c}^T \mathbf{x} + \boldsymbol{\lambda}^T (\mathbf{b} - \mathbf{Ax}) - \boldsymbol{\mu}^T \mathbf{x} \tag{4.96}$$

We then apply the KKT conditions to obtain:

$$\mathbf{A}^T \boldsymbol{\lambda} + \boldsymbol{\mu} = \mathbf{c} \tag{4.97a}$$

$$\mathbf{Ax} = \mathbf{b} \tag{4.97b}$$

$$\boldsymbol{\mu} \geq \mathbf{0} \tag{4.97c}$$

$$\mu_i x_i = 0, \quad i = 1, 2, \ldots, n \tag{4.97d}$$

We return now to the standard LP Formulation (4.95) and define the following logarithmic barrier function B($\mathbf{x}$, π):

$$\beta_\ell(\mathbf{x}, \pi) = \mathbf{c}^T \mathbf{x} - \pi \sum_{i=1}^{n} \ln(x_i) \tag{4.98}$$

with $\pi > 0$. As the barrier function ensures that $\mathbf{x} \geq \mathbf{0}$ will be satisfied, we can drop the $\mathbf{x} \geq \mathbf{0}$ constraint, and Problem (4.95) is equivalent to the following problem:

$$\min_{\mathbf{x}} \beta(\mathbf{x}, \ \pi)$$

$$\text{s.t.} \tag{4.99}$$

$$\mathbf{Ax} = \mathbf{b}$$

We now define the modified Lagrangian function:

$$\mathcal{L}(\mathbf{x}, \boldsymbol{\lambda}, \pi) = \mathbf{c}^T \mathbf{x} - \pi \sum_{i=1}^{n} \ln(x_i) + \boldsymbol{\lambda}^T (\mathbf{b} - \mathbf{Ax}) \tag{4.100}$$

We will be using the following definitions:

$$\mathbf{X} = \text{diag}\{x_i\} = \begin{bmatrix} x_1 & 0 & \cdots & 0 \\ 0 & x_2 & \ddots & \vdots \\ \vdots & \ddots & \ddots & 0 \\ 0 & \cdots & 0 & x_n \end{bmatrix} \tag{4.101}$$

$$\mathbf{e} = \begin{bmatrix} 1 \\ 1 \\ \vdots \\ 1 \end{bmatrix} \tag{4.102}$$

Note that $\mathbf{x} = \mathbf{X}\mathbf{e}$. We now express the KKT conditions for (4.100):

$$\mathbf{A}^T \boldsymbol{\lambda} + \pi \mathbf{X}^{-1} \mathbf{e} = \mathbf{c} \tag{4.103a}$$

$$\mathbf{A}\mathbf{x} = \mathbf{b} \tag{4.103b}$$

We define $\mathbf{s} = \pi \mathbf{X}^{-1} \mathbf{e}$, and we finally obtain:

$$\mathbf{A}^T \boldsymbol{\lambda} + \mathbf{s} = \mathbf{c} \tag{4.104a}$$

$$\mathbf{A}\mathbf{x} = \mathbf{b} \tag{4.104b}$$

$$\mathbf{X}\mathbf{S}\mathbf{e} = \pi \mathbf{e} \tag{4.104c}$$

Note again that $\mathbf{s} = \mathbf{S}\mathbf{e}$. It is important to note the similarities with the KKT conditions for the standard LP problem (Eq. (4.97)) and to realize that we can recover them by allowing $\pi \rightarrow 0$. Based on the equivalence of the KKT conditions of the standard LP formulation and Formulation (4.99), we may develop a methodology for solving (4.99) iteratively with the additional requirement that we also need to ensure that $\pi \rightarrow 0$ (i.e., decreasing the value of π in every iteration).

One way to achieve that is to solve at each iteration a quadratic approximation of the Problem (4.99). To this end, note that the gradient $\nabla \beta_\ell$ and the hessian $\nabla^2 \beta_\ell$ of the logarithmic barrier function are given by:

$$\nabla \beta_\ell(\mathbf{x}, \pi) = \mathbf{c} - \pi \mathbf{X}^{-1} \mathbf{e} \tag{4.105a}$$

$$\nabla^2 \beta_\ell(\mathbf{x}, \pi) = \pi \mathbf{X}^{-2} \tag{4.105b}$$

The quadratic approximation for determining the search direction $\mathbf{p}$ is as follows:

$$\min_{\mathbf{p}} \quad \nabla^T \beta_l \mathbf{p} + \frac{1}{2} \mathbf{p}^T \nabla^2 \beta_l \mathbf{p}$$
$$\text{s.t.} \tag{4.106}$$
$$\mathbf{A}(\mathbf{x} + \mathbf{p}) = \mathbf{b}$$

Assuming that $\mathbf{A}\mathbf{x} = \mathbf{b}$, it follows that $\mathbf{A}\mathbf{p} = \mathbf{0}$. This is classical "Newton" problem for which we may express the KKT conditions through the following matrix equations:

$$\begin{bmatrix} \nabla^2 \beta_\ell & \mathbf{A}^T \\ \mathbf{A} & \mathbf{0} \end{bmatrix} \begin{bmatrix} -\mathbf{p} \\ \lambda \end{bmatrix} = \begin{bmatrix} \nabla^T \beta_\ell \\ \mathbf{0} \end{bmatrix} \tag{4.107}$$

The algorithmic implementation is relatively simple, and we start the algorithm with an initial feasible $\mathbf{x}^0$ and π^0. We then, at iteration k, solve (4.107) to determine $\mathbf{p}^k$ and λ^k. We solve the one-dimensional problem $\min_a \beta_\ell(\mathbf{x}^k, \pi^k)$, where $\mathbf{x}^k = \mathbf{x}^{k-1} + a\mathbf{p}^k$, s.t. $\mathbf{A}\mathbf{x}^k = \mathbf{b}$. We increase the counter, update the multiplier used $\pi^k = \rho\pi^{k-1}, \rho \in (0, 1)$, and check for convergence. If convergence has not been achieved, we return to solving Problem (4.107).

This has been a very brief presentation of the vast field of interior point methods. The interested reader is referred to the literature given at the end of this book for further study of this subject.

Learning Summary

In this chapter, we studied an important special case of constrained optimization problems in which the objective function and all constraints are linear leading to a linear programming (LP) problem:

$$\min_{\mathbf{x}} \mathbf{c}^T\mathbf{x}$$
$$\text{s.t.}$$
$$\mathbf{A}\mathbf{x} = \mathbf{b}$$
$$\mathbf{C}\mathbf{x} \leq \mathbf{d}$$
$$\mathbf{x}_L \leq \mathbf{x} \leq \mathbf{x}_U$$

where $\mathbf{A}$ is a m-by-n constant matrix, $\mathbf{C}$ is a p-by-n constant matrix, $\mathbf{b}$ is a m dimensional constant vector, and $\mathbf{d}$ is a p dimensional constant vector. The simplex algorithm assumes that the LP problem has been defined (or transformed) in the *standard form*:

$$\min_{\mathbf{x}} \mathbf{c}^T\mathbf{x}$$
$$\text{s.t.}$$
$$\mathbf{A}\mathbf{x} = \mathbf{b}$$
$$\mathbf{0} \leq \mathbf{x}$$

The simplex method is based on successive movements between neighboring extreme points (by exchanging one variable between the basic and nonbasic variables) and moving across the boundary of the feasible space. The interior point methods, on the other hand, approach the solution from the interior of the feasible space by applying Newton's method to a linearly constrained, nonlinear (barrier) objective function. Linear programming models have been developed for a multitude of problems involving, among others, network models, scheduling models, cutting stock models, and statistical estimation models.

Terms and Concepts

You must be able to discuss the concept of:

Cutting Stock problem
Extreme point of feasible space in LP problems
Facility location model
Initial feasible solution
Interior point methods
Maximum flow
Network Models
Production and employee scheduling problems
Shortest Route problem
Simplex method
Standard LP problem
Statistical estimation model
Transshipment problem
Transportation problem
Trim Loss problem
Vertex solution

Problems

4.1 In Table P4.1, a sample of the saturation properties of water in the range 273.16–473.15 K is presented. Use these data to develop an equation that determines the saturation pressure of water as a function of absolute temperature.

4.2 Use the data from Problem 4.1 to develop an equation that determines the enthalpy of evaporation of water as a function of the absolute temperature.

4.3 Solve graphically the following LP problem, and determine the Lagrange multipliers:

Table P4.1 Saturation properties of water

i	T (K)	p^{sat} (kPa)	Δh_{vap} (kJ/kg)
1	273.16	0.6117	2 500.9
2	293.15	2.3392	2 453.5
3	313.15	7.3851	2 406.0
4	333.15	19.947	2 357.7
5	353.15	47.416	2 308.0
6	373.15	101.42	2 256.4
7	393.15	198.67	2 202.1
8	413.15	361.53	2 144.3
9	433.15	618.23	2 082.0
10	453.15	1002.8	2 014.2
11	473.15	1 554.9	1 939.8

$$\max_{x_1,\ x_2} \quad f = 3x_1 + 2x_2$$

s.t.

$$2x_1 + x_2 \le 8$$
$$x_1 + 3x_2 \le 15$$
$$0 \le x_1,\ x_2$$

4.4 Repeat Problem 4.3 using the simplex method. Validate your solution using MATLAB.

4.5 Your company operates three coal mines A, B, and C which provide 800, 1000, and 1400 t of coal, respectively, per week. Five customers (1, 2, 3, 4, and 5) have ordered 1000, 800, 600, 600, and 1200 t of coal, respectively, for next week. Transportation cost are shown in the table that follows:

c_{ij} in $/t	1	2	3	4	5
A	6	24	2	24	21
B	27	15	12	18	18
C	9	2	6	19	3

The penalty for not satisfying the demand for customers 1 and 3 is 8 $/t and for customers 2, 4, and 5 is 6 $/t. Determine the weekly shipping schedule that minimizes the total cost.

4.6 A power plant (see Fig. P4.6) uses an extraction turbine that receives $m_S = 200$ t/h of superheated steam at 454 °C and 60 atm.

The electrical power generated is given by the following equation:

$$P_{el}(\text{MW}) = 0.4m_S - 0.38m_{HP} - 0.35m_{LP}$$

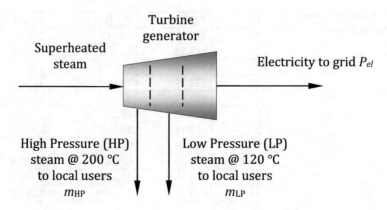

Fig. P4.6 Turbine generator for combined heat and power generation

where m_{HP} is the production rate of high-pressure steam (in t/h) and m_{LP} is the production rate of low-pressure steam (in t/h). The operating constraints are the following:

(a) At least 15 t/h of LP steam must be produced.
(b) HL steam production to LP steam production ration must be at least 4:3.
(c) The maximum amount of LP and HP steam that can be produced is 50% of the saturated steam fed to the unit.

    The selling price of electricity is 0.1 $/kWh. LP steam can be sold for 20 $/t and HP steam for 25 $/t. Maximize the profit of the operation.

4.7 You are responsible for the production scheduling of five different products (say P1, P2,..., P5) which can be produced using three machines (say M1, M2, and M3). Products P1 and P3 can be sold for 10 $/unit and products P2, P4, and P5 for 8 $/unit. Labor cost is $8/h for all machines, while raw material costs are 4$ for products P1 and P3 and $2 for P2, P4, and P5. All machines are available for 5 days per week (120 h/week or 7200 min/week). The processing times in min for each product and machine are the following:

pt_{ij} (min)	M1	M2	M3	Selling price ($/unit)	Raw mat. cost ($/unit)
P1	18	12	8	10	4
P2	10	14	15	8	2
P3	12	6	10	10	4
P4	15	0	5	8	2
P5	10	16	3	8	2

    Formulate a LP problem to determine the optimal production schedule, and use MATLAB to obtain the solution.

Fig. P4.8 A water
management network

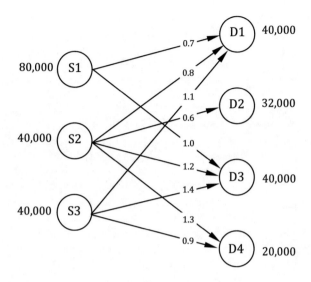

4.8 In Fig. P4.8, a network is shown which involves water supplies (water wells) on the left and major residential areas on the right (water consumers). The amounts given in the water sources (water supplies) or sinks (water demands) are in m^3/d, while the weights on the arcs are the costs in $\$/m^3$. Develop a LP formulation for the minimum cost satisfaction of the water demand, and solve the problem using MATLAB.

4.9 One of the most well-known problems in optimal control is known as the students problem which can be described as follows:

$$\min_{u(t)} f(u(t)) = \int_0^T u(t)dt$$

s.t.

$$\frac{dx(t)}{dt} = -ax(t) + bu(t), \ x(0) = 0$$

$$x(T) = x_f$$

$$0 \le u(t) \le u^U$$

Assume that $a = 0.1$, $b = 0.5$, $T = 15$, $x_f = 50$, and $u^U = 15$; use finite differences to approximate the solution of the differential equation, and prove that the discrete problem is a LP problem. Solve the problem in MATLAB.

Chapter 5
Integer and Mixed Integer Programming Problems

5.1 Introduction

In all chapters up to this point we have studied problems in which all variables were continuous. In this chapter, we will study optimization formulations and problems in which either all or some of the variables obtain integer or binary (either 0 or 1) values. This is arguably the most important class of optimization problems in engineering as it encompasses the modeling of decisions. We will first present a number of example problems in which integer (mostly binary) variables appear naturally in the mathematical formulation. We will then present some solution strategies and the solver available in MATLAB®.

5.2 Examples of Integer Programming Formulations

We will present a number of representative examples of the use of integer programming (IP) models. The aim is for the reader to become familiar with their development and to gradually build some basic modeling skills in developing IP models.

Example 5.1 *A Knapsack Problem*
    Assume that you have managed, over the recent years, to save $70,000 and that you are looking for a combination of investments that will maximize the net present worth (NPW) of your capital. Your financial advisor tells you that there are four investments available that match the risk criteria that you have stated. A summary of these investments is presented in Table 5.1. Note that if one invests 35 k$ in investment 1, then the expected return has a NPW of 55 k$, while if one invests 25 k$ in investment 2, then the expected return has a NPW

I. K. Kookos, *Practical Chemical Process Optimization*, Springer Optimization and Its Applications 197, https://doi.org/10.1007/978-3-031-11298-0_5

Table 5.1 Summary of investment opportunities

	Investment 1	Investment 2	Investment 3	Investment 4
Cash outflow (k$)	35	25	20	15
NPW (k$)	55	40	30	20

of 40 k$, etc. Any combination of investment opportunities is permissible provided that the cash outflow does not exceed the available capital of 70 k$.

The first thing to think about is what the optimization variables are in this example problem. We realize that what we are looking for is whether to select or not a specific investment opportunity or a specific combination of investment opportunities. Our decision that needs to be modeled is a yes/no decision related to the selection (yes) or to the rejection (no) of each one of the available investment opportunities. This leads us to define the binary (1 or 0) variables y_i which obtain the value of 1 (or yes decision) if investment i is made and the value of 0 (or no decision) if investment i is not made (for $i = 1, 2, 3, 4$). This is normally stated as follows:

$$y_i = \begin{cases} 1, & \text{if investment } i \text{ is made} \\ 0, & \text{otherwise} \end{cases}, \quad i = 1, 2, 3, 4 \qquad (5.1)$$

Our aim is to make a combined selection of investments so as to maximize the return on our investment. If we select investment 1, then the NPW is equal to 55 k$, while if we do not select investment 1, then the NPW due to investment 1 is zero. This can be modeled by the product $55y_i$ for which we note that:

$$55y_1 = \begin{cases} 55, & \text{if investment } i = 1 \text{ is made or } y_1 = 1 \\ 0, & \text{otherwise} \end{cases} \qquad (5.2)$$

The objective function follows directly if we consider all available investments:

$$\max_{y_1 y_2 y_3 y_4} 55y_1 + 40y_2 + 30y_3 + 20y_4 \qquad (5.3)$$

Following the same reasoning if investment 1 is made, then 35 k$ should be invested, while if investment 2 is made, then 25 k$ should be invested, etc. The constraint that we face is that at most, 70 k$ can be invested, or:

$$35y_1 + 25y_2 + 20y_3 + 15y_4 \leq 70 \qquad (5.4)$$

The complete formulation is therefore:

$$\min_{y_1,y_2,y_3,y_4} -55y_1 - 40y_2 - 30y_3 - 20y_4$$
$$\text{s.t.} \tag{5.5}$$
$$35y_1 + 25y_2 + 20y_3 + 15y_4 \le 70$$
$$y_i = 0 \text{ or } 1, \quad i = 1,2,3,4$$

The mathematical problem given by Formulation (5.5) is of the general form:

$$\max_{y_i} \sum_{i=1}^{n} v_i y_i$$
$$\text{s.t.} \tag{5.6}$$
$$\sum_{i=1}^{n} w_i y_i \le W$$
$$y_i = 0 \text{ or } 1, \quad i = 1,2,\cdots,n$$

where v_i is the value of item i, w_i is the weight of item i, and W is the maximum capacity. This is the famous *knapsack problem* in combinatorial optimization: a set of items is given, each with a weight w_i and a value v_i, and we seek to determine the number of items to include in a collection so as to maximize the collective value, while the total weight is less than or equal to a given capacity W. Its name comes from the problem faced by anyone who is constrained by a fixed-size knapsack and must fill it with the most valuable (non-divisible) items.

After contacting the financial adviser to let her know that you will be interested in investing your savings (but you have to wait until you have finished this chapter so as to determine what is the best combination of the investment opportunities), she informs you on some additional constraints that you have to take into consideration:

(a) You must invest in at least two investment opportunities.
(b) As there is a lot of interest in investments 1 and 2, you cannot select both at the same time.
(c) If you decide to invest in investment plan 1, then you also need to invest in investment plan 4.
(d) You may increase the amount invested in investment opportunity 3 without limit, and each dollar you invest above the minimum of 20 k$ has a NPW of 1.58 $.

Clearly, you need to amend the formulation in order to account for these additional complications.

In order to express mathematically condition a, we note that the sum $y_1 + y_2 + y_3 + y_4$ is equal to the number of investment opportunities selected. In order to select at least two investment opportunities, the sum must be greater than 2:

$$y_1 + y_2 + y_3 + y_4 \geq 2 \tag{5.7}$$

or:

$$2 - \sum\nolimits_{i=1}^{4} y_i \leq 0 \tag{5.8}$$

For condition b, we can express the constraint as follows:

$$y_1 + y_2 \leq 1 \tag{5.9}$$

Note that if you select both investment opportunity 1 and investment opportunity 2, then $y_1 = y_2 = 1$ and $y_1 + y_2 = 2$, and the inequality is violated (infeasible solution). If you select investment opportunity 1 and not investment opportunity 2, then $y_1 = 1$ and $y_2 = 0$ and $y_1 + y_2 = 1$, and the inequality is satisfied as equality. If you select investment opportunity 2 and not investment opportunity 1, then $y_1 = 0$ and $y_2 = 1$ and $y_1 + y_2 = 1$, and the inequality is satisfied as equality. If you do not select any of the investment opportunities 1 and 2, then $y_1 = 0$ and $y_2 = 0$ and $y_1 + y_2 = 0$, and the inequality is satisfied. We conclude that constraint (5.9) forbids the simultaneous selection of investment opportunities 1 and 2 but allows all permissible combinations (only one of the two or none).

Condition c can be expressed through the following inequality constraint:

$$y_1 \leq y_4 \tag{5.10}$$

or:

$$y_1 - y_4 \leq 0 \tag{5.11}$$

Note that if investment opportunity 1 is selected, then $y_1 = 1$. From (5.10), it then follows that $1 \leq y_4 \leq 1$, or $y_4 = 1$, and investment opportunity 4 is also selected. If investment opportunity 1 is not selected, then $y_1 = 0$. From (5.10), it then follows that $0 \leq y_4 \leq 1$, or y_2 can take any permissible value.

Condition d is more complicated and requires the introduction of the continuous and non-negative variable x to denote the amount invested in investment opportunity 3 more than the minimum of 20 k\$. This variable must take the value of zero if investment opportunity 3 is not selected and any value between the minimum investment of 0 k\$ and maximum of

(70–20=) 50 k\$ if investment opportunity 3 is selected. These conditions can be satisfied simultaneously by using the following inequality constraints:

$$0 \leq x \leq 50y_3 \qquad (5.12)$$

Note that if $y_3 = 0$, then $0 \leq x \leq 0$ or $x = 0$, while if $y_3 = 1$, then $0 \leq x \leq 50$. However, we are not finished yet as we need to modify both the objective function (5.3) to account for any amount invested in investment opportunity 3 more than the minimum:

$$\max_{y_1,y_2,y_3,y_4} 55y_1 + 40y_2 + 30y_3 + 1.58x + 20y_4 \qquad (5.13)$$

and constraint (5.4) that guarantees that we do not invest more that 70 k\$:

$$35y_1 + 25y_2 + 20y_3 + x + 15y_4 \leq 70 \qquad (5.14)$$

The overall formulation is as follows:

$$\min_{x,y_1,y_2,y_3,y_4} -1.58x - 55y_1 - 40y_2 - 30y_3 - 20y_4$$
$$\text{s.t.} \qquad (5.15)$$
$$x + 35y_1 + 25y_2 + 20y_3 + 15y_4 \leq 70$$
$$-y_1 - y_2 - y_3 - y_4 \leq -2$$
$$y_1 - y_4 \leq 0$$
$$x - 50y_3 \leq 0$$
$$-x \leq 0$$
$$y_i \in \{0, 1\}$$

Note that while (5.5) is a linear formulation with binary variables only (an integer linear programming or ILP problem), Formulation (5.15) is also linear but involves both continuous and integer variables (a mixed integer linear programming or MILP problem).

Example 5.2 *Set Covering Problem*

In Fig. 5.1, a geographical area consisting of ten counties (denoted by a, b, $c, . . ., j$) is shown. There are eight available locations (denoted by $k = 1, 2, . . ., 8$) at which certain major health service facilities can be located. A service facility k can serve a county provided that the service facility is located at its border. Given this restriction, we may note that county a can be served by service facilities 1 and 2, county b can be served by service facilities 2 and 3, ..., and county j can be served by service facilities 4, 5, 6, 7, and 8.

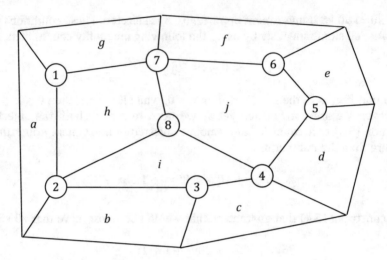

Fig. 5.1 Representative set covering problem

Our objective is to locate (build) the minimum number of service centers (facility locations) to "cover" all available counties. This is a special case of the famous *set covering problem* in which each member of a given set (set of counties) must be "covered" by a member of some other related set (potential service locations) with the aim to minimize the number of elements of the second set that are required to achieve the stated objective.

To develop the mathematical description of the problem, we define the binary variables y_k, $k \in \{1,2,3,\ldots,8\}$ to denote the location of a service facility at position k:

$$y_k = \begin{cases} 1, & \text{if service station is to be located at point } k \\ 0, & \text{otherwise} \end{cases} \quad , \quad k \in \{1, 2, 3, \ldots, 8\}$$

$$(5.16)$$

The objective function follows immediately as our aim is to minimize the number of service facilities to build:

$$\min_{y_1 y_2 y_3 y_4 y_5 y_6 y_7 y_8} \sum_{k=1}^{8} y_k = y_1 + y_2 + y_3 + y_4 + y_5 + y_6 + y_7 + y_8 \quad (5.17)$$

The constraints follow from the requirement to "cover" every county. As county a is "covered" if we place a service center at location 1 or at location 2 or at both locations, we may write:

$$\text{county a}: \quad y_1 + y_2 \geq 1 \quad (5.18)$$

In a similar way:

$$\text{county b}: \quad y_2 + y_3 \geq 1 \tag{5.19}$$

$$\text{county c}: \quad y_3 + y_4 \geq 1 \tag{5.20}$$

$$\text{county d}: \quad y_4 + y_5 \geq 1 \tag{5.21}$$

$$\text{county e}: \quad y_5 + y_6 \geq 1 \tag{5.22}$$

$$\text{county f}: \quad y_6 + y_7 \geq 1 \tag{5.23}$$

$$\text{county g}: \quad y_1 + y_7 \geq 1 \tag{5.24}$$

$$\text{county h}: \quad y_1 + y_2 + y_7 + y_8 \geq 1 \tag{5.25}$$

$$\text{county i}: \quad y_2 + y_3 + y_4 + y_8 \geq 1 \tag{5.26}$$

$$\text{county j}: \quad y_4 + y_5 + y_6 + y_7 + y_8 \geq 1 \tag{5.27}$$

$$y_k \in \{0, 1\}, \quad \forall k \tag{5.28}$$

Example 5.3 *Integer Programming of Recreational Problems: The Magic Square Problem and the Sudoku Puzzle Problem as ILP Problems*

In Fig. 5.2, a famous engraving by the German Renaissance artist Albrecht Dürer is shown. The engraving contains (upper right corner just below the bell) a solution to the well-known recreational *problem of the magic square* (MSP). A magic square is a *n*-by-*n* ($n \geq 3$) square grid filled with the positive integers $1, 2, \ldots, n^2$ such that each cell contains a different integer and the sum of the integers in each row, column, and main diagonal is equal to $M = n(n^2 + 1)/2$ (known as the *magic constant*). This is not an example of great practical importance but is very interesting and challenging (in terms of solving the problem as ILP problem for $n > 6$ or 7). An example of a solution to the MSP for the case of $n = 3$ is the following:

4	9	2
3	5	7
8	1	6

Note that as $n = 3$, each row, column, and the two main diagonals sum up to $M = 3 \cdot (3^2 + 1)/2 = 15$.

Assume that we are given n and we want to solve the MSP by integer linear programming. The integers that we can place into the $n \times n$ cells are the integers $k = 1, 2, \ldots, n^2$. Let's use index $i = 1, 2, \ldots, n$ to denote the row and index $j = 1, 2, \ldots, n$ to denote the column and also define the binary variables y_{ijk} to denote the placement of integer k into the i–j cell or not:

Fig. 5.2 Melencolia I is a 1514 engraving by the German Renaissance artist Albrecht Dürer, and its central subject is an enigmatic and gloomy winged female figure thought to be a personification of melancholia. The engraving features a 4-by-4 magic square!. (From Wikipedia/Wikipedia domain/public domain work of art)

$$y_{ijk} = \begin{cases} 1, & \text{if integer } k \text{ is placed in the cell at row } i \text{ and column } j \\ 0, & \text{otherwise} \end{cases} \tag{5.29}$$

We also define the continuous variable x_{ij} to be equal to the value of the integer that has been placed into the i–j cell. x_{ij} can be obtained from y_{ijk} by the following equation:

$$x_{ij} = \sum_{k=1}^{n^2} k \cdot y_{ijk}, \quad \forall i,j \tag{5.30}$$

To satisfy the restrictions set by the recreational problem, the sum of each row must be equal to the magic number:

$$M = \sum_{j=1}^{n} x_{ij}, \quad \forall i \tag{5.31}$$

and the same must hold true for each column:

$$M = \sum_{i=1}^{n} x_{ij}, \quad \forall j \tag{5.32}$$

The problem statement also requires that the two diagonals have the same sum:

$$x_{11} + x_{22} + \cdots + x_{kk} + \cdots x_{nn} = M \tag{5.33}$$

$$x_{n1} + x_{n-1,2} + \cdots + x_{n-(k-1),k} + \cdots x_{1n} = M \tag{5.34}$$

For a complete formulation, we must ensure that each integer k is used only once:

$$\sum_{i=1}^{n} \sum_{j=1}^{n} y_{ijk} = 1, \quad \forall k \tag{5.35}$$

and that each cell contains one integer only:

$$\sum_{k=1}^{n^2} y_{ijk} = 1, \quad \forall i,j \tag{5.36}$$

The equations that have been presented so far define what the feasible space of the problem is using only linear constraints in terms of the binary and continuous variables. To develop a complete ILP formulation, we need to devise an appropriate objective function. The statement of the MSP does not include any objective function as its aim is to define a feasibility problem. Other classes of MSP have additional objectives. We can, therefore, define any arbitrary objective function such as maximize the difference between the integers placed in cell 1–1 and cell n–n:

$$\min_{y_{ijk}, x_{ij}} x_{11} - x_{nn} \tag{5.37}$$

A similar recreational problem is that of the sudoku puzzle problem in which you are asked to fill a 9-by-9 grid with integers from 1 to 9 so that each integer appears only once in each row, column, and each major 3-by-3 square. The initial grid is partially filled with integer numbers, and the task is to fill in the unoccupied cells. An example sudoku puzzle is the following (taken from https://sudoku.game/):

					7	1		8
2			8		1	9	3	
	8		3			4		
		1		8			4	
	7			9			2	
	5			7		8		
		2			8		1	
	4	8	7		2			9
3			5	4				

To express the restrictions involved in filling in the sudoku cells, we need to define the following indices:

$i = 1, 2, 3$ to denote the rows within an internal 3-by-3 grid
$j = 1, 2, 3$ to denote the columns within an internal 3-by-3 grid
$m = 1, 2, 3$ to denote the row of the external 3-by3 grid
$n = 1, 2, 3$ to denote the column of the external 3-by-3 grid
$k = 1, 2, 3,\ldots, 9$ to denote the possible integers that can be used to fill in the cells

To clarify the exact use of these indices, the following can be consulted:

	n=1			n=2			n=3		
m=1	ij=11	ij=12	ij=13						
	ij=21	ij=22	ij=23		i-j			i-j	
	ij=31	ij=32	ij=33						
m=2				ij=11	ij=12	ij=13			
		i-j		ij=21	ij=22	ij=23		i-j	
				ij=31	ij=32	ij=33			
m=3							ij=11	ij=12	ij=13
		i-j			i-j		ij=21	ij=22	ij=23
							ij=31	ij=32	ij=33

We also define the binary variables y_{mnijk} as follows:

$$y_{mnijk} = \begin{cases} 1, & \text{if integer } k \text{ is placed in the cell with external row and column } m, n \\ & \text{\&internal row/column } i, j \\ \\ 0, & \text{otherwise} \end{cases}$$

$$(5.38)$$

An example from the sudoku given in the previous page is $y_{23124} = 1$. We can now express the sudoku number placement restrictions using these binary variables. The first one is that each cell contains a single integer:

$$\sum_{k=1}^{9} y_{mnijk} = 1, \quad \forall m,n,i,j \qquad (5.39)$$

To understand this equation, it is important to note that each combination of the first four indices m,n,i,j corresponds to a unique cell of the sudoku. There are as many binary variables defined for this cell as there are potential integers that can be placed in this cell. However, one of these nine binary variables must be equal to 1 (indicating that we have placed the corresponding integer into the cell), and all others must be equal to 0.

The second restriction is that each integer appears only once in each column:

$$\sum_{m=1}^{3}\sum_{i=1}^{3} y_{mnijk} = 1, \quad \forall n,j,k \tag{5.40}$$

Note that by enumerating all combinations of m and i, we are actually enumerating all cells that belong to a particular column that is selected by assigning values to n and j.

The third restriction is that each integer appears only once in each row:

$$\sum_{n=1}^{3}\sum_{j=1}^{3} y_{mnijk} = 1, \quad \forall m,i,k \tag{5.41}$$

Finally, each integer appears only once in each internal 3-by-3 grid:

$$\sum_{i=1}^{3}\sum_{j=1}^{3} y_{mnijk} = 1, \quad \forall m,n,k \tag{5.42}$$

Again, as sudoku puzzle problem is defined as a feasibility problem and any arbitrary objective function is acceptable.

Example 5.4 *ILP Model for Truck Load Determination in Liquid Fuels Distribution*

In Fig. 5.3, a characteristic example of a tank truck, used by oil companies to distribute liquid fuels such as gasoline and diesel, is shown. These tank trucks have several compartments for loading simultaneously several different fuels. The orders received by the customers are initially processed and organized according to the geographical area of the customers and then sent to the appropriate department for generating a preliminary dispatching schedule. In this example, we consider the case where several tank trucks are available for satisfying the orders shown in Table 5.2. Each order might correspond to a single customer or customers aggregated according to their geographical proximity. The aim is to satisfy each order by minimizing the potential deviations from the initial order called "quantity adjustments." These adjustments

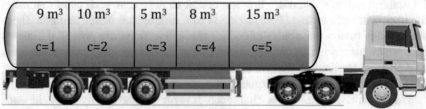

Fig. 5.3 Tank trucks used in liquid fuel distribution by oil companies have several compartments for loading products with different quality specifications

Table 5.2 Orders to be delivered by the tank truck shown in Fig. 5.3

Volumes in m^3	Order 1	Order 2	Order 3
Liquid fuel 1	16	12	12
Liquid fuel 2	3	15	16
Liquid fuel 3	29	22	24
Total	48	49	52

are deemed necessary as it is impossible to "fit" all potential orders into the available tank trucks. In this example, we only consider negative quantity adjustments (delivering less than the volume ordered) at a cost which is proportional to the adjustment made and which is usually determined at managerial level. Our aim is to propose a mathematical formulation that can be used to solve the problem.

To obtain some understanding of the problem, we consider order 1. To solve the problem, we decide to satisfy first the fuel with the largest volume. We may use compartments 2, 3, and 5 to load all 29 m^3 of liquid fuel 3. Then we use compartment 1 for liquid fuel 1 (but we will deliver $16 - 9 = 7$ m^3 less than the order) and finally use compartment 4 to load liquid fuel 2. Even though the total available volume in the tank truck is 47 m^3 and the order is 48 m^3, we will be delivering only 41 m^3, i.e., 15% quantity adjustment relative to the volume of the order. This is clearly not very satisfactory, and we need to find out whether we can do any better. We may keep on trying until we find a satisfactory solution, but it will be much better if we can always find the best solution using a computer to do all the hard work.

We first define the set $c = \{1, 2, 3, 4, 5\}$ to denote the compartments of the tank truck and the set $f = \{\text{fuel 1, fuel 2, fuel 3}\}$ to denote the liquid fuels included in the orders. We then define the binary (decision) variables $y_{f,c}$ as follows:

$$y_{f,c} = \begin{cases} 1, & \text{if fuel } f \text{ is placed in compartment } c \\ 0, & \text{otherwise} \end{cases} \tag{5.43}$$

The first requirement is that only one fuel is loaded in a compartment, i.e.:

$$\sum_{f=1}^{3} y_{f,c} \le 1, \quad \forall c \tag{5.44}$$

It should also be noted that a fuel in a specific order can be loaded in more than one compartment to satisfy the demand, and all fuels must be loaded in at least one compartment (i.e., we may deliver smaller volumes, but we must include all fuels in the order):

$$\sum\nolimits_{c=1}^{5} y_{f,c} \geq 1, \quad \forall f \tag{5.45}$$

As it is unlikely that we will satisfy all demands, we define the non-negative continuous variables s_f to denote the quantity (volume) adjustments (volume of fuel not "fit" to the tank truck). The volume of the fuel loaded to the tank truck plus the volume adjustment must always be equal to the volume of the fuel in the order d_f:

$$\sum\nolimits_{c=1}^{5} V_{f,c} + s_f = d_f, \quad \forall f \tag{5.46}$$

$$0 \leq s_f \leq S_f, \quad \forall f \tag{5.47}$$

where S_f is the maximum permissible volume adjustment for fuel f.

Finally, each compartment has a finite minimum and maximum capacity:

$$V_{c,min} y_{f,c} \leq V_{f,c} \leq V_{c,max} y_{f,c}, \forall f, c \tag{5.48}$$

The objective function to be minimized is the penalty due to volume adjustments:

$$\min_{y_{f,c}, s_f, V_{f,c}} \sum\nolimits_{f=1}^{3} c_f s_f \tag{5.49}$$

where c_f is the penalty that corresponds to volume adjustment for fuel type f. The complete formulation is the following:

$$\min_{y_{f,c}, s_f, V_{f,c}} \sum\nolimits_{f=1}^{3} c_f s_f$$

$$\text{s.t.} \tag{5.50}$$

$$\sum\nolimits_{f=1}^{3} y_{f,c} \leq 1, \quad \forall c$$

$$\sum\nolimits_{c=1}^{5} y_{f,c} \geq 1, \quad \forall f$$

$$V_{c,min} y_{f,c} \leq V_{f,c} \leq V_{c,max} y_{f,c}, \forall f, c$$

$$\sum\nolimits_{c=1}^{5} V_{f,c} + s_f = d_f, \quad \forall f$$

$$0 \leq s_f \leq S_f, \quad \forall f$$

$$y_{f,c} \in \{0, 1\}, \quad \forall f, c$$

Example 5.5 *ILP Model for Scheduling the Operation of Thermal Power Units (the Unit Commitment Problem)*

Thermal power units are complex production units with significant costs associated with the starting up or the shutting down of a unit. There are also important constraints related to the rate at which the power production can be increased or decreased while the plant is in operation. Any thermal power unit can be characterized by the following parameters of the unit:

$C_{up,i}$:   the startup cost of unit i (in $)
$C_{dn,i}$:   the shutdown cost of unit i (in $)
$C_{fx,i}$:   is the fixed cost of operation of unit i (in $/h)
$C_{vr,i}$:   is the unit variable cost of operation of unit i (in ($/h)/MW)
$P_{M,i}$:   is the maximum output power of unit i (in MW)
$P_{m,i}$:   is the minimum output power of unit i (in MW)
$R_{up,i}$:   is the maximum power increment or ramp-up limit (in MW/h) of unit i
$R_{dw,i}$:   is the maximum power decrement ramp-down limit (in MW/h) of unit i

The problem is to decide when to start up and when to shut down a set of power plants $i = \{1, 2, \ldots, N\}$ so as to satisfy a varying demand D_t, where t is a time period within the time horizon of interest $t = \{1, 2, \ldots, 24\}$ (i.e., one day divided in 1 h time periods or any other appropriate selection of time periods $t = \{1, 2, 3, \ldots, H\}$).

To evaluate the operating cost, we need to define several binary variables to indicate the state of each power unit at each time period. All events are assumed to take place at the beginning of each time interval and to remain at the same state until the beginning of the next time interval. It is important to note that each power unit can be:

1. In normal operation
2. In shutdown state (not in operation)
3. In starting up state
4. In shutting down state

Let's define the binary variable $y_{i,t}$ as follows:

$$y_{i,t} = \begin{cases} 1, & \text{if plant } i \text{ is in normal operation at time interval } t \\ 0, & \text{otherwise} \end{cases} \tag{5.51}$$

For any unit to be in normal operation at time interval t, a start-up event must have taken place at any time interval $t' < t$, and no shutdown event has occurred in the meantime. We need to define the binary variable $u_{i,t}$ to denote a start-up event and the binary variable $d_{i,t}$ to denote a shutdown event:

$$u_{i,t} = \begin{cases} 1, & \text{if plant } i \text{ starts up at time interval } t \\ 0, & \text{otherwise} \end{cases} \quad (5.52)$$

$$d_{i,t} = \begin{cases} 1, & \text{if plant } i \text{ shuts down at time interval } t \\ 0, & \text{otherwise} \end{cases} \quad (5.53)$$

The binary variables are related through the following logical equation:

$$y_{i,t} - y_{i,t-1} = u_{i,t} - d_{i,t}, \quad \forall i,t \quad (5.54)$$

where $y_{i,0}$ are known. Careful consideration is needed in order to understand the way that Eq. (5.54) works and "connects" the start-up and shutdown events with the state of each power unit.

If plant i is in normal operating state, then its power output at time interval t can be any number between the minimum and maximum power:

$$P_{m,i} y_{i,t} \le P_{i,t} \le P_{M,i} y_{i,t}, \quad \forall i,t \quad (5.55)$$

The power output from all available units must satisfy the demand D_t at time interval t:

$$\sum_{i=1}^{N} P_{i,t} \ge D_t, \quad \forall t \quad (5.56)$$

In addition, in normal operating state, the output power can vary at each time interval, but the increase or decrease in the output power between consecutive time intervals is limited by the ramp-up and ramp-down limits:

$$R_{dw,i} y_{i,t} \le P_{i,t} - P_{i,t-1} \le R_{up,i} y_{i,t}, \quad \forall i,t \quad (5.57)$$

where $P_{i,0}$ are known.

The objective function is the following:

$$\min_{y_{i,t}, u_{i,t}, d_{i,t}, P_{i,t}} \sum_{i=1}^{N} \sum_{t=1}^{H} \left(C_{fx,i} y_{i,t} + C_{vr,i} P_{i,t} + C_{up,i} u_{i,t} + C_{dn,i} d_{i,t} \right) \quad (5.58)$$

The first term accounts for the fixed costs and the second one for the variable costs, and both contribute only when power plant i is in operation ($y_{i,t} = 1$). The third and fourth terms account for the cost of start-up (when $u_{i,t} = 1$) and shutdown (when $d_{i,t} = 1$) of plant i.

Example 5.6 *ILP Model for the Single-Machine Scheduling Problem*

Scheduling is the field of study concerned with the optimal allocation of resources (machines or processors) over time, to a set of activities or tasks. Scheduling problems are classified according to the characteristics of the processors and the characteristics of the tasks. In this example, we will study the case of a single machine problem which can be used to process N jobs without any sequence-dependent constraints. To make the example more concrete, let's consider the case where we need to satisfy the demand for three different types of products: paints of different colors, namely, white (W), red (R), and blue (B) paint. Data for the problem are presented in Table 5.3. The white (W) paint needs 12 h of machine operation (preparation and operation of a mixing tank with addition of ingredients and subsequent processing including packaging), and it must be delivered (due time) in 60 h (zero time is understood to be the time when the first paint production commences). In the case of delayed delivery, there is a penalty of 15 $/h of delay (usually estimated at managerial level). The red (R) paint must be delivered in 52 h, its processing time is 48 h, and the penalty for delayed delivery is 10 $/h. The blue (B) paint has a processing time of 36 h, a due time of 84 h, and a penalty of 30 $/h.

Let's develop a possible solution and investigate its characteristics. We will adopt the rule that we schedule the tasks according to their due time (the earliest to be delivered is processed first). According to this rule, paint R is produced first followed by paint W, and finally, paint B is produced. The proposed schedule is shown in Fig. 5.4. Paint R is ready for delivery 4 hours in advance of the due time, and paint W becomes available exactly at the time of its delivery. However, the production of paint B is completed at $t = 96$ h, and there is a delay in its delivery $\tau_B = 96 - 84 = 12$ h. The resulting penalty for the delay in delivering paint B is $\tau_B \times w_B = 12$ h $\times$ 30 $/h $= 360$ $. Our aim is to develop an ILP formulation that can be used to develop automatically the best possible sequence of paint production.

Several different formulations have been proposed over the years for modeling the single machine scheduling problem, and we will present one of them which treats the time as a continuous variable. The most important element of the problem is the order at which the tasks are processed from the processor or machine. Let's define the binary variables $y_{i,j}$, where i, $j = \{W, R, B\}$ are the tasks, through which the ordering of the processing of tasks can be inferred:

Table 5.3 Data for the paint production scheduling problem

Paint color (i)	Processing time pt_i in h	Due time dt_i in h	Penalty w_i in $/h
W	12	60	15
R	48	52	10
B	36	84	30

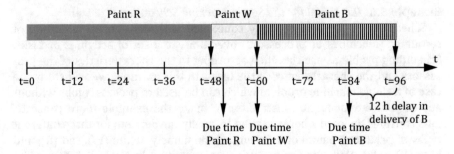

Fig. 5.4 A potential solution to the paint production scheduling problem

$$y_{i,j} = \begin{cases} 1, & \text{if task } i \text{ is performed before task } j \\ 0, & \text{otherwise} \end{cases} \tag{5.59}$$

In the solution shown in Fig. 5.4, $y_{R,W} = y_{R,B} = y_{W,B} = 1$, and all other $y_{i,j} = 0$. We also define the continuous variable st_i to denote the time at which task i starts processing in the machine. It then follows that the time at which processing of task i is completed, denoted by ft_i, is given by:

$$ft_i = st_i + pt_i, \quad \forall i \tag{5.60}$$

where pt_i is the processing time for task i. To determine whether there is a time delay in completing task i, a non-negative delay time τ_i is used:

$$ft_i - \tau_i \le dt_i, \quad \forall i \tag{5.61}$$

For the solution shown in Fig. 5.4, it follows that $st_R = 0$, $st_W = 48$, and $st_B = 60$ and, using (5.60), $ft_R = 0 + 48 = 48$, $ft_W = 48 + 12 = 60$, and $ft_B = 60 + 36 = 96$. We now substitute the numbers in (5.61) to obtain:

$$\text{Paint R} \quad 48 - \tau_R \le 52 \tag{5.62}$$

$$\text{Paint W} \quad 60 - \tau_W \le 60 \tag{5.63}$$

$$\text{Paint B} \quad 96 - \tau_B \le 84 \tag{5.64}$$

It then follows that the solution is $\tau_R = \tau_W = 0$ and $\tau_B = 12$ h (it is reminded that we seek for the minimum non-negative τ_R, τ_W, and τ_B so that the inequalities are satisfied).

The most complicating part of the formulation is the requirement that there is no overlapping of the tasks. If task i starts at time st_i and task j at time st_j and task i precedes (is performed before) task j (i.e. $y_{i,j} = 1$), then:

$$st_i + pt_i \leq st_j, \quad \forall i,j \tag{5.65}$$

However, if task j precedes task i (i.e., $y_{i,j} = 0$), then:

$$st_j + pt_j \leq st_i, \quad \forall i,j \tag{5.66}$$

Inequalities (5.65) and (5.66) cannot hold simultaneously, but we need a method to select the "valid" one based on the value of the binary variables $y_{i,j}$. This can be achieved by using the following formulation:

$$\left. \begin{aligned} st_i + pt_i &\leq st_j + M\left(1 - y_{i,j}\right) \\ st_j + pt_j &\leq st_i + M y_{i,j} \end{aligned} \right\}, \quad \forall i,j \tag{5.67}$$

where M is a sufficiently large positive constant. This is a classical modeling trick to handle cases where only one of two constraints can be active. Note that if $y_{i,j} = 1$, the first constraint is active, and the second is redundant (because its right-hand side will include M, which is much larger than any combination of st_j and pt_j). If $y_{i,j} = 0$, the first constraint is redundant (as M appears in the right-hand side), and the second is active. As we aim at minimizing the penalty incurring due to the delayed completion of the tasks, the following overall formulation follows:

$$\min_{y_{i,j}, st_i, \tau_i} \sum_{i \in \{R, W, B\}} w_i \tau_i$$

$$\text{s.t.} \tag{5.68}$$

$$st_i + pt_i - \tau_i \leq dt_i, \quad \forall i$$

$$\left. \begin{aligned} st_i + pt_i &\leq st_j + M\left(1 - y_{i,j}\right) \\ st_j + pt_j &\leq st_i + M y_{i,j} \end{aligned} \right\}, \quad \forall i,j : j > i$$

$$st_i, \tau_i \geq 0, \quad \forall i$$

$$y_{i,j} \in \{0, 1\}, \quad \forall i,j : j > i$$

Note that in the final formulation, only the combinations i,j for which $j > i$ holds true (i.e., R-W, R-B, and W-B) are used. This is correct as if we have assigned value to $y_{i,j}$, then knowledge of $y_{j,i}$ is redundant (as $y_{i,j} + y_{j,i} = 1$ must always hold true).

5.3   Solving Integer Programming Problems Using the Branch and Bound Method

In the previous section, we have presented several IP formulations which may involve continuous as well as binary variables. If we think of what the general formulation is, we will conclude that for mixed, integer (practically binary) linear programming (MILP) problems, a general mathematical formulation is the following:

$$\min_{x,y} f = \mathbf{c}^T \mathbf{x} + \mathbf{d}^T \mathbf{y}$$

$$\text{s.t.} \tag{5.69}$$

$$\mathbf{Ax} + \mathbf{By} = \mathbf{b}$$

$$\mathbf{Cx} + \mathbf{Dy} \leq \mathbf{e}$$

$$-\mathbf{x} \leq \mathbf{0}, \mathbf{y} \in \{0, 1\}^n$$

where $\mathbf{c}$, $\mathbf{d}$, $\mathbf{b}$, and $\mathbf{e}$ are constant vectors; $\mathbf{A}$, $\mathbf{B}$, $\mathbf{C}$, and $\mathbf{D}$ are constant matrices; $\mathbf{x}$ is the vector of continuous variables; and $\mathbf{y}$ is the vector of binary variables.

The first question to ask is how we can prove that a promising solution $(f^*, \mathbf{x}^*, \mathbf{y}^*)$ is the optimal solution for Problem (5.69). One popular method for proving optimality is to construct an algorithm that generates, in a clever way, a decreasing sequence of upper bounds:

$$\bar{f}^1 > \bar{f}^2 > \cdots > \bar{f}^k \geq f^* \tag{5.70}$$

(k denotes the iteration counter) and at the same time an increasing sequence of lower bounds:

$$\underline{f}^1 < \underline{f}^2 < \cdots < \underline{f}^k \leq f^* \tag{5.71}$$

If the lower- and upper-bound converge (in an acceptable number of iterations k) so that, for a suitably chosen small positive constant ε, the following holds true:

$$\bar{f}^k - \underline{f}^k \leq \varepsilon \tag{5.72}$$

then we can conclude that the optimal solution of (5.69) has been obtained. The important question, from the practical point of view, is how one can generate the decreasing sequence of upper bounds and the increasing sequence of lower bounds.

For MILP problems of the form given by (5.69), it is important to note that if, in the k-th iteration of the algorithm, we assign constant values to the binary variables $\mathbf{y} = \mathbf{y}^k$, then the problem is transformed to a LP problem in terms of $\mathbf{x}$:

$$\min_{\mathbf{x}} f = \mathbf{c}^T \mathbf{x} + \mathbf{d}^T \mathbf{y}^k$$

s.t. $\qquad\qquad\qquad\qquad$ (5.73)

$$\mathbf{Ax} = \widetilde{\mathbf{b}} = \left(\mathbf{b} - \mathbf{By}^k \right)$$

$$\mathbf{Cx} \leq \widetilde{\mathbf{e}} = \left(\mathbf{e} - \mathbf{Dy}^k \right)$$

$$-\mathbf{x} \leq 0$$

As we have seen in the previous chapter, efficient algorithms are available for solving LP problems today. Assume that an optimal solution of (5.73) exists and is given by $(f^k, \mathbf{x}^k, \mathbf{y}^k)$. We may conclude that f^k is always an upper bound on the optimal solution of (5.69), i.e., for a fixed vector of binary variables $\mathbf{y}^k$:

$$f^k \geq f^* \qquad\qquad\qquad\qquad (5.74)$$

The equality holds true only if $\mathbf{y}^k = \mathbf{y}^*$. It then follows that by fixing the binary variables to any arbitrary values, the solution of the resulting LP problem is an upper bound to the optimal solution. This also suggests a method for generating upper bounds: solving the initial problem for a constant vector of the binary variables. As the algorithm proceeds, the best upper bound is updated:

$$\bar{f}^k = \min \left\{ \bar{f}^{k-1}, f^k \right\} \geq f^* \qquad\qquad\qquad (5.75)$$

It is important to note that if the solution of (5.73) is feasible and optimal, then it is also feasible for Problem (5.69) (provided that $\mathbf{y}^k$ is also a feasible set of binary variables). The need is not to find any feasible solution but to find "good" feasible solutions that provide tight upper bounds on the optimal solution.

This general idea of producing upper bounds based on fixing the values of the binary variables and solving the resulting LP problem to generate a feasible solution is common to most MILP solution algorithms. They differ in the methodology used to generate lower bounds. Here we will be presenting a methodology known as LP-based *Branch and Bound* (B&B) methodology. In this methodology, the generation of increasing lower bounds is based on the *linear programming relaxation* idea. To understand the methodology, we need first to define the term relaxation.

When we face a "difficult" problem, we can attack the problem by "relaxing" the characteristics that make the problem difficult to solve and ensure that the relaxed problem offers a lower bound on the objective function of the initial problem. When it comes to binary programming problems, the complicating factor is the requirement that the y variables are binary, i.e., $y_i \in \{0,1\}$. We may relax the requirement that the y_i variables obtain the values of either zero or one by the requirement that y_i are continuous variables in the interval between zero and one, i.e., $y_i \in [0,1]$. We practically enlarge the feasible set over which we seek to minimize the objective function. This is the reason why the relaxed problem offers a lower bound. As we have more options (wider feasible space) in trying to minimize the objective function, it is guaranteed that the objective function will always be better (smaller) than the objective function obtained when the complicating restrictions are taken into consideration.

We consider the general MILP problem given by Formulation (5.69). We consider the following relaxed problem by replacing the $y_i \in \{0,1\}$ requirement with the $y_i \in [0,1]$ requirement:

$$\min_{\mathbf{x},\mathbf{y}} f = \mathbf{c}^T \mathbf{x} + \mathbf{d}^T \mathbf{y}$$

$$\text{s.t.} \hspace{6cm} (5.76)$$

$$\mathbf{Ax} + \mathbf{By} = \mathbf{b}$$

$$\mathbf{Cx} + \mathbf{Dy} \leq \mathbf{e}$$

$$-\mathbf{x} \leq \mathbf{0}, y_i \in [0, 1], \forall i$$

Formulation (5.76) is a LP problem which can be solved using widely available software.

Assume that at the first iteration of the algorithm, the solution obtained for the relaxed problem (5.76) is $(f_R{}^0, \mathbf{x}^0, \mathbf{y}^0)$. If there is no feasible solution for the relaxed problem, then there is no feasible solution for the initial problem (5.69), and the algorithm terminates. If all y_i obtain binary values, then the optimal solution for (5.76) is feasible for the initial problem (5.69) and therefore optimal for the initial problem. These two options terminate the algorithm at the initialization step but are not common in problems of interest.

In the most common case, in the optimal solution of the relaxed problem (5.76) (we will refer to (5.76) as R^0 problem), some of the relaxed binary variables will obtain non-integral values, i.e., they are between zero and one (and not exactly zero or one). The value of the objective function of the relaxed problem $f_R{}^0$ is a lower bound for the optimal solution of (5.69). How do we proceed? Remember that what we are trying to do is to avoid facing the integrality constraints. We can, and this is one of the options available, select among the binary variables the one that is closest to be either 0 or 1 and "branch" on this variable. In this way, we generate two new subproblems, the one corresponding to $y_k = 1$ (node R^1) and the other to $y_k = 0$ (node R^2) as shown in Fig. 5.5. The next step is to solve the corresponding relaxed problems R^1 and R^2 (with y_k equal to the binary variable denoted in Fig. 5.5). There are three potential outcomes:

1. The relaxed problem is infeasible.
2. The relaxed problem has an optimal solution with integer values for the **y** (binary) variables.
3. The relaxed problem has an optimal solution with non-integral values for the **y** (binary) variables.

Fig. 5.5 Partial B&B tree

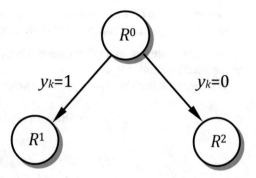

In case 1, the branch is *fathomed* (is not developed any further) as imposing any additional constraints will not alter the infeasibility. In case 2, the branch is also fathomed as imposing additional constraints will only worsen the objective function (and therefore the optimal solution corresponding to this branch has already been found).

We use integer solutions (case 2 above) to generate upper bounds on the objective function. The currently available best integer solution is called the *incumbent* and is denoted by f^*:

$$f* = \text{incumbent}$$
$$= \text{value of the objective function at the best feasible(binary)}$$
$$\text{solution that has been found so far}$$

In the case we discover a new feasible solution with all y_i obtaining binary values, we compare the value of the objective function with the incumbent and update the incumbent if an improved integer solution (with lower objective function) has been discovered.

In case 3 we only update the lower bound for the descendants of this subproblem, and we then generate a new branch by selecting a new branching variable. It is interesting to note that subsequent "generations" created by the branching of a particular subproblem must always have a larger value for the solution of the relaxed problems as they are produced by adding new constraints on the binary variables (increasing lower bound on a branch of the tree).

The procedure continues by selecting any active (not dismissed or fathomed) node and solving the relaxed problem. We continue doing this until we have fathomed all nodes and the incumbent corresponds to the optimal solution. If no incumbent has been found, then the initial problem is infeasible. We summarize the fathoming criteria:

1. The relaxed problem is infeasible.
2. The relaxed problem has a binary solution (for the **y** variables).

3. The relaxed problem does not have an integer solution, but the objective function is greater than the currently available upper bound (incumbent).

We will now solve an extended case study to clarify the implementation details of the B&B methodology.

We consider the Knapsack problem studied in Example 5.1:

$$\min_{y_1 y_2 y_3 y_4} \; -55y_1 - 40y_2 - 30y_3 - 20y_4$$

$$\text{s.t.} \tag{5.77}$$

$$35y_1 + 25y_2 + 20y_3 + 15y_4 \le 70$$

$$y_i = 0 \text{ or } 1, \quad i = 1,2,3,4$$

to demonstrate the application of the LP-based B&B methodology. We start by relaxing the requirement that $y_i \in \{0,1\}$, $i = 1, 2, 3, 4$ with the requirement that $y_i \in [0,1]$, $i = 1, 2, 3, 4$ and solve the resulting LP problem. This can be done in MATLAB® as follows:

```
>> clear all
>> c=[-55 -40 -30 -20];
>> A=[35 25 20 15];
>> b=70;
>> LB=zeros(4,1);
>> UB=ones(4,1);
>> [y,f,flag]=linprog(c,A,b,[],[],LB,UB)


Optimal solution found.

y = 1.0000
    1.0000
    0.5000
         0
f =   -110
flag =   1
```

We note that the solution $\mathbf{y} = [1\ 1\ 0.5\ 0]^T$ is not integral and that the objective function is $f^0 = -110$. This is a lower bound on the objective function (i.e., the optimal solution of (5.77) cannot be better/lower than -110). We branch on the only non-integral variable which is y_3. We create two nodes: node R^1 with $y_3 = 1$ and node R^2 with $y_3 = 0$. This is also shown in Fig. 5.6. We solve the relaxed problems in these two nodes to obtain the following results:

Node R^1: $\mathbf{y} = [0.7143\ 1\ 1\ 0]^T$ which is not integral with objective function $f^1 = -109.3$.
Node R^2: $\mathbf{y} = [1\ 1\ 0\ 0.6667]^T$ which is not integral with objective function $f^2 = -108.3$.

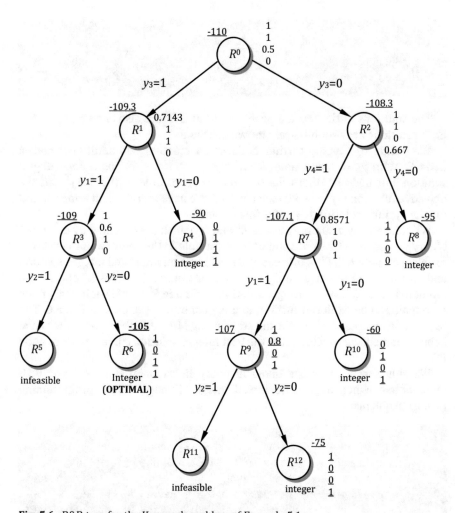

Fig. 5.6 B&B tree for the Knapsack problem of Example 5.1

This can be achieved in MATLAB, for the case of R^1, as follows:

```
>> Aeq=[0 0 1 0];
>> beq=1;
>> [y,f,flag]=linprog(c,A,b,Aeq,beq,LB,UB)

Optimal solution found.

y = 0.7143
    1.0000
```

(continued)

```
    1.0000
         0
 f = -109.2857
 flag =   1
```

The solutions of the relaxed problems R^1 and R^2 are also shown in Fig. 5.6. Both solutions are non-integral and we continue.

We decide to develop further node R^1 by creating two additional nodes: node R^3 with $y_1 = 1$ and node R^4 with $y_1 = 0$. Node R^3 has a non-integral solution, but node R^4 offers the first binary solution $\mathbf{y} = [0\ 1\ 1\ 1]^T$ and the objective function is $f^4 = -90$ (see Fig. 5.6). We have therefore an upper bound for the optimal solution $\bar{f} = -90$ (incumbent).

We continue by examining nodes R^5 and R^6 which are obtained by assigning binary values to y_2 (the only nonintegral variable in the solution of the relaxed problem at node R^3). R^5 is infeasible, and R^6 gives an improved integer solution and an improved upper bound $\bar{f} = -105$ (incumbent). Nodes R^4 and R^6 are fathomed: node R^6 has an integer solution, and node R^5 is infeasible. Only node R^2 remains to be explored further, and we continue as shown in Fig. 5.6. The reader is encouraged to repeat the steps using MATLAB. When all nodes have been fathomed, we conclude that the best integer solution corresponds to node R^6.

We summarize the steps taken at every iteration of the LP-based B&B methodology when applied to mixed integer (binary) linear programming (MILP) problems:

1. Select one subproblem (node) from the remaining ones (the most recently created or the most promising) and among the binary variables that have a non-integral value in the solution (the first in the natural ordering or the closest to be either 0 or 1). This is the *branching variable*, and use it to create two new subproblems: the one corresponds to the value of zero and the other to the value of one for the *branching variable*.
2. Solve the LP relaxation of the two new subproblems, and obtain the value of the objective function.
3. Apply the fathoming tests: the node is fathomed if:

   (a) The LP relaxation is infeasible.
   (b) The LP relaxation has binary solution (for the $\mathbf{y}$ variables).
   (c) The LP relaxation has an optimal value that is worse (larger) than the incumbent.

   In case b above, if the objective function is better (smaller) than the incumbent, then update the incumbent before checking case c.

(continued)

4. If no remaining nodes/subproblems exist, then the algorithm has terminated, and the incumbent is the solution. If there is no incumbent, then the problem is infeasible.

As we have already discussed, the algorithm is initialized by setting the upper bound to $+\infty$, the lower bound to $-\infty$, and solving the LP relaxation of the initial problem. The solution then offers an improved lower bound (unless it is infeasible or has an integer solution in which cases, the algorithm terminates). The solution of the relaxed problems at subsequent nodes can generate an improved incumbent or global upper bound (in the case that an improved integer solution is obtained) or an improved lower bound (in the case a feasible but non-integral solution is obtained) for the descendants of the particular node. When this lower bound becomes at any node greater than the current incumbent, then this node is fathomed as no improved upper bound can be obtained from this node. The importance of discovering a "good" incumbent is therefore extremely important in speeding up the B&B algorithm.

We will now consider an additional example and apply the LP-based B&B methodology. We will consider the MILP formulation of the paint production or single-machine scheduling problem presented in Example 5.6. The complete mathematical formulation is the following (see (5.68)):

$$\min_{y_{i,j}, st_i, \tau_i} \quad w_W \tau_W + w_R \tau_R + w_B \tau_B$$

$$\text{s.t.} \tag{5.78}$$

$$st_W - \tau_W \le dt_W - pt_W$$
$$st_R - \tau_R \le dt_R - pt_R$$
$$st_B - \tau_B \le dt_B - pt_B$$
$$st_W - st_R + My_{W,R} \le M + pt_W$$
$$st_W - st_B + My_{W,B} \le M + pt_W$$
$$st_R - st_B + My_{R,B} \le M + pt_R$$
$$st_R - st_W + My_{W,R} \le -pt_R$$
$$st_B - st_W + My_{W,B} \le -pt_B$$
$$st_B - st_R + My_{R,B} \le -pt_B$$
$$st_W, st_R, st_B, \tau_W, \tau_R, \tau_B \ge 0$$
$$y_{W,R}, y_{W,B}, y_{R,B} \in \{0, 1\}$$

The optimization variables are the starting times (st_W, st_R, and st_B), the delay time (τ_W, τ_R, τ_B), and the binary variables that denote the ordering of the production sequence ($y_{W,R}$, $y_{W,B}$, $y_{R,B}$). The due times dt_i, the processing times

pt_i, and the penalties for delayed delivery are all given in Table 5.3. To facilitate the solution in MATLAB, we write the inequality constraints in matrix form:

$$
\begin{bmatrix}
1 & 0 & 0 & -1 & 0 & 0 & 0 & 0 & 0 \\
0 & 1 & 0 & 0 & -1 & 0 & 0 & 0 & 0 \\
0 & 0 & 1 & 0 & 0 & -1 & 0 & 0 & 0 \\
1 & -1 & 0 & 0 & 0 & 0 & M & 0 & 0 \\
1 & 0 & -1 & 0 & 0 & 0 & 0 & M & 0 \\
0 & 1 & -1 & 0 & 0 & 0 & 0 & 0 & M \\
-1 & 1 & 0 & 0 & 0 & 0 & -M & 0 & 0 \\
-1 & 0 & 1 & 0 & 0 & 0 & 0 & -M & 0 \\
0 & -1 & 1 & 0 & 0 & 0 & 0 & 0 & -M
\end{bmatrix}
\cdot
\begin{bmatrix}
st_W \\ st_R \\ st_B \\ \tau_W \\ \tau_R \\ \tau_B \\ y_{W,R} \\ y_{W,B} \\ y_{R,B}
\end{bmatrix}
\leq
\begin{bmatrix}
dt_W - pt_W \\ dt_R - pt_R \\ dt_B - pt_B \\ M + pt_W \\ M + pt_W \\ M + pt_R \\ -pt_R \\ -pt_R \\ -pt_B
\end{bmatrix}
$$

$$(5.79)$$

Using the data on Table 5.3 and the compact Formulation (5.79), we can solve the LP relaxation of the overall formulation in MATLAB as shown below (script file):

```
clear all
% W R B
pt=[12 48 36];
dt=[60 52 84];
w =[15 10 30];
c=[zeros(3,1);w';zeros(3,1)];
M=100;
A=[1  0  0 -1  0  0  0  0  0;
   0  1  0  0 -1  0  0  0  0;
   0  0  1  0  0 -1  0  0  0;
   1 -1  0  0  0  0  M  0  0;
   1  0 -1  0  0  0  0  M  0;
   0  1 -1  0  0  0  0  0  M;
  -1  1  0  0  0  0 -M  0  0;
  -1  0  1  0  0  0  0 -M  0;
   0 -1  1  0  0  0  0  0 -M];
b=[dt(1)-pt(1);
   dt(2)-pt(2);
   dt(3)-pt(3);
   M-pt(1);
   M-pt(1);
   M-pt(2);
   -pt(2);
   -pt(3);
   -pt(3)];
```

(continued)

```
LB=zeros(9,1);

UB=[M; M; M; M; M; M; 1; 1; 1];

[x,f,flag]=linprog(c,A,b,[],[],LB,UB)
```

```
>> introBnB3
Optimization terminated.

x = 21.2886
     1.5376
    18.5970
     0.0000
     0.0000
     0.0000
     0.4856
     0.5894
     0.6091

f =  1.5010e-11

flag = 1
```

Note that the solution of the initial LP relaxation gives a zero lower bound for the objective function, and the binary variables obtain non-integer values. At the initial node, we select the first binary variable $y_{W,R}$ as the branching variable and create two new nodes R^1 ($y_{W,R} = 1$) and R^2 ($y_{W,R} = 0$). We then proceed and solve the LP-relaxed subproblems. The results are shown in Fig. 5.7. The results at node R^1 can be used to obtain the improved lower bound of 80. No node can be fathomed, and we repeat. We select node R^1 and $y_{W,B}$ as the branching variable and generate nodes R^3 and R^4 as shown in Fig. 5.7. R^3 and R^4 give non-integer solutions, and we select R^3 and $y_{R,B}$ as the branching variable and generate nodes R^5 and R^6 which both have integer solutions with equal objective functions. The common value of the objective function becomes the first incumbent which is equal to 440. Using this incumbent, R^4 can be fathomed, and R^2 remains the only available node. We continue to finally discover that the optimal solution corresponds to node R^9 and the optimal solution is identical to the one shown in Fig. 5.4. Note that B&B has not been very efficient in this case study as we had to examine 13 nodes out of the potential 15 nodes before we can guarantee optimality.

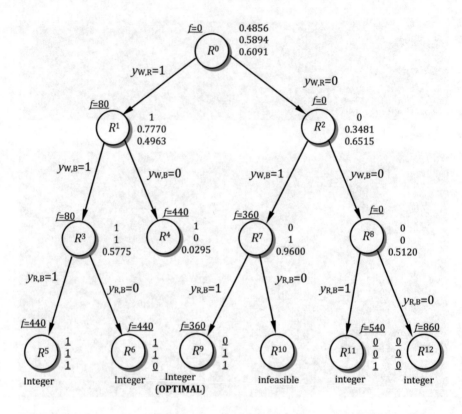

Fig. 5.7 B&B tree for the single-machine scheduling problem of Example 5.6

5.4 Solving MILP Problems in MATLAB®

The MILP solver available in MATLAB® is `intlinprog` and solves the problem given by (5.69). Its general structure is the following:

```
[xopt,fopt,flag,output]
    =intlinprog(c,INTEGERS,A,b,Aeq,beq,LB,UB)
```
attempts to solve the linear programming problem given by:

```
min c'*x subject to: A*x <= b, Aeq*x=bed & LB<x<
UB

x(i) integer
```
where i is in the index vector INTEGERS (integer
constraints)
Set A=[] and B=[] if no inequalities exist. Set Aeq=[] and
Beq=[] if no equalities exist. Use empty matrices for LB and UB if
no bounds exist. xopt is the optimal solution and fopt the
optimum value of the objective function. When flag=1 the global
solution has been found.

Note that `intlinprog` solves the more general mixed integer linear programming where the integers can obtain any integer value and not just binary variables. In this book, we restrict attention to binary variables. Note however that any integer Y can be expressed as a function of binary variables y_i:

$$Y = 2^0 y_1 + 2^1 y_2 + \cdots + 2^{k-1} y_k + \cdots + 2^{N-1} y_N \qquad (5.80)$$

Note that for $N = 10$, we may express all integers up to $Y = 1023$ as a function of ten binary variables (note that $N =$ the smaller integer larger than $\log_2(Y)$, $\log_2(1023) = 9.9986$, and $N = 10$, for instance).

We can now solve the Knapsack problem in MATLAB as follows:

```
>> clear all
>> c=[-55 -40 -30 -20];
>> A=[35 25 20 15];
>> b=70;
>> LB=zeros(4,1);
>> UB=ones(4,1);

>> INTEGERS=[1 2 3 4];
>> [x,f,flag]=intlinprog(c,INTEGERS,A,b,[],[],LB,UB)
```

```
Optimal solution found.

Intlinprog stopped at the root node because the objective value is
within a gap tolerance of the optimal value,
options.AbsoluteGapTolerance = 0 (the default value).
The intcon variables are integer within tolerance,
options.IntegerTolerance = 1e-05 (the default value).

x = 1.0000
         0
    1.0000
    1.0000

f = -105.0000

flag =  1
```

This agrees well with the solution that it was found by our "manual" application of the B&B methodology summarized in Fig. 5.6.

We will now also solve the single-machine scheduling problem in MATLAB:

```
clear all
% W R B
pt=[12 48 36];
dt=[60 52 84];
w =[15 10 30];
c=[zeros(3,1);w';zeros(3,1)];
M=100;
A=[1  0  0 -1  0  0  0  0  0;
   0  1  0  0 -1  0  0  0  0;
   0  0  1  0  0 -1  0  0  0;
   1 -1  0  0  0  0  M  0  0;
   1  0 -1  0  0  0  0  M  0;
   0  1 -1  0  0  0  0  0  M;
  -1  1  0  0  0  0 -M  0  0;
  -1  0  1  0  0  0  0 -M  0;
   0 -1  1  0  0  0  0  0 -M];
b=[dt(1)-pt(1);
   dt(2)-pt(2);
   dt(3)-pt(3);
   M-pt(1);
   M-pt(1);
   M-pt(2);
   -pt(2);
   -pt(3);
   -pt(3)];
LB=zeros(9,1);
UB=[M; M; M; M; M; M; 1; 1; 1];

INTEGERS=[7 8 9]
[x,f,flag,output]=intlinprog(c,INTEGERS,A,b,[],[],LB,UB)
```

```
Branch and Bound:

   nodes       total    num int         integer       relative
explored    time (s)   solution            fval        gap (%)
       5        0.01           3   3.600000e+02   7.756233e+01
      11        0.01           3   3.600000e+02   0.000000e+00

Optimal solution found.

Intlinprog stopped because the objective value is within a gap
tolerance of the optimal value, options.AbsoluteGapTolerance =
0 (the default value).
The intcon variables are integer within tolerance,
options.IntegerTolerance = 1e-05 (the default value).
```

(continued)

```
x = 48.0000
        0
   60.0000
        0
        0
   12.0000
        0
    1.0000
    1.0000

f = 360.0000

flag = 1
```

This solution agrees well with the solution obtained by "manual" application of the B&B methodology and summarized in Fig. 5.7, as expected.

5.5 Solving MINLP Problems Using the B&B and Outer Approximation

When the objective function or the constraints of an optimization problem that involves integer (binary) variables are nonlinear, then we obtain the more general class of optimization problems studied in this book. These problems are called mixed integer nonlinear programming (MINLP) problems, and their general structure is the following:

$$\min_{\mathbf{x},\mathbf{y}} f(\mathbf{x}, \mathbf{y})$$

$$\text{s.t.} \tag{5.81}$$

$$\mathbf{h}(\mathbf{x}, \mathbf{y}) = \mathbf{0}$$
$$\mathbf{g}(\mathbf{x}, \mathbf{y}) \leq \mathbf{0}$$
$$\mathbf{x} \in X \subseteq R^n, \quad \mathbf{y} \in \{0, 1\}^m$$

MILP problems is a special case of MINLP problems where the objective function, equality, and inequality constraints are linear in both the continuous variables $\mathbf{x}$ and binary variables $\mathbf{y}$. When specific values $\mathbf{y}^k$ are assigned to the complicating binary variables, then the MINLP problem (5.81) is simplified to an NLP problem (it is reminded that MILP problems simplify to LP problems for fixed $\mathbf{y}^k$). The solution of the resulting NLP is always an upper bound to the optimal solution of the MINLP problem (5.81). In the same way, a relaxation can be defined by relaxing the constraints $y_i \in \{0,1\}$ to $y_i \in [0,1]$, and as is the case for the MILP problems, the relaxation offers a lower bound to the solution of

(5.81). With these observations in mind, we can extend the B&B algorithm that has been presented for MILP problems directly to MINLP problems. However, in the case of MILP problems, we have the distinct advantage that for the LP-based relaxations, we can always find the global minimum as they are convex problems. This is, unfortunately, not the case of the general relaxed MINLP (or RMINLP) problems as convexity is scarce in realistic problems. To guarantee convergence, we need to ensure that we can locate, in each iteration of the B&B algorithm, the global minimum of the relaxed problem. In most cases, we are satisfied by using a local NLP solver and finally discover a near-optimal solution or simply a good solution.

An algorithm that has been quite successful in solving MINLP problems is the outer approximation algorithm. It is reminded that for a convex function $f(\mathbf{x})$, the tangent line approximation at each point has the following property:

$$f(\mathbf{x}) \geq f(\mathbf{x}^k) + \nabla f^T(\mathbf{x}^k) \cdot (\mathbf{x} - \mathbf{x}^k) \tag{5.82}$$

i.e., the linear approximation of the objective function is always an underestimation of the objective function. Consider now the special form of an inequality constrained MINLP problem which is linear and separable in the binary variables:

$$\min_{\mathbf{x},\mathbf{y}} f(\mathbf{x}) + \mathbf{c}^T\mathbf{y}$$
$$\text{s.t.} \tag{5.83}$$
$$\mathbf{g}(\mathbf{x}) + \mathbf{B}\mathbf{y} \leq \mathbf{0}$$
$$\mathbf{x} \in X \subseteq R^n, \quad \mathbf{y} \in \{0, 1\}^m$$

We also assume that the feasible space defined by the inequality constrains is convex. Then, as shown in Fig. 5.8 for the case of a convex feasible region defined by three inequalities, there are finite points $\mathbf{x}^i$, $i \in F$ such that:

$$\mathbf{g}(\mathbf{x}^i) + \nabla \mathbf{g}^T(\mathbf{x}^i) \cdot (\mathbf{x} - \mathbf{x}^i) \leq \mathbf{0}, \quad i \in F \tag{5.84}$$

defines an overestimation of the initial convex feasible region. Putting these two ideas together, we can propose that the solution of the following problem:

$$\min_{\mathbf{x},\mathbf{y},\mu} \mu + \mathbf{c}^T\mathbf{y}$$
$$\text{s.t.} \tag{5.85}$$
$$\left.\begin{array}{l} f(\mathbf{x}^i) + \nabla f^T(\mathbf{x}^i) \cdot (\mathbf{x} - \mathbf{x}^i) \leq \mu \\ \mathbf{g}(\mathbf{x}^i) + \nabla \mathbf{g}^T(\mathbf{x}^i) \cdot (\mathbf{x} - \mathbf{x}^i) + \mathbf{B}\mathbf{y} \leq \mathbf{0} \end{array}\right\}, \quad \forall i \in F$$
$$\mathbf{x} \in X \subseteq R^n, \quad \mathbf{y} \in \{0, 1\}^m$$

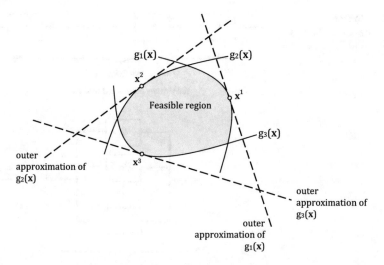

Fig. 5.8 The outer approximation of a convex region overestimates the feasible region

offers a lower bound on the optimal solution of problem (5.83). Note that (5.85) is a MILP problem that can be solved to global optimality. The outer approximation (5.85) involves an overestimation of the feasible region and an underestimation of the objective function, and its solution is, therefore, a lower bound to (or better than) the solution of the initial MINLP problem (5.83). Furthermore, the solution of (5.85) at iteration k offers a new set of values for the binary variables $\mathbf{y}^k$, and the resulting NLP problem:

$$\min_{\mathbf{x}} \ f(\mathbf{x}) + \mathbf{c}^T \mathbf{y}^k$$
$$\text{s.t.} \qquad\qquad\qquad\qquad (5.86)$$
$$\mathbf{g}(\mathbf{x}) + \mathbf{B}\mathbf{y}^k \le \mathbf{0}$$
$$\mathbf{x} \in X \subseteq R^n, \quad y \in \{0, 1\}^m$$

offers an upper bound on the solution of the initial MINLP (5.83). This is the basic idea on which the outer approximation methodology has been developed. To avoid reexamining a specific combination of values for the binary variables, we add the following *integer cut* constraints to Formulation (5.85):

$$\sum_{i \in B^k} y_i - \sum_{i \in N^k} y_i \le |B^k| - 1 \qquad\qquad (5.87)$$

where B^k consists of all i for which $y_i^k = 1$, N^k consists of all i for which $y_i^k = 0$, and $|S|$ denotes the cardinality (number of elements) of S.

Fig. 5.9 System
consisting of N systems (**a**)
in series and (**b**) in parallel

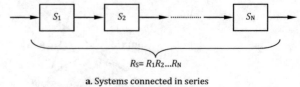

$R_S = R_1 R_2 ... R_N$

a. Systems connected in series

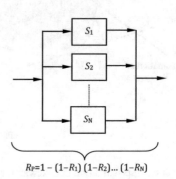

$R_P = 1 - (1-R_1)(1-R_2)... (1-R_N)$

Example 5.7 *MINLP Model for the Reliability Optimization of a Series-Parallel System and Solution Using B&B*

Reliability of a system is defined as the probability that the system will perform its operation (not fail) throughout a prescribed operating period. Systems are connected in several complex configurations, and special techniques are required to determine their reliability. The simplest configurations are the series and parallel configurations:

- Series (see Fig. 5.9a): system performs satisfactorily if all components are fully functional.
- Parallel (see Fig. 5.9b): system performs its operation if any one component remains operational.

A critical computer-controlled system (CCS) in a process consists of three systems in series (a sensor, a computer, and a pump), and each system can be composed by up to three identical systems in parallel (see Fig. 5.10). The sensor (S), the computer (C), and the pump (P) have different reliability characteristics and different costs. Our aim is to select:

- How many identical sensors to install
- How many identical computer systems to install
- How many identical pumps to install

to achieve a minimum acceptable reliability of the overall system. If w_i, $i = \{S, C, P\}$ is the cost of installed equipment and R_i is the reliability of each type of individual equipment, then the optimization problem is:

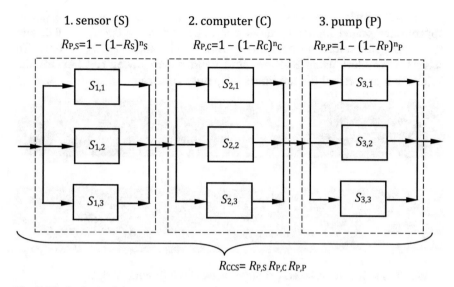

1. sensor (S)   **2. computer (C)**   **3. pump (P)**

$R_{P,S} = 1 - (1 - R_S)^{n_S}$ $R_{P,C} = 1 - (1 - R_C)^{n_C}$ $R_{P,P} = 1 - (1 - R_P)^{n_P}$

$S_{1,1}$ $S_{2,1}$ $S_{3,1}$

$S_{1,2}$ $S_{2,2}$ $S_{3,2}$

$S_{1,3}$ $S_{2,3}$ $S_{3,3}$

$$R_{CCS} = R_{P,S}\, R_{P,C}\, R_{P,P}$$

Fig. 5.10 Analysis of the computer control system under study

$$\min_{n_S, n_C, n_P} \quad w_S n_S + w_C n_C + w_P n_P$$

$$\text{s.t.} \tag{5.88}$$

$$\left[1 - (1 - R_S)^{n_S}\right] \cdot \left[1 - (1 - R_C)^{n_C}\right] \cdot \left[1 - (1 - R_P)^{n_P}\right] \geq R_{CCS,\min}$$

$$n_S, n_C, n_P \in \{1, 2, 3\}$$

where $R_{CCS,\min}$ is the minimum acceptable reliability of the overall system and n_S, n_C, and n_P are the number of sensors, computers, and pumps, respectively. This can be transformed to a MINLP with binary-only variables where the binary variables appear linearly:

$$\min_{\substack{x_S, x_C, x_P \\ y_{0,S}, y_{0,C}, y_{0,P}, y_{1,S}, y_{1,C}, y_{1,P}}} \quad w_S x_S + w_C x_C + w_P x_P$$

$$\text{s.t.} \tag{5.89}$$

$$R_{CCS,\min} - \left[1 - Q_S^{x_S}\right] \cdot \left[1 - Q_C^{x_C}\right] \cdot \left[1 - Q_P^{x_P}\right] \leq 0$$

$$x_S = 2^0 y_{0,S} + 2^1 y_{1,S}$$

$$x_C = 2^0 y_{0,C} + 2^1 y_{1,C}$$

$$x_P = 2^0 y_{0,P} + 2^1 y_{1,P}$$

$$y_{0,S} + y_{1,S} \geq 1$$

$$y_{0,C} + y_{1,C} \geq 1$$

$$y_{0,P} + y_{1,P} \geq 1$$

$$x_S, x_C, x_P \in [1, 3]$$

$$y_{0,S}, y_{0,C}, y_{0,P}, y_{1,S}, y_{1,C}, y_{1,P} \in \{0, 1\}$$

where we have introduced the unavailability $Q_i = 1 - R_i$.

We will consider the case where $w_S = 8$, $w_C = 6$, and $w_P = 10$ (all in appropriate scaled monetary units per unit) and $R_S = 0.9$, $R_C = 0.8$, and $R_P = 0.7$ with minimum acceptable overall reliability $R_{CCS,min} = 0.65$. To solve the problem using B&B, we first build the following m-files for the objective function and the nonlinear inequality constraint:

```
function f=ObjFun(x)
f=8*x(1)+6*x(2)+10*x(3);
```

```
function [g,h]=NonLin(x)
h=[];
g=0.65-(1-0.1^x(1))*(1-0.2^x(2))*(1-0.3^x(3));
```

We then solve the relaxed problem using the following script:

```
clear all
%X = [ xS xC xP y0S y1S y0C y1C y0P y1P]
x0 =[ 3 3 3   1  1  1  1  1  1];
LB =[ 1 1 1   0  0  0  0  0  0];
UB =[ 3 3 3   1  1  1  1  1  1];
A =[ 0 0 0 -1 -1  0  0  0  0;
         0 0 0  0  0 -1 -1  0  0;
         0 0 0  0  0  0  0 -1 -1];
b =-ones(3,1);
Aeq=[ 1 0 0 -1 -2  0  0  0  0;
         0 1 0  0  0 -1 -2  0  0;
         0 0 1  0  0  0  0 -1 -2];
beq=[0;0;0];
[xopt,fopt,flag]=fmincon(@ObjFun,x0,A,b,Aeq,beq,LB,UB,
@NonLin)
```

```
>> reliabilityOptimMINLP

Local minimum found that satisfies the constraints.

xopt = 1.0220   1.4226   1.3337   0.9875   0.0172   0.8128   0.3049
0.8247   0.2545


fopt = 30.0483


flag =    1
```

The value obtained by the binary variables is shown in Fig. 5.11. We note that the objective function of R^0 node is a lower bound on the objective function. As the first binary variable (in the ordering used) is closest to be integer, we branch on $y_{0,S}$ to obtain nodes R^1 ($y_{0,S} = 0$) and R^2 ($y_{0,S} = 1$). The solutions of the relaxed problem at the two nodes do not give a binary solution, and we continue by branching first on $y_{0,C}$ and then on $y_{0,P}$. The first integer solution is obtained at node R^5 with an objective function of 35 (incumbent). We continue to develop the B&B tree which is shown in Fig. 5.11. Nodes R^1, R^8, R^9, and R^{10} are fathomed for the following reasons:

R^1:   the objective function of the relaxed problem is greater than the incumbent.

R^8:   the objective function of the relaxed problem is greater than the incumbent.

R^9:   the relaxed problem is infeasible.

R^{10}:   the relaxed problem has a binary solution.

Expanding node R^3 is left as an exercise to the reader (to show that no improved integer solution can be obtained).

The optimal solution corresponds the solution at node R^5 and is shown in Fig. 5.12. It involves the use of one sensor, one controller, and two pumps with an overall cost of 34 monetary units and an overall reliability that is slightly greater than the minimum acceptable overall reliability.

Example 5.8 *A Numerical MINLP Example and Solution Using B&B*

We will consider the following numerical example of an MINLP problem:

$$\min_{y_1,y_2,y_3,x_1,x_2} (x_1^2 + x_2^2) + (y_1 + 1.5y_2 + 0.5y_3)$$

$$\text{s.t.}$$

$$(x_1 - 2)^2 - x_2 \leq 0$$

$$-x_1 + 2y_1 \leq 0$$

$$x_1 - x_2 - 4(1 - y_2) \leq 0$$

$$-x_1 + (1 - y_1) \leq 0$$

$$-x_2 + y_2 \leq 0$$

$$-x_1 - x_2 + 3y_3 \leq 0$$

$$-y_1 - y_2 - y_3 + 1 \leq 0$$

$$0 \leq x_1 \leq 4$$

$$0 \leq x_2 \leq 4$$

$$y_1, y_2, y_3 \in \{0, 1\}$$

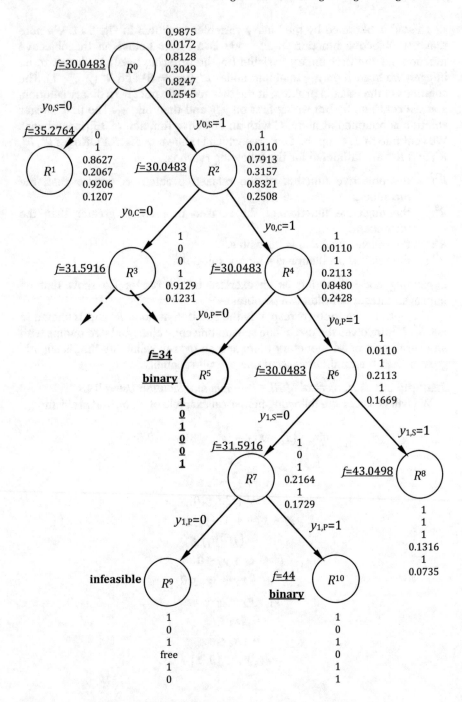

Fig. 5.11 B&B tree for the reliability-optimization problem of Example 5.7

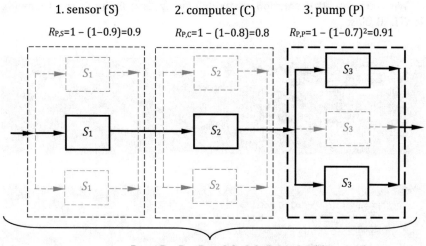

$$R_{CCS} = R_{P,S}\, R_{P,C}\, R_{P,P} = 0.9 \cdot 0.8 \cdot 0.91 = 0.6552 > 0.65$$

Fig. 5.12 Solution of the reliability optimization problem

To demonstrate how we can solve this numerical example efficiently in MATLAB, we note that the problem at hand can be written as follows:

$$\min_{y_1, y_2, y_3, x_1, x_2} f(\mathbf{x}) + \mathbf{c}^T \mathbf{y}$$

$$\text{s.t.}$$

$$g(\mathbf{x}) \le 0$$

$$\mathbf{Ax} + \mathbf{By} \le \mathbf{d}$$

$$0 \le x_1 \le 4$$

$$0 \le x_2 \le 4$$

$$y_1, y_2, y_3 \in \{0, 1\}$$

where $f(\mathbf{x}) = x_1^2 + x_2^2$, $g(\mathbf{x}) = (x_1 - 2)^2 - x_2$, and:

$$\mathbf{c} = \begin{bmatrix} 1 \\ 1.5 \\ 0.5 \end{bmatrix}, \ \mathbf{A} = \begin{bmatrix} -1 & 0 \\ +1 & -1 \\ -1 & 0 \\ 0 & -1 \\ -1 & -1 \\ 0 & 0 \end{bmatrix}, \ \mathbf{B} = \begin{bmatrix} +2 & 0 & 0 \\ 0 & +4 & 0 \\ -1 & 0 & 0 \\ 0 & +1 & 0 \\ 0 & 0 & +3 \\ -1 & -1 & -1 \end{bmatrix}, \ \mathbf{d} = \begin{bmatrix} 0 \\ +4 \\ -1 \\ 0 \\ 0 \\ -1 \end{bmatrix}$$

We start the B&B solution methodology by first building the following MATLAB files:

```
function obj=minlp_obj(X)
y=X(1:3); % binary variables
x=X(4:5); % continuous variables
obj = (x(1)^2+x(2)^2) + [1; 1.5; 0.5]'*y;
end
```

```
function [g,h]=minlp_constr(X)
y=X(1:3); % binary variables
x=X(4:5); % continuous variables
h=[]; g = (x(1)-2)^2-x(2);
end
```

We then solve the relaxed problem using the following script:

```
clear all
A=[-1  0;
   +1 -1;
   -1  0;
    0 -1;
   -1 -1;
    0  0];
B=[+2  0  0;
    0 +4  0;
   -1  0  0;
    0 +1  0;
    0  0 +3;
   -1 -1 -1];
d=[0; +4; -1; 0; 0; -1];
ub=[1;1;1;4;4];
lb=[0;0;0;0;0];
x0=[1;0;1;1;0];
options = optimoptions('fmincon');
options.Display='iter';
[xopt,fopt,flag]=fmincon(@minlp_obj,x0,[B A],d,[],[],lb,ub,
@minlp_constr,options)


>> runNumericalMINLP
```

(continued)

Iter F-count		f(x)	Feasibility	Steplength	Norm of step	First-order optimality
0	6	2.500000e+00	2.000e+00		2.000e+00	
1	12	2.686736e+00	1.132e-04	1.000e+00	1.265e+00	1.021e+00
2	18	2.494607e+00	3.453e-02	1.000e+00	4.284e-01	5.403e-01
3	24	2.531718e+00	4.995e-04	1.000e+00	7.831e-02	1.285e-02
4	30	2.532344e+00	1.192e-06	1.000e+00	1.741e-03	1.699e-03
5	36	2.532346e+00	5.142e-09	1.000e+00	1.440e-04	3.296e-05
6	42	2.532346e+00	1.030e-12	1.000e+00	2.031e-06	5.415e-08

Local minimum found that satisfies the constraints.

```
xopt =
    0.3767
         0
    0.6233
    1.1539
    0.7159


fopt =   2.5323


flag =   1
```

We therefore note that the solution of the relaxed problem is not an integer solution and the value of the objective function (+2.5323) is a lower bound on the optimal solution of the MINLP problem. We branch on the first binary variable (y_1) to obtain two nodes: node R^1 ($y_1 = 0$) and node R^2 ($y_1 = 1$). We solve the new relaxed problems:

```
>> Aeq=[1 0 0 0 0]; beq=1; % node R2
>> [xopt,fopt,flag]=fmincon(@minlp_obj,x0,[B A],d,Aeq,beq,lb,
ub, @minlp_constr,options)
```

Iter F-count		f(x)	Feasibility	Steplength	Norm of step	First-order optimality
0	6	2.500000e+00	2.000e+00		2.000e+00	
1	12	5.250000e+00	0.000e+00	1.000e+00	1.118e+00	1.000e+00
2	18	5.000000e+00	0.000e+00	1.000e+00	5.000e-01	4.000e-01
3	24	5.000000e+00	0.000e+00	1.000e+00	0.000e+00	2.220e-16

Local minimum found that satisfies the constraints.

```
xopt =   1
         0
         0
         2
         0
```

(continued)

```
fopt =    5

flag =    1

>> Aeq=[1 0 0 0 0]; beq=0; % node R1
>> [xopt,fopt,flag]=fmincon(@minlp_obj,x0,[B A],d,Aeq,beq,lb,
ub, @minlp_constr,options)
```

					Norm of	First-order
Iter	F-count	f(x)	Feasibility	Steplength	step	optimality
0	6	2.500000e+00	2.000e+00		2.000e+00	
1	12	2.833333e+00	0.000e+00	1.000e+00	1.491e+00	1.000e+00
2	18	2.697676e+00	1.995e-02	1.000e+00	3.228e-01	3.886e-01
3	24	2.720044e+00	1.968e-06	1.000e+00	1.973e-02	1.704e-02
4	30	2.720046e+00	1.291e-07	1.000e+00	7.252e-04	9.292e-04
5	36	2.720046e+00	8.461e-09	1.000e+00	1.852e-04	2.779e-07

```
Local minimum found that satisfies the constraints.

xopt =  0
   0.3740
   0.6260
   1.1424
   0.7355

fopt =   2.7200

flag =   1
```

We note that the solution at node R^1 does not give an integer solution (see Fig. 5.13) but the solution at node R^2 is integer, and this becomes the incumbent solution. Node R^1 is the only active node and we branch on y_2. The relaxed problems at nodes R^3 ($y_1 = y_2 = 0$) and R^4 ($y_1 = 0, y_2 = 1$) have integer solutions, and the algorithm terminates as there is no active node. Node R^4 corresponds to the new incumbent and is also the optimal solution to the MINLP problem. The complete enumeration tree is shown in Fig. 5.13. Note that in this case study, all nodes are fathomed due to the fact that the relaxed problems have integer solution.

```
>> Aeq=[1 0 0 0 0;0 1 0 0 0]; beq=[0;1]; % node R4
>> [xopt,fopt,flag]=fmincon(@minlp_obj,x0,[B A],d,Aeq,beq,lb,
ub, @minlp_constr,options)
```

					Norm of	First-order
Iter	F-count	f(x)	Feasibility	Steplength	step	optimality
0	6	2.500000e+00	2.000e+00		2.000e+00	
1	12	3.750000e+00	0.000e+00	1.000e+00	1.803e+00	1.000e+00
2	18	3.500000e+00	0.000e+00	1.000e+00	5.000e-01	4.615e-01
3	24	3.500000e+00	0.000e+00	1.000e+00	0.000e+00	0.000e+00

(continued)

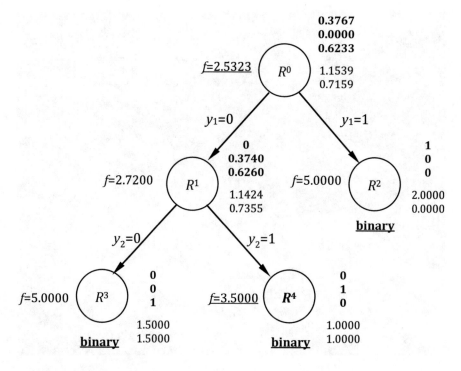

Fig. 5.13 B&B tree for the numerical MINLP problem of Example 5.8

```
xopt =

   0
   1
   0
   1
   1

fopt =

   3.5000

flag =

   1

>> Aeq=[1 0 0 0 0;0 1 0 0 0]; beq=[0;0]; % node R3
>> [xopt,fopt,flag]=fmincon(@minlp_obj,x0,[B A],d,Aeq,beq,lb,
ub, @minlp_constr,options)
```

(continued)

Iter	F-count	f(x)	Feasibility	Steplength	Norm of step	First-order optimality
0	6	2.500000e+00	2.000e+00		2.000e+00	
1	12	5.500000e+00	0.000e+00	1.000e+00	2.236e+00	2.000e+00
2	18	5.031250e+00	0.000e+00	1.000e+00	8.839e-01	6.250e-01
3	24	5.000000e+00	0.000e+00	1.000e+00	1.768e-01	2.645e-08

```
xopt =


         0
         0
    1.0000
    1.5000
    1.5000


fopt =

    5.0000


flag =

    1
```

Learning Summary

In this chapter, we studied an important class of optimization formulations that include integer (mostly binary) variables in order to model decisions related to the topology or structure of the system under study or to model "logical" constraints. The models studied include the mixed integer linear programming problems:

$$\min_{\mathbf{x},\mathbf{y}} f = \mathbf{c}^T \mathbf{x} + \mathbf{d}^T \mathbf{y}$$

$$\text{s.t.}$$

$$\mathbf{Ax} + \mathbf{By} = \mathbf{b}$$

$$\mathbf{Cx} + \mathbf{Dy} \leq \mathbf{e}$$

$$-\mathbf{x} \leq \mathbf{0}, \mathbf{y} \in \{0, 1\}^n$$

and the mixed integer nonlinear programming models:

$$\min_{x,y} f(\mathbf{x}, \mathbf{y})$$

$$\text{s.t.}$$

$$\mathbf{h}(\mathbf{x}, \mathbf{y}) = \mathbf{0}$$

$$\mathbf{g}(\mathbf{x}, \mathbf{y}) \leq \mathbf{0}$$

$$\mathbf{x} \in X \subseteq R^n, \quad y \in \{0, 1\}^m$$

A systematic and general methodology for solving optimization problems involving integer variables is the branch and bound methodology. In this chapter, we presented the application of the branch and bound methodology to linear and nonlinear problems involving binary variables.

Terms and Concepts

You must be able to discuss the concept of:

Branch and bound (B&B) methodology
Branching variable
Fathoming of a node in the B&B tree
Incumbent solution
Knapsack problem
Magic square problem
Outer approximation
Relaxation
Reliability optimization
Scheduling (single machine) problem
Set covering problem
Su Doku problem
Unit commitment problem

Problems

5.1. Solve Example 5.2 using the B&B methodology. Validate your results using MATLAB.

5.2. Use MATLAB and `intlinprog` to find all integer solutions in Example 5.2 that correspond to the same optimal value of the objective function.

5.3. Solve the problem presented in Example 5.4 for Orders 1, 2, and 3. Consider the case where our aim is to minimize quantity adjustments.

5.4. You are considering the inclusion of a new product in the production line of your company. The forecast for the demand of the new product for the coming year is as follows (in t/month):

Jan	Feb	Mar	Apr	May	Jun	Jul	Aug	Sep	Oct	Nov	Dec
800	800	1600	1600	2400	2500	2600	2800	2200	1600	1200	600

There is an inventory of 1000 t available currently. The production of the new product has a setup cost of 50 k$ and a unit production cost of 0.1 k$/t. The setup cost must be taken into consideration when production is scheduled for a month but was not scheduled for the previous month. The storage of any inventory costs 0.05 k$/t of product stored. Find the cost optimal solution for producing the new product. Develop a mathematical formulation, and solve the problem in MATLAB.

5.5. The traveling salesman problem (TSP) is among the most famous optimization problems. A set of nodes (cities) is given and directed or undirected arcs (roads) connecting the nodes. The constraints of the problem are:

(a) Each city must be entered exactly once.
(b) Each city must be exited exactly once.
(c) No subtours (closed tours involving a subset of the nodes) are allowed.

The objective is to minimize the distance traveled (or a related objective). Develop a model for the TSP using binary variables y_{ij} to denote movement of the salesman between cities i and j ($y_{ij} = 1$) or not ($y_{ij} = 0$). Apply your formulation to the problem summarized in Fig. P5.5 without taking constraint c into consideration. Then use integer cuts to eliminate solutions that do not satisfy constraint c until you obtain a feasible solution.

Fig. P5.5 Simple traveling salesperson problem

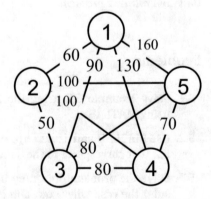

5.6. You need to produce a product with the following specifications:

Ingredient	Minimum (%)	Maximum (%)
A	3.0	3.5
B	0.3	0.5
C	1.4	1.7
D	2.7	3.0

The raw materials available are the following:

Raw material	Cost (Relative)	% A	% B	% C	% D	Amount available
RM1	1.5	4.0	0.0	0.9	2.3	$+\infty$
RM2	3.5	0.0	10.0	4.5	15.0	$+\infty$
RM3	3.0	0.0	0.0	0.0	40.0	$+\infty$
RM4	5.0	0.0	0.0	60.0	18.0	$+\infty$
RM5	1.1	0.4	0.0	1.0	0.0	0.1 t
RM6	1.05	0.1	0.0	0.3	0.0	0.1 t
RM7	1.0	0.1	0.0	0.3	0.0	0.1 t

You have been asked to produce 1 t of product to minimize the quality specifications at the minimum cost. Develop and solve an appropriate mathematical programming formulation to solve the problem. Repeat the problem when any three out of the four quality specifications need to be satisfied (i.e., one can be violated).

5.7. There are seven major cities in a county, and you need to determine where to locate an emergency service so as to ensure that there is at least one emergency service within 20 minutes' driving distance from each major city. The driving distances between the cities are given in the table that follows:

td_{ij} (min) from	to City A	City B	City C	City D	City E	City F	City G
City A	0	15	25	30	40	22	29
City B	15	0	22	28	31	17	22
City C	25	22	0	16	33	24	27
City D	30	28	16	0	19	23	33
City E	40	31	33	19	0	16	21
City F	22	17	24	23	16	0	33
City G	29	22	27	33	21	33	0

Formulate an appropriate binary linear programming problem, and find the solution using MATLAB.

5.8. This problem refers to Example 2 of the paper Kocis and Grossmann, *Industrial and Engineering Chemistry Research*, 27, pp. 1407–1421, 1988. Solve Example 2 of the paper using B&B, and compare your results with the results reported in the paper. The mathematical description of the problem is as follows:

$$\min_{x_1,x_2,y_1,y_2,y_3} \; (y_1 + 2y_2 + 3y_3 - y_4)(2y_1 + 5y_2 + 3y_3 - 6y_4)$$

s.t.

$$y_1 + 2y_2 + y_3 + 3y_4 \geq 4$$
$$y_1,y_2,y_3,y_4 \in \{0, 1\}$$

The global solution is $y = [0\ 0\ 1\ 1]^T, f = -6$.

Repeat the same for the following problem (Example 3 from the same publication):

$$\min_{x_1,x_2,y_1,y_2,y_3} \; 2x_1 + 3x_2 + 1.5y_1 + 2y_2 - 0.5y_3$$

s.t.

$$x_1^2 + y_1 = 1.25$$
$$x_2^{1.5} + 1.5y_2 = 3$$
$$x_1 + y_1 \leq 1.6$$
$$1.333x_2 + y_2 \leq 3$$
$$-y_1 - y_2 + y_3 \leq 0$$
$$x_1,x_2,x_3 \geq 0, \quad y_1,y_2,y_3 \in \{0, 1\}$$

The global solution is $y = [0\ 1\ 1]^T, x = [1.118\ 1.310]^T, f = 7.667$.

5.9. This problem refers to Example 1 of the paper Duran and Grossmann, *Mathematical Programming*, 36, pp. 307–339, 1986. Solve Example 1 of the paper using B&B, and compare your results with the results reported in the paper. The mathematical description of the problem is as follows:

$$\min_{x_1,x_2,y_1,y_2,y_3} \; 10 + (5y_1 + 6y_2 + 8y_3) + (10x_1 - 7x_3)$$
$$- 18 \ln(1 + x_2) - 19.2 \ln(1 + x_1 - x_2)$$

s.t.

$$- 0.8 \ln(1 + x_2) - 0.96 \ln(1 + x_1 - x_2) + 0.8x_3 \leq 0$$
$$-2 - \ln(1 + x_2) - 1.2 \ln(1 + x_1 - x_2) + x_3 + 2y_3 \leq 0$$
$$x_2 - x_1 \leq 0$$
$$x_2 - 2y_1 \leq 0$$
$$x_1 - x_2 - 2y_2 \leq 0$$
$$y_1 + y_2 - 1 \leq 0$$
$$y_1,y_2,y_3 \in \{0, 1\}, 0 \leq x_1,x_2 \leq 2, 0 \leq x_3 \leq 1$$

The global solution is $y = [0\ 1\ 0]^T, x = [1.301\ 0\ 1] f = -6.0098$.

5.10. Solve Example 5.8:

$$\min_{y_1,y_2,y_3,x_1,x_2} \left(x_1^2 + x_2^2\right) + \left(y_1 + 1.5y_2 + 0.5y_3\right)$$

s.t.

$$(x_1 - 2)^2 - x_2 \leq 0$$
$$-x_1 + 2y_1 \leq 0$$
$$x_1 - x_2 - 4(1 - y_2) \leq 0$$
$$-x_1 + (1 - y_1) \leq 0$$
$$-x_2 + y_2 \leq 0$$
$$-x_1 - x_2 + 3y_3 \leq 0$$
$$-y_1 - y_2 - y_3 + 1 \leq 0$$
$$0 \leq x_1 \leq 4$$
$$0 \leq x_2 \leq 4$$
$$y_1,y_2,y_3 \in \{0, 1\}$$

with the outer approximation methodology (OA). It is reminded that the problem can be stated as:

$$\min_{y_1,y_2,y_3,x_1,x_2,\mu_{OA}} \mu_{OA} + \mathbf{c}^T\mathbf{y}$$

s.t.

$$f(\mathbf{x}) \leq \mu_{OA}$$
$$\mathbf{Ax} + \mathbf{By} \leq \mathbf{d}$$
$$0 \leq x_1 \leq 4$$
$$0 \leq x_2 \leq 4$$
$$y_1,y_2,y_3 \in \{0, 1\}$$

where $f(\mathbf{x}) = x_1^2 + x_2^2$, $g(\mathbf{x}) = (x_1 - 2)^2 - x_2$, and:

$$\mathbf{c} = \begin{bmatrix} 1 \\ 1.5 \\ 0.5 \end{bmatrix}, \ \mathbf{A} = \begin{bmatrix} -1 & 0 \\ +1 & -1 \\ -1 & 0 \\ 0 & -1 \\ -1 & -1 \\ 0 & 0 \end{bmatrix}, \ \mathbf{B} = \begin{bmatrix} +2 & 0 & 0 \\ 0 & +4 & 0 \\ -1 & 0 & 0 \\ 0 & +1 & 0 \\ 0 & 0 & +3 \\ -1 & -1 & -1 \end{bmatrix}, \ \mathbf{d} = \begin{bmatrix} 0 \\ +4 \\ -1 \\ 0 \\ 0 \\ -1 \end{bmatrix}$$

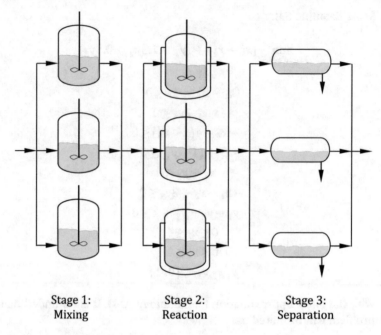

Stage 1:   Stage 2:   Stage 3:
Mixing   Reaction   Separation

Fig. P5.10 Description of the batch process

5.11. This problem refers to Example 4 of the paper Kocis and Grossmann, *Industrial and Engineering Chemistry Research*, 27, pp. 1407–1421, 1988. Study the mathematical formulation (use also the paper by Grossmann & Sargent from the same journal, 18, pp. 343–348, 1978 for further information). Solve the case study presented below using B&B. Use MIPB1 formulation as it features a smaller number of binary variables. Show clearly the development of the B&B tree (Fig. P5.10).

Data:
Products: product A and product B
Stages: stage 1: mixing, stage 2: reaction, stage 3: separation
Cost data:

Mixing vessels: $C_1 = 250 \cdot V_1^{0.6}$, V in L, C in \$
Reaction vessels: $C_2 = 500 \cdot V_2^{0.6}$
Separatores: $C_3 = 350 \cdot V_3^{0.6}$

Processing time:

t_{ij} (h)	Stage (j)		
	Mixing	Reaction	Separation
Product (i)	1	2	3
A	8	20	4
B	10	12	3

Size factor (Volume per unit product – L/g):

S_{ij} (h)	Stage (j)		
	Mixing	Reaction	Separation
Product (i)	1	2	3
A	2	3	4
B	4	6	3

Production capacity:

Product A: $Q_A = 200{,}000$ kg/y
Product B: $Q_B = 150{,}000$ kg/y

Time horizon: $H = 6000$ h/y
Other data: equipment volumes are restricted between 250 L and 2500 L

Up to 2 units per stage are allowed.

Mathematical formulation:

$$\min_{y_j, V_j, M_i, CT_i} n_1 250 V_1^{0.6} + n_2 500 V_2^{0.6} + n_3 350 V_2^{0.6}$$

s.t.

$$\frac{Q_A}{M_A} \cdot CT_A + \frac{Q_B}{M_B} \cdot CT_B \leq H$$

$$S_{ij} \cdot M_i \leq V_j, \forall i,j$$

$$t_{ij} \leq n_j \cdot CT_i, \forall i,j$$

$$n_j = 1 + y_j, \forall j$$

$$250 \leq V_j \leq 2500$$

$$M_j^L \leq M_j \leq M_j^U$$

$$\max_j \left\{ \frac{t_{ij}}{2} \right\} \leq CT_i \leq \max_j \left\{ t_{ij} \right\}$$

$$1 \leq n_j \leq 2$$

$$y_j \in \{0, 1\}$$

where the optimization variables are as follows:

n_j are the number of identical units for stage j.

y_j is a binary variable denoting the existence of a second unit of type j.

V_j is the volume in L of unit at stage j.

M_i is the mass of product i produced in one batch in g.

CT_i is the cycle time for product i in h.

and the model parameters (constants):

Q_i is the annual production of product i in g/y.

H is the operating time in h/y.

S_{ij} are the size factors in L/g.

t_{ij} are the processing times in h.

Chapter 6
Solving Optimization Problems in GAMS®

6.1 Introduction

In this chapter, the General Algebraic Modelling System or GAMS® will be introduced. GAMS is a modeling environment for mathematical programming and optimization purposes, which simplifies the model development process. Arguably, the best way to introduce GAMS is through examples. We have followed this approach with MATLAB®, and we will do the same for GAMS. We will start with a notorious mathematical problem, the circle packing in a square problem. We have presented a closely related example in Chap. 3 (Sect. 3.4). Circle packing in a square is a problem in applied mathematics, where the aim is to pack n unit circles into the smallest possible square (with side length L). The mathematical formulation is a nonlinear programming problem with inequality only constraints:

$$\min_{x_i, y_i, L} L$$

$$\text{s.t.} \tag{6.1}$$

$$(x_i - x_j)^2 + (y_i - y_j)^2 \geq (2r)^2, \forall i, j, j > i$$

$$\left.\begin{array}{l} r \leq x_i \leq L - r \\ r \leq y_i \leq L - r \end{array}\right\} \forall i = 1, 2, \ldots, n$$

where (x_i, y_i) are the coordinates of the center of circle i, $i \in \{1, 2, \ldots, n\}$. r is the radius of the circles and for unit circles $r = 1$.

I. K. Kookos, *Practical Chemical Process Optimization*, Springer Optimization and Its Applications 197, https://doi.org/10.1007/978-3-031-11298-0_6

6.2   Elements of a GAMS® Model

Each GAMS model consists of the following main elements:

Sets: SETS are used to define the indices in the algebraic representations of
models.

For example, in the unit circle packing in a square problem, we have the
following set consisting of the circles considered and the index i:

$$i \in \{\, 1, 2, \ldots, n \,\}$$

or, in a more descriptive form:

$$i \in \{\, Circle_1, Circle_2, \ldots, Circle_n \,\}$$

Data: The input data of each GAMS model can be SCALARS, PARAMETERS, or
TABLES. SCALARS are single-value quantities (constants), while PARAME-
TERS and TABLES are defined over the sets and constitute vector and
matrix quantities. In the unit circle packing in a square problem, we need
to define the circle radius r, π ($=3.14159\ldots$), and two integers constants
Nmin and Nmax such that $\text{Nmin}^2 \leq \text{card(i)} \leq \text{Nmax}^2$ (card(A), where A is any
set, is the "number of elements" in the set, in the case of the unit circle
packing in a square problem card$(i) = n$).

Variables: The decision variables of the optimization model are defined as
VARIABLES in the GAMS modelling language. VARIABLES can obtain any
value (from $-\infty$ to $+\infty$), while we can also define POSITIVE VARIABLES
(from 0 to $+\infty$), NEGATIVE VARIABLES (from $-\infty$ to 0), BINARY VARI-
ABLES (0 or 1), and INTEGER VARIABLES (0, 1, 2,..., 100).

Equations: The equations describe the mathematical model. Equations can be
defined over the SETS and must be declared and defined in separate
statements. First comes the keyword, EQUATIONS in this case, followed
by the name, domain, and text of one or more groups of equations or
inequalities being declared. In the case of the unit circle packing in a square
problem, we have three inequalities, the first is defined for every i and j with
$j > i$, and the other two are defined for every i.

Model and Solve Statements: The model is defined as a set of equations which
contain an objective function. The solve statement asks GAMS to solve the
model.

Lines beginning with an asterisk (*) are ignored by GAMS as they are
comments. With this introduction in place, we present our first GAMS model
in Table 6.1. The model consists of six parts. In the first part, the set of circles is
defined:

Table 6.1 GAMS model for the unit circle packing in a square problem

```
* SECTION 1: DEFINE SETS AND INDICES .......................
SET i define number of circles /Circle1*Circle5/;
ALIAS (i,j);


* SECTION 2: DEFINE SCALARS, PARAMETERS & TABLES ............
SCALAR  pi /3.141592654/;
SCALAR  r radius of circle i in m /1/;
SCALARS Nmax,Nmin;
Nmax = CEIL(SQRT(CARD(i)));
Nmin = FLOOR(SQRT(CARD(i)));


* SECTION 3: DEFINE VARIABLES ..............................
POSITIVE VARIABLES x(i),y(i);
VARIABLE L;


* SECTION 4: DEFINE EQUATIONS ..............................
EQUATIONS Ineq1,Ineq2,Ineq3;
Ineq1(i,j)$(ORD(i)<ORD(j)).. POWER(x(i)-x(j),2)+POWER(y(i)-
                                  y(j),2) =G= (2*r)**2;
Ineq2(i).. x(i)+r =L= L;
Ineq3(i).. y(i)+r =L= L;


* SECTION 5: DEFINE BOUNDS AND INITIAL VALUES ...............
x.lo(i)=r;              x.up(i)=2*r*Nmax-r;      x.l(i)=r;
y.lo(i)=r;              y.up(i)=2*r*Nmax-r;      y.l(i)=r;
L.lo    =2*r*Nmin;      L.up    =2*r*Nmax;       L.l=L.up;


* SECTION 6: MODEL & SOLVE STATEMENTS ......................
MODEL  CirclePackingInSquare /ALL/;
OPTION NLP=KNITRO;
SOLVE  CirclePackingInSquare USING NLP MINIMIZING L;
```

```
SET i define number of circles /Circle1*Circle5/;
ALIAS (i,j);
```

Although GAMS is case insensitive (SET, Set, and set are the same for GAMS), we will be using capital letters for reserved word in GAMS. First the keyword SET is given followed immediately by the index i and the optional comment "define number of circles". You should note the typographical differences between the GAMS format and the usual mathematical format for listing the elements of a set. GAMS uses slashes "/ /" rather than curly braces "{ }" to delineate the set. A convenient feature to use when you are assigning members to a set is the asterisk. It applies to cases when the elements follow a sequence, i.e.:

/Circle1, Cyrcle2, Cyrcle3, Circle4, Circle5/

is equivalent to:

/Circle1*Circle5/

It is important to emphasize the presence of the semicolon at the end of the first line. Without it, the GAMS compiler would attempt to interpret both lines as parts of the same statement.

The second line uses the reserved word ALIAS which is used when it is necessary to have more than one name for the same set. In this way, the index j is just another name (or an alias) for the index i. After the use of the command ALIAS(i,j), i and j are completely equivalent (undisguisable).

In the second part of the GAMS program, we declare and assign values to SCALARS or matrix parameters. In the particular program, only scalar quantities are defined in the way shown. The reserved word SCALAR is given first, followed by the declaration of the scalar, optional comments, and the number that is assigned to be the value of the scalar quantity in slashes "/ /":

```
SCALAR pi /3.141592654/;
SCALAR r radius of circle i in m /1/;
SCALARS Nmax,Nmin;
Nmax = CEIL(SQRT(CARD(i)));
Nmin = FLOOR(SQRT(CARD(i)));
```

Table 6.2 Most commonly encountered GAMS function

abs(x)	Absolute value of x, $|x|$
ceil(x)	Ceiling of x, smallest integer $\geq x$
cos(x)	cos of x, x in radians
exp(x)	Exponential of x, e^x
floor(x)	Floor of x, larger integer $\leq x$
log(x)	Natural logarithm, $\log_e x$
log10(x)	Common logarithm, $\log_{10} x$
max(x,y,z,...)	Maximum value among arguments
min(x,y,z,...)	Minimum value among arguments
power(x,n)	Integer power of x, x^n
sin(x)	sin of x, x in radians
sqrt(x)	Square root of x, $\sqrt{x}$

We can avoid using the SCALAR keyword multiple times in the way shown below:

```
SCALARS pi /3.141592654/,
    r radius of circle i in m /1/,
    Nmax, Nmin;
Nmax = CEIL(SQRT(CARD(i)));
Nmin = FLOOR(SQRT(CARD(i)));
```

CEIL, SQRT, and FLOOR are functions in GAMS. The most used functions in GAMS are summarized in Table 6.2. CARD is used to determine the number of elements in a set. In the particular case where $i \in \{\text{Circle}_1, \text{Circle}_2, \ldots, \text{Circle}_5\}$, CARD(i) = 5 (i.e., the set has five elements).

In the third part of the program, we declare the variables appearing in the model. The unit circle packing in a square problem has as variables the x-y coordinates of the center of each circle (x_i, y_i) which are always positive variables (POSITIVE VARIABLES that obtain values between 0 and $+\infty$). The other variable appearing in the model is the length of the side of the square, but, as this is also our objective function, we are forced to define L as an unrestricted variable (VARIABLE which obtains values from $-\infty$ to $+\infty$):

```
POSITIVE VARIABLES x(i),y(i);
VARIABLE L;
```

The fourth part is arguably the most important as it is the part where we define the set of equality and inequality constraints of the mathematical model. It starts with the keyword EQUATIONS followed by a list of equation names separated by commas. The names can be descriptive or arbitrary but always start with a letter and can only contain up to 63 alphanumeric characters. The equation name can be followed by optional text to describe the equation. The declaration of the equations is followed by their definition:

```
EQUATIONS Ineq1,Ineq2,Ineq3;

Ineq1(i,j)$(ORD(i)<ORD(j))..
POWER(x(i)-x(j),2) + POWER(y(i)-y(j),2) =G= (2*r)**2;
Ineq2(i).. x(i)+r =L= L;
Ineq3(i).. y(i)+r =L= L;
```

In the definition of any equation, the name of the equation is stated first followed by the domain of its definition. `Ineq1`, for instance, is defined for all pairs of `i` and `j` for which `j > i`. To achieve that, we first declare, immediately after the name, the domain of its definition, i.e., `(i,j)`. To enforce the condition `j > i`, we use the dollar operator `$` which is a logical operator. The `ORD` operator is used to determine the order of an element in a set (relative position of a member of a set). Apparently if i precedes j, then ord(i) < ord(j), and the condition `$(ORD(i) < ORD(j))` is true. In this way, when five circles are considered, only the pairs 1–2, 1–3, 1–4, 1–5, 2–3, 2–4, 2–5, 3–4, 3–5, and 4–5 are used when equations are generated. `Ineq2` and `Ineq3` are both defined for each element of the set i. Note the two dots "`..`" that are always used between the name and domain definition and the start of the algebra. Finally, equations are defined using common mathematical operations, variables, and parameters and consist of the left-hand-side expression (`lhs_expression`) and the right-hand-side expression (`rhs_expression`):

$$\texttt{lhs\_expression} \left\{ \begin{array}{l} = E = \\ = G = \\ = L = \end{array} \right\} \texttt{rhs\_expression}$$

where $= E =$ denotes that the `lhs_expression` must be equal to `rhs_expression` (equality constraint), $=G=$ denotes that the `lhs_expression` must be greater or equal to the `rhs_expression`, and $= L =$ that the `lhs_expression` must be less or equal to the `rhs_expression` (inequality constraints).

The lines that follow:

```
x.lo(i)=r;          x.up(i)=2*r*Nmax-r;  x.l(i)=r;
y.lo(i)=r;          y.up(i)=2*r*Nmax-r;  y.l(i)=r;
L.lo =2*r*Nmin;     L.up =2*r*Nmax;      L.l=L.up;
```

are used to define lower bounds (`*.lo`), upper bounds (`*.up`), and initial values (`*.l`) for all variables appearing in the model. This part is optional but greatly improves our chances of finding a solution to our problem and is highly recommended.

Up to this point, we have defined the set and indices, the parameters, and the constants of the model and the equations. What remains to be done is to define the equations that form our mathematical model and then solve the problem. The MODEL statement is used to collect equations into a group and to label them so that they can be solved. The simplest form of the model statement uses the keyword ALL, i.e., the model consists of all equations that have been declared:

```
MODEL CirclePackingInSquare /ALL/;
```

The model name "CirclePackingInSquare" in our case can be any valid name (must start with a letter followed by more letters or digits; it can only contain up to 63 alphanumeric characters). Instead of using the keyword ALL, we could have used the names of all equations that we wish to include in the model in slashes "/ /":

```
MODEL CirclePackingInSquare /Ineq1, Ineq2, Ineq3/;
```

What remains to be done is to select a specific solver:

```
OPTION NLP=KNITRO;
```

and, finally, solve the model using the SOLVE ... USING ... MINIMIZING/ MAXIMIZING reserved words:

```
SOLVE CirclePackingInSquare USING NLP MINIMIZING L;
```

The remaining elements of this last line are self-evident. The specific problem is a nonlinear programming problem or NLP. Other types of problems are:

LP   Linear programming (proposed solver CPLEX)
MIP   Mixed integer linear programming (proposed solver CPLEX)
MINLP   Mixed integer nonlinear programming (proposed solvers DICOPT, SBB, BARON)

The variable that is optimized, as we have already mentioned, must always be an unrestricted variable.

At this point, we have completed building our first model in GAMS, and we may press F9 or the run GAMS button in the GAMS environment (see Fig. 6.1) to solve the problem. GAMS activates a new window that supplies information about the progress in completing the task, and when finished it generates a

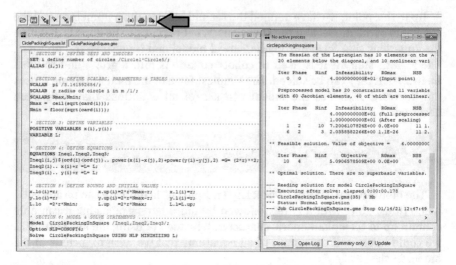

Fig. 6.1 GAMS programming environment after solving the unit circle packing in a square problem

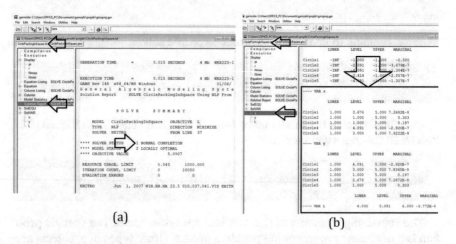

Fig. 6.2 GAMS output for the solution of the unit circle packing in a square problem: (**a**) solution report and (**b**) variables at the solution

short report (see window on the right of Fig. 6.1) and an *.1st file with the solution of the problem and more information that can be useful. The reader is referred to the GAMS manual for more information.

The reader needs to study carefully Fig. 6.2 and to make sure that she/he always check the SOLVE SUMMARY in the *.1st file to make sure that the SOLVER STATUS is NORMAL COMPLETION and MODEL STATUS is LOCALLY OPTIMAL (or OPTIMAL for the case of LP or MILP models). If this is not the case, then the results are not of any value as the solver has apparently failed to provide a solution. If a valid solution has been obtained, then we can use the results reported in the `*.1st` file and in the `SolVAR` section to recover all details about the solution obtained.

For the case study under study, the results obtained when five circles are considered are given by the length of the side of the square $L = 5.093$ and the coordinates of the centers of the five circles (what follows is an extract from the `*.1st` file):

```
---- VAR x
             LOWER   LEVEL   UPPER    MARGINAL
    Circle1  1.000   2.676   5.000   7.2682E-8
    Circle2  1.000   1.000   5.000       0.303
    Circle3  1.000   1.000   5.000       0.197
    Circle4  1.000   4.091   5.000   -2.920E-7
    Circle5  1.000   3.000   5.000   7.9222E-9
---- VAR y
             LOWER   LEVEL   UPPER    MARGINAL
    Circle1  1.000   4.091   5.000   -2.920E-7
    Circle2  1.000   3.000   5.000   7.9365E-9
    Circle3  1.000   1.000   5.000       0.197
    Circle4  1.000   2.676   5.000   7.2672E-8
    Circle5  1.000   1.000   5.000       0.303
```

Note that four values are reported:

LOWER: lower bound provided by the user
LEVEL: value at the solution
UPPER: upper bound provided by user
MARGINAL: marginal value

The single dots in the output represent zeroes. The entry EPS, which stands for "epsilon," means very small but nonzero. The solution for the case of five circles is also shown in Fig. 6.3. The best known solution for the problem of packing five unit circles in a square is $L = 2(1 + \sqrt{2}) \approx 4.828427$ (see Fig. 6.4). Several solvers that are available in GAMS are used, and the results are summarized as follows:

NLP solver	5 circles Best solution	Comput. time (s)	6 circles Best solution	Comput. time (s)	7 circles Best solution	Comput. time (s)
MSNLP	4.828427	0.36	5.328201	0.21	5.732051	0.56
BARON	4.828424	25.93	5.328197	10,492.47	5.732051	>18,000
KNITRO	5.090662	0.09	5.328201	0.17	5.906508	0.16
CONOPT4	5.090658	0.05	5.328201	0.05	5.763523	0.06
MINOS5	Infeasible		Infeasible		Infeasible	

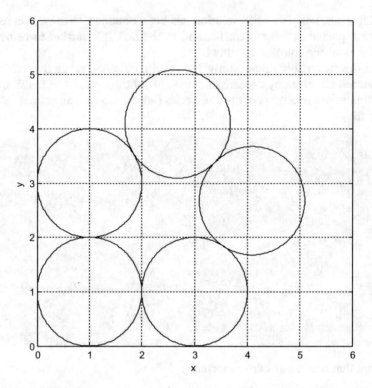

Fig. 6.3 GAMS generated local solution for the unit circle packing in a square problem using KNITRO solver (5 circles) and all circles initially placed at (r, r) point in x–y space

The solution obtained using either MSNLP or BARON is as follows:

```
---- VAR x
           LOWER    LEVEL    UPPER    MARGINAL
Circle1    1.000    1.000    5.000    0.500
Circle2    1.000    2.414    5.000    .
Circle3    1.000    3.828    5.000    .
Circle4    1.000    3.828    5.000    .
Circle5    1.000    1.000    5.000    4.528E-13
---- VAR y
           LOWER    LEVEL    UPPER    MARGINAL
Circle1    1.000    1.000    5.000    0.500
Circle2    1.000    2.414    5.000    .
Circle3    1.000    3.828    5.000    .
Circle4    1.000    1.000    5.000    .
Circle5    1.000    3.828    5.000    .
```

Note the dots appearing in the marginal values (denoting zero values).

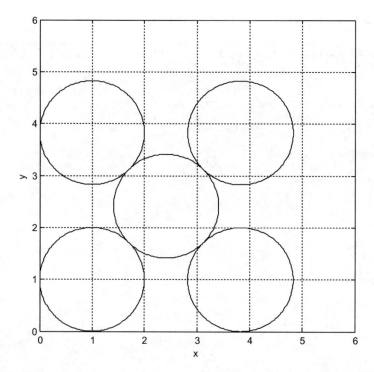

Fig. 6.4 GAMS generated local solution for the unit circle packing in a square problem using MSNLP or BARON solvers (five circles) and all circles initially placed at (r, r) point in x–y space

It is important to note that different solvers have different characteristics and different performance. However, no general conclusion can be drawn from this single case study. The GAMS user must always compare the performance of different solvers for the particular case study under investigation. Some solvers can be better on a specific case study, and some others can be better on other case studies.

We will now present the solution of another problem in order to introduce some additional characteristics of GAMS. We will consider the cutting stock problem studied in Example 4.3. We first present the GAMS model in Table 6.3. Note the definition of vector and matrix data using the PARAMETER and the TABLE keywords:

```
PARAMETER D(j) demand of rods in demands
         / rod5cm 150
           rod3cm 100/;
TABLE Y(j,k)
          pattern1 pattern2 pattern3 pattern4 pattern5 pattern6
rod5cm 2     1        0        1        0        0
rod3cm 0     1        3        0        2        1;
```

Table 6.3 GAMS model for the cutting stock problem

```
$ TITLE Cutting Stock Problem Section 4.3
* SECTION 1: DEFINE SETS AND INDICES .........................
SETS i available rods  /rod10cm, rod7cm, rod4cm/,
     j rods in demand  /rod5cm, rod3cm/,
     k cutting pattern /pattern1*pattern6/;
* SECTION 2: DEFINE SCALARS, PARAMETERS & TABLES ..............
SCALAR Ccut cost of one cut /0.1/;

PARAMETER c(i) cost of available rods
          /rod10cm  2,
           rod7cm  1.5,
           rod4cm   1/;

PARAMETER S(i) supply of available rods
          /rod10cm  50,
           rod7cm   80,
           rod4cm  100/;

PARAMETER kappa(k) cost of available rods
          /pattern1  1,
           pattern2  2,
           pattern3  3,
           pattern4  1,
           pattern5  2,
           pattern6  1/;

PARAMETER D(j) demand of rods in demands
          / rod5cm 150
            rod3cm 100/;

TABLE Y(j,k)
       pattern1 pattern2 pattern3 pattern4 pattern5 pattern6
rod5cm 2        1        0        1        0        0
rod3cm 0        1        3        0        2        1;

TABLE a(i,k)
        pattern1 pattern2 pattern3 pattern4 pattern5 pattern6
rod10cm 1        1        1        0        0        0
 rod7cm 0        0        0        1        1        0
 rod4cm 0        0        0        0        0        1;
* SECTION 3: DEFINE VARIABLES .................................
POSITIVE VARIABLES x(k) number of times pattern k is cut;
VARIABLE Cost;
* SECTION 4: DEFINE EQUATIONS .................................
EQUATIONS Ineq1(i), Eq1(j), ObjFun;
Ineq1(i).. SUM(k,a(i,k)*x(k)) =L= S(i);
Eq1(j)..   SUM(k,Y(j,k)*x(k)) =E= D(j);
ObjFun..   Cost =E=
SUM(i,c(i)*SUM(k,a(i,k)*x(k)))+Ccut*SUM(k,kappa(k)*x(k));
* SECTION 6: MODEL & SOLVE STATEMENTS .....................
MODEL CuttingStock /ALL/;
SOLVE CuttingStock USING LP MINIMIZING Cost;
```

The most important new element available in GAMS that is introduced is the summation operation used in both equations. The first inequality constraint:

```
Ineq1(i).. SUM( k, a(i,k)*x(k) ) =L= S(i);
```

implements the following inequality constraint:

$$\sum_{k\in\{1,\,2,\,..,\,6\}} a_{ik}x_k \leq S_i, \ \forall i \in \{1, 2, 3\}$$

The equivalence between the equality constraint:

$$\sum_{k\in\{1,\,2,\,..,\,6\}} Y_{jk}x_k = D_j, \ \forall j \in \{1, 2\}$$

and its implementation in GAMS:

```
Eq1(j)..   SUM(k,Y(j,k)*x(k)) =E= D(j);
```

is also self-evident. The summation operation available in GAMS is among the strongest characteristics that makes programming of nonlinear programming models in GAMS particularly easy. There is the option of automatic double summation, as it is the case for the following mathematical operation (see, for instance, formulation (4.61)):

$$\sum_{i=1}^{m}\sum_{j=1}^{n} c_{ij}x_{ij} = \text{Cost}$$

which can be implemented in GAMS as follows:

```
Eq..  SUM( (i,j), c(i,j)*x(i,j) ) =E= Cost;
```

We also present a case of a mixed integer problem. We will consider the paint production scheduling problem studied in Example 5.6 and summarized in Formulation (5.68). It is reminded that the relevant data are given in Table 5.3. Table 6.4 presents the GAMS file for the solution of the paint production scheduling problem. There are several new GAMS elements that are introduced in this GAMS program. The first is the use of the $TITLE

Table 6.4 GAMS model for the paint production scheduling problem

```
$ TITLE Paint Production Problem
$ontext
Example 5.6 of the book, Data given in Table 5.3
$offtext
* SECTION 1: DEFINE SETS AND INDICES ..........................
SET    i different paints  /Red, White, Blue/;
ALIAS (i,j);


* SECTION 2: DEFINE SCALARS, PARAMETERS & TABLES ..............
SCALAR    M large number /1000/;
PARAMETER w(i) penalty for delayed delivery in $ per h
          /Red   15, White 10, Blue  30/;
PARAMETER dt(i) due time for paint i
          /Red   60, White 52, Blue  84/;
PARAMETER pt(i) processing time for paint i
          /Red   12, White 48, Blue  36/;
* SECTION 3: DEFINE VARIABLES .................................
POSITIVE VARIABLES st(i)  starting time for processing paint i,
                   tau(i) delay time in delivering paint i;
VARIABLE          Cost;
BINARY VARIABLES  y(i,j);
* SECTION 4: DEFINE EQUATIONS .................................
EQUATIONS Ineq1(i), Ineq2(i,j), Ineq3(i,j), ObjFun;
Ineq1(i)..                st(i)+pt(i) =L= tau(i)+dt(i);
Ineq2(i,j)$(ord(i)<ord(j))..   st(i)+pt(i)   =L=   st(j)+M*(1-
y(i,j));
Ineq3(i,j)$(ord(i)<ord(j)).. st(j)+pt(j) =L= st(i)+M*y(i,j);
ObjFun..                  Cost =E= sum( i, w(i)*tau(i) );
* SECTION 5: DEFINE BOUNDS AND INITIAL VALUES .................
* SECTION 6: MODEL & SOLVE STATEMENTS .........................
MODEL   PaintProductionSchedulling /ALL/;
OPTION MIP=CPLEX;
SOLVE  PaintProductionSchedulling USING MIP MINIMIZING Cost;

* Declare the output file and name it
FILE out /output.dat/;
* Make 'out' the active output file
PUT  out;
* Put some text in the first lines ...
PUT   'Paint production Scheduling Problem' / ;
PUT 'Number of Paints=',card(i):<3:0 /;
* Write the results is formated columns
PUT '|Paint||Start Time||Delay Time||Penatly|'/;
PUT '1234567123456789012121234567890121234567889'/;
LOOP(i,
    PUT ORD(i)        :<>7:0,
        st.l(i)       :<12:1,
        tau.l(i)      :>12:2,
        w(i)          :>9:3  /
    );
PUTCLOSE out;
```

keyword which is used to assign a title to our program (in addition, the $TITLE statement causes the subsequent text to be printed at the top of each page of output). In the GAMS program of Table 6.4, binary variables are defined and used:

```
BINARY VARIABLES  y(i,j);
```

The most important new element used in the program of Table 6.4 is the *put writing facility*. The put writing facility generates documents automatically when GAMS is executed. A document is written to an external file sequentially, a single page at a time. In the first line of the put writing facility, the internal file name out is defined and connected to the external file out.dat using the keyword FILE:

```
* Declare the output file and name it
FILE out /output.dat/;
```

The second line of this put writing facility example assigns the file out.dat as the current file, that is, the file which is currently available to be written to.

```
* Make 'out' the active output file
PUT out;
```

In the third line of the writing to the document begins using a put statement with a textual item:

```
PUT  'Paint production Scheduling Problem' / ;
```

Notice that the text is quoted. The slashes following the quoted text represent carriage returns. The example continues with another textual item followed by the scalar card(i). Notice that these output items are separated with commas:

```
PUT 'Number of Paints=',CARD(i):<3:0 /;
```

Some more textual items follow:

```
* Write the results is formated columns
PUT '|Paint||Start Time||Delay Time||Penatly|'/;
PUT '12345671234567890121234567890121234567890'/;
```

Writing the results into the output file selected requires varying the index in a systematic way. This is achieved in this case using the LOOP statement which has the general format:

```
LOOP( i,
      ....statements...
   ) ;
```

Within the LOOP statement, we use again the PUT statement to write into the out file the elements that are of interest to us:

```
PUT ORD(i)     :<>7.0,
    st.l(i)    :<12:1,
    tau.l(i)   :>12:2,
    w(i)       :>9:3 /
```

Note that: <>7:0 is used to format the way ORD(i) is written into the output file. This is an example of the local format feature. The syntax of this feature is as follows:

```
item:{<>}width:decimals
```

The item is followed by a justification symbol, the field width, and the number of decimals to be displayed. The specification of the number of decimals is only valid for numeric output. The following local justification symbols are applicable:

>   Right justified
<   Left justified
<>   Center justified

Omitting any of the components causes their corresponding global format settings to be used. The item width and decimals are delimited with colons as shown above. The output that is automatically generated (out.dat) after execution is the following:

```
Paint production Scheduling Problem
Number of Paints= 3
|Paint||Start Time||Delay Time||Penatly|
123456712345678901212345678901212345678
1.00 48.0          0.00 15.000
2.00  0.0          0.00 10.000
3.00 60.0         12.00 30.000
```

The keyword PUTCLOSE is used to close a file during the execution of a GAMS program.

Up to this point, we have introduced most of the commonly used features of a GAMS program, and we can, based on these features, solve particularly complex optimization models. We will devote the rest of the chapter to solve some optimization problems in GAMS in order to explore the capabilities of the software, introduce some more advanced elements, and help the reader improve his/her GAMS modelling ability.

6.3   Two Recreational Problems Solved in GAMS®

In this section, we will introduce further elements of GAMS through two recreational problems. Note that in this presentation, our aim is not to present the most compact or the more efficient program in GAMS but to present them with the aim to incorporate interesting features available in GAMS. In the first problem, we will study the famous "River Crossing Problem" (RCP): on his way home, a farmer came to the bank of a river and rented a boat, but crossing the river by boat, the farmer could carry only himself and only one of his purchases: the wolf, the goat, or the cabbage. If left unattended together, the wolf would eat the goat, or the goat would eat the cabbage. The farmer's challenge is to carry himself and his purchases to the far bank of the river, leaving each purchase intact (see Fig. 6.5). Our aim in this section is to propose an integer formulation of the RCP.

There is a thorough analysis of this interesting recreational problem in the literature (Borndörfer et al., Alcuin's Transportation Problems and Integer Programming, In Butzer, P L, Ed, *Charlemagne and His Heritage: 1200 Years of Civilization and Science in Europe*. Brepols, Belgium, 1998, pp. 379–409).

There are several different approaches to attack this challenging problem. We will follow an approach that has a direct generalization to problems that are of interest to chemical engineering. We first define the set of "items", $i \in \{W, G, C, M\}$, where W denotes the wolf, G denotes the goat, C denotes the cabbage, and F denotes the farmer. We also use the index $t \in \{0, 1, 2, \ldots, n\}$ to denote the time (or state) after a river crossing has taken place. Clearly $t = 0$ corresponds to the initial state or at zero time and $t = 1$ the time after the first crossing, etc. We furthermore define the binary variables $z_{i,t}$ and $y_{i,t}$ as follows:

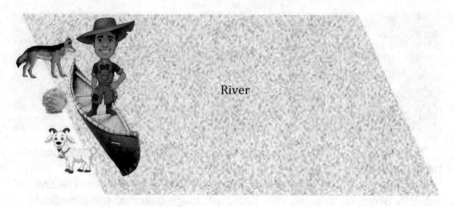

River

Fig. 6.5 The "River Crossing Problem"

$$
z_{i,t} = \begin{cases} 1, & \text{if } i \text{ is on the near (left) bank of the river at time } t \\\\ 0, & \text{if } i \text{ is not on the near (left) bank of the river at time } t \end{cases}
$$

$$
y_{i,t} = \begin{cases} 1, & \text{if } i \text{ is on the far (right) bank of the river at time } t \\\\ 0, & \text{if } i \text{ is not on the far (right) bank of the river at time } t \end{cases}
$$

Having defined the basic variables of the model, we then define the parameter direction_t which shows when the crossing at t is from the near bank to the far bank ($= -1$) or from the far bank to the near bank ($= +1$):

$$
\text{direction}_t = (-1)^{1+t}
$$

We now present the equations through which we can determine whether an item can be found in the near or far bank of the river at $t + 1$ given its state at t:

$$
z_{i,t+1} = z_{i,t} + \text{direction}_t \cdot u_{i,t}, \quad \forall i,t
$$

$$
y_{i,t+1} = y_{i,t} - \text{direction}_t \cdot u_{i,t}, \quad \forall i,t
$$

where we have defined the following "control variables":

$$
u_{i,t} = \begin{cases} 1, & \text{if } i \text{ is crossing the river between the time intervals } t \& t + 1 \\\\ 0, & \text{otherwise} \end{cases}
$$

As the boat can only carry the farmer and up to one additional item, it follows that:

$$\sum_{i \in \{W,\, G,\, C,\, F\}} u_{i,t} \leq 2, \ \forall t$$

$$u_{i,t} \leq u_{F,t}, \ \forall i \in \{W, G, C\}, t$$

The second constraint ensures that any item can cross the river if and only if the farmer is on the boat (i.e., the wolf, the goat, or the cabbage cannot cross the river alone). We may simplify the equations by using the fact that an item can be at either the left bank or the right bank:

$$z_{i,t} + y_{i,t} = 1, \ \forall i, t$$

Finally, we need to exclude the "forbidden" states at which the wolf and the goat or the goat and the cabbage are on the same bank of the river and also take into consideration the fact that these states are allowable if the farmer is on the same bank of the river (the items are under supervision, and they cannot eat each other):

$$\left.\begin{aligned}
z_{W,t} + z_{G,t} &\leq 1 + z_{F,t} \\
y_{W,t} + y_{G,t} &\leq 1 + y_{F,t} \\
z_{G,t} + z_{C,t} &\leq 1 + z_{F,t} \\
y_{G,t} + y_{C,t} &\leq 1 + y_{F,t}
\end{aligned}\right\} \forall t$$

To have a complete model, we need to define an indicator integer variable AllCrossed_t that becomes equal to 1 when all items have crossed the river intact. To determine this variable, we note that all items have crossed the river when $z_{i,t} = 0, \ \forall i$ and $y_{i,t} = 1, \ \forall i$. If any of the $y_{i,t}$ is equal to zero, then not all items have crossed the river. We may therefore write:

$$\text{AllCrossed}_t \leq y_{i,t}, \ \forall i, t$$

Finally, we use the following objective function with the aim to enforce the fastest possible completion of the crossing:

$$\max \sum_{t=0}^{n} \text{AllCrossed}_t$$

The complete mathematical formulation contains binary variables only and is as follows:

$$\max_{z_{i,t},\ y_{i,t},u_{i,t},\text{AllCrossed}_t} \sum_{t=0}^{n} \text{AllCrossed}_t$$

$$\text{s.t.}$$

$$\left.\begin{aligned}
z_{i,t+1} &= z_{i,t} + \text{direction}_t \cdot u_{i,t} \\[4pt]
y_{i,t+1} &= 1 - z_{i,t} \\[4pt]
\text{AllCrossed}_t &\le y_{i,t}
\end{aligned}\right\} \forall i,t$$

$$\left.\begin{aligned}
\sum_{i\in\{W,G,C,F\}} u_{i,t} &\le 2 \\[4pt]
u_{i,t} &\le u_{F,t}, \quad \forall i \in \{W,\ G,\ C\} \\[4pt]
z_{W,t} + z_{G,t} &\le 1 + z_{F,t} \\[4pt]
y_{W,t} + y_{G,t} &\le 1 + y_{F,t} \\[4pt]
z_{G,t} + z_{C,t} &\le 1 + z_{F,t} \\[4pt]
y_{G,t} + y_{C,t} &\le 1 + y_{F,t}
\end{aligned}\right\} \forall t$$

$$z_{i,t}, y_{i,t}, u_{i,t}, \text{AllCrossed}_t \in \{0,1\}, \forall i,t$$

The formulation can be greatly simplified as there are several redundant constraints in the formulation and not all variables need to be defined as binary variables. The interested reader is referred to the GAMS MODEL LIBRARY (model cross.gms) for an alternative, very clever, and efficient formulation.

After solving the model, we obtain the following extract from the *.lst file (as always, we first check the SOLVER STATUS and the MODEL STATUS to ensure that the information provided in the *.lst file corresponds to an optimal solution):

```
---- VAR u control variable: which item is on the boat at t

          LOWER    LEVEL    UPPER    MARGINAL

W.2         .      1.000    1.000    EPS
G.0         .      1.000    1.000    EPS
G.3         .      1.000    1.000    EPS
G.6         .      1.000    1.000    EPS
C.4         .      1.000    1.000    EPS
F.0         .      1.000    1.000    EPS
F.1         .      1.000    1.000    EPS
F.2         .      1.000    1.000    EPS
F.3         .      1.000    1.000    EPS
F.4         .      1.000    1.000    EPS
F.5         .      1.000    1.000    EPS
F.6         .      1.000    1.000    EPS
```

Table 6.5 GAMS model for the river crossing problem

```
$TITLE river crossing problem

* SECTION 1: DEFINE SETS AND INDICES ........................
SETS i /W,G,C,F/,
     t /0*10/;

* SECTION 2: DEFINE SCALARS, PARAMETERS & TABLES ..............
PARAMETER direction(t);
* -1 leaving left bank, +1 leaving right bank
direction(t)=POWER(-1,ord(t));

* SECTION 3: DEFINE VARIABLES ................................
BINARY VARIABLES y(i,t) items at the right (far) bank at t,
                 z(i,t) items at the left (near) bank at t,
                 AllCrossed(t) indicator: all items have
crossed,
                 u(i,t) control variable: item i on the boat
at t;
VARIABLE FinalTime;

* SECTION 4: DEFINE EQUATIONS ................................
EQUATION
LeftBankDynamics(i,t)  balance of items on the left ,
RightBankDynamics(i,t) balance of items on the right,
Up2TwoPassengers(t)    no more that 2 items on the boat,
ManPassenger(i,t)      farmer on the boat for crossing,
ForbiddenLeft1(t)      wolf cannot eat goat-left bank,
ForbiddenLeft2(t)      goat cannot eat cabbage-left bank,
ForbiddenRight1(t)     wolf cannot eat goat-right bank,
ForbiddenRight2(t)     goat cannot eat cabbage-right bank,
AllHaveCrossed(i,t)    all have crossed the river intact,
objfun                 a dummy objective function;

LeftBankDynamics(i,t+1).. z(i,t+1) =E= z(i,t) +
                                      direction(t)*u(i,t);
RightBankDynamics(i,t).. y(i,t) =E= 1-z(i,t);
Up2TwoPassengers(t)..    SUM(i,u(i,t)) =L= 2;
ManPassenger(i,t)$(ord(i)<4).. u('F',t) =G= u(i,t) ;
ForbiddenLeft1(t)..      z('W',t)+z('G',t) =L= 1+z('F',t);
ForbiddenLeft2(t)..      z('C',t)+z('G',t) =L= 1+z('F',t);
ForbiddenRight1(t)..     y('W',t)+y('G',t) =L= 1+y('F',t);
ForbiddenRight2(t)..     y('C',t)+y('G',t) =L= 1+y('F',t);
AllHaveCrossed(i,t)..    AllCrossed(t) =l= y(i,t);
objfun..                 FinalTime=E= SUM(t,AllCrossed(t));

* SECTION 5: DEFINE BOUNDS AND INITIAL VALUES ................
* initial state: all items on the near bank of the river
z.fx(i,t)$(ord(t)=1) = 1;

* SECTION 6: MODEL & SOLVE STATEMENTS ........................
MODEL RiverCrossingProblem /ALL/;
OPTION MIP=CPLEX
SOLVE RiverCrossingProblem USING MIP MaxIMIZING FinalTime;
```

We have only kept the nonzero elements for saving space. We note that at the first crossing u(G.0) = 1 & u(F.0) = 1 and the farmer crosses the river from the near to the far bank together with the goat. In the second crossing u (F.1) = 1 only and returns to the near bank alone. In the third crossing, u (W.2) = 1 & u(F.2) = 1, and the farmer crosses the river from the near to the far bank together with the wolf. Then the farmer returns to the near bank together with the goat as u(G.3) = 1 & u(F.3) = 1 (fourth crossing) and then leaves the goat on the near bank and crosses the river carrying the cabbage as u(C.4) = 1 & u(F.4) = 1 (fifth crossing). Then the farmer returns alone to the near left bank (sixth crossing) and finally crosses the river for seventh and final time together with the goat as u(G.6) = 1 & u (F.6) = 1. There is a second completely equivalent solution in which in the third crossing, instead of crossing the river carrying the wolf, the farmer can select to carry the cabbage, then returns with the goat, and finally crosses the river from the near bank to the far bank two additional times, initially carrying the wolf and finally the goat. Again, seven crossings are necessary, and the solution is equivalent.

We now move to the second problem that we will study in this section: the Traveling Salesman (or Salesperson) Problem or simply TSP. In the TSP problem, we consider n cities, and the salesman wants to start from city 1, visit all cities exactly one time, and finally return to city 1 by minimizing the "cost" of travelling. We will assume that the cost of travelling between city i and city j is c_{ij}, which is always a positive constant. We will also restrict attention to the so-called symmetric TSP problem in which the cost of traveling between j and i is exactly the same as the cost of traveling between i and j. We will also consider that there are direct connections between all cities, and if this is not the case, then we can always set $d_{ij} = M$, where M is a very large number, to make the particular selection unattractive.

Fig. 6.6 TSP example problem

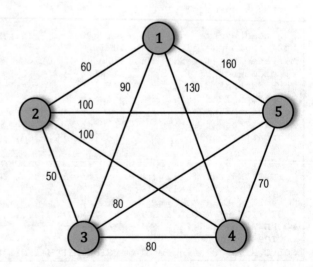

An example TSP problem is shown in Fig. 6.6. The cost of travelling between any two cities is independent of the direction of the movement and is given in Fig. 6.6. To develop the mathematical model, we define the binary variables y_{ij} that are equal to 1 if the salesperson travels from city i to city j, $\forall i, j \in \{1, 2, \ldots, 5\}$. The mathematical formulation is easy to derive as it follows easily from the general assignment problem:

$$\min_{y_{ij}} \sum_i \sum_j c_{ij} y_{ij}$$

$$\text{s.t.}$$

$$\sum_{j,j \neq i} y_{ji} = 1, \forall i$$

$$\sum_{j,j \neq i} y_{ij} = 1, \forall i$$

$$y_{ij} \in \{0, 1\}, \forall i,j$$

The first equality constraint ensures that the salesman arrives at each city and the second one that the salesman departs from each city. Is this formulation a complete representation of the TSP problem? The following is a GAMS file that implements the formulation:

```
$ TITLE TSP Example Problem with 5 cities

SET i cities /1*5/; ALIAS (i,j)

TABLE c(i,j) cost of travelling between i&j
     1    2    3    4    5
1         60   90  130  160
2   60         50  100  100
3   90   50         80   80
4  130  100   80         70
5  160  100   80   70        ;

BINARY VARIABLE x(i,j);
VARIABLE cost;

EQUATIONS Arrive(i), Depart(i), ObjFun;
Arrive(i) .. SUM(j,x(j,i)) =E= 1;
Depart(i) .. SUM(j,x(i,j)) =E= 1;
ObjFun..   SUM((i,j),c(i,j)*x(i,j)) =E= Cost;
x.fx(i,i)=0;

MODEL TSPexample /ALL/;
OPTIONS MIP=CPLEX;
SOLVE TSPexample USING MIP MINIMIZING Cost;
```

When we solve the model, we obtain the following solution (extract from the
*.1st file, only the nonzero elements are given to save space):

```
---- VAR x
      LOWER   LEVEL    UPPER    MARGINAL
1.2    .      1.000    1.000    60.000
2.3    .      1.000    1.000    50.000
3.1    .      1.000    1.000    90.000
4.5    .      1.000    1.000    70.000
5.4    .      1.000    1.000    70.000
                LOWER    LEVEL    UPPER    MARGINAL
---- VAR cost   -INF    340.000  +INF       .
```

This solution is also shown in Fig. 6.7.

What is immediately observed is that the solution is not a complete closed
tour but involves two subtours that are not connected: subtour $1 \rightarrow 2 \rightarrow 3 \rightarrow 1$
and subtour $4 \rightarrow 5 \rightarrow 4$. This demonstrates clearly that the formulation is not a
complete one, and additional constraints are necessary to make the formation
of subtour infeasible. These constraints are known as *subtour elimination
constraints*.

One way to include subtour elimination constraints is to enumerate all
potential subsets of the initial set of cities or vertices $i \in V = \{1, 2, \ldots, n\}$
with more than two elements, i.e., all v, with $2 < \mathrm{card}(v) < n - 1$ (there cannot
be a subtour with 1 city while the subtours with n cities are feasible solutions).
For a subtour to be formed involving all elements of V', there must be as many
nonzero x_{ij} in this set as it is the number of elements in the same set (see
Fig. 6.7, for instance, for the case of a subset with three elements and a subset
of two elements), i.e., the following must hold true:

Fig. 6.7 TSP example
problem initial solution
that involves subtours

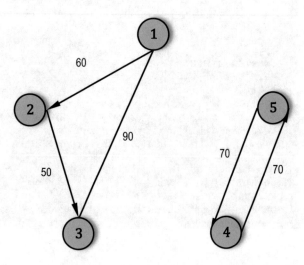

$$\sum_{i \in v} \sum_{j \in v} x_{ij} = \text{card}(v)$$

To eliminate this, we may add the following constraints:

$$\sum_{i \in v} \sum_{j \in v} x_{ij} \leq \text{card}(v) - 1, \quad \text{for all } v \subset V : 2 \leq \text{card}(v) < \text{card}(V)$$

In the case study of Fig. 6.6, we need to apply these inequalities for all subsets with two elements, three elements, and four elements. How many 2-element subsets do we have? The answer is that the number of 2-element subsets is equal to the ways that we may select two objects from a set of five objects: $n!/(n - m)!m! = 5!/3!2! = 10$. We can therefore write the following inequality constraints:

Set elements	Inequality constraint
1–2	$x_{12} + x_{21} \leq 1$
1–3	$x_{13} + x_{31} \leq 1$
1–4	$x_{14} + x_{41} \leq 1$
1–5	$x_{15} + x_{51} \leq 1$
2–3	$x_{23} + x_{32} \leq 1$
2–4	$x_{24} + x_{42} \leq 1$
2–5	$x_{25} + x_{52} \leq 1$
3–4	$x_{34} + x_{43} \leq 1$
3–5	$x_{35} + x_{53} \leq 1$
4–5	$x_{45} + x_{54} \leq 1$

There are also ten 3-element subsets:

Set elements	Inequality constraint
1–2–3	$x_{12} + x_{13} + x_{23} + x_{21} + x_{31} + x_{32} \leq 2$
1–2–4	$x_{12} + x_{14} + x_{24} + x_{21} + x_{41} + x_{42} \leq 2$
1–2–5	$x_{12} + x_{15} + x_{25} + x_{21} + x_{51} + x_{52} \leq 2$
1–3–4	$x_{13} + x_{14} + x_{34} + x_{31} + x_{41} + x_{43} \leq 2$
1–3–5	$x_{13} + x_{15} + x_{35} + x_{31} + x_{51} + x_{53} \leq 2$
1–4–5	$x_{14} + x_{15} + x_{45} + x_{41} + x_{51} + x_{54} \leq 2$
2–3–4	$x_{23} + x_{24} + x_{34} + x_{32} + x_{42} + x_{43} \leq 2$
2–3–5	$x_{23} + x_{25} + x_{35} + x_{32} + x_{52} + x_{53} \leq 2$
2–4–5	$x_{24} + x_{25} + x_{45} + x_{42} + x_{52} + x_{54} \leq 2$
3–4–5	$x_{34} + x_{35} + x_{45} + x_{43} + x_{53} + x_{54} \leq 2$

Finally, there are five 4-element subsets:

Set elements	Inequality constraint
1–2–3–4	$x_{12} + x_{13} + x_{14} + x_{23} + x_{24} + x_{34} + x_{21} + x_{31} + x_{41} + x_{32} + x_{42} + x_{43} \leq 3$
1–2–3–5	$x_{12} + x_{13} + x_{15} + x_{23} + x_{25} + x_{35} + x_{21} + x_{31} + x_{51} + x_{32} + x_{52} + x_{53} \leq 3$
1–2–4–5	$x_{12} + x_{14} + x_{15} + x_{24} + x_{25} + x_{45} + x_{21} + x_{41} + x_{51} + x_{42} + x_{52} + x_{54} \leq 3$
1–3–4–5	$x_{13} + x_{14} + x_{15} + x_{34} + x_{35} + x_{45} + x_{31} + x_{41} + x_{43} + x_{51} + x_{53} + x_{54} \leq 3$
2–3–4–5	$x_{23} + x_{24} + x_{25} + x_{34} + x_{35} + x_{45} + x_{32} + x_{42} + x_{43} + x_{52} + x_{53} + x_{54} \leq 3$

Clearly, despite the fact that many of these constraints are redundant, the number of additional constraints that we need to integrate in our formulation is significant. GAMS, however, makes this easy.

Instead of adding all these constraints in our formulation, we can follow a step-by-step procedure where we initially ignore all of them, and we then add them by increasing the cardinality of the sets considered in each step. In our case, we start by adding all subtour elimination constraints that correspond to 2-element subsets:

```
$ TITLE TSP Example Problem with 5 cities
SET i cities /1*5/; ALIAS (i,j);

TABLE c(i,j) cost of travelling between i & j
       1   2   3    4    5
1          60  90  130  160
2   60         50  100  100
3   90  50         80   80
4  130 100  80         70
5  160 100  80   70           ;

BINARY VARIABLE x(i,j);
VARIABLE cost;

EQUATIONS Arrive(i), Depart(i), Subtour(i,j), ObjFun;
Arrive(i).. SUM(j,x(j,i)) =E= 1;
Depart(i).. SUM(j,x(i,j)) =E= 1;
ObjFun..   SUM((i,j),c(i,j)*x(i,j)) =E= Cost;
Subtour(i,j).. x(i,j)+x(j,i) =L= 1;
x.fx(i,i)=0;

MODEL TSPexample /ALL/;
OPTIONS MIP=CPLEX;
SOLVE TSPexample USING MIP MINIMIZING Cost;
```

The solution obtained is the following:

```
---- VAR x
       LOWER   LEVEL    UPPER    MARGINAL
1.2      .     1.000    1.000     60.000
2.3      .     1.000    1.000     50.000
3.5      .     1.000    1.000     80.000
4.1      .     1.000    1.000    130.000
5.4      .     1.000    1.000     70.000
                        LOWER    LEVEL     UPPER    MARGINAL
---- VAR cost           -INF     390.000   +INF       .
```

This solution is the optimal solution of the TSP case study and is shown in Fig. 6.8. As there is no subtour involved in this solution, we do not need to add any additional constraints. If a solution with disconnected subtours had been obtained, then we would have added the subtour elimination constraints that correspond to all subsets involving three elements, etc.

An alternative approach that is efficient in practice is to solve the relaxed problem, i.e., the problem without the subtour elimination constraints, and then if subtours are discovered, then integer cut constraints are added to eliminate only the subtours that have already discovered. This can be achieved in the case study we investigate by first solving the relaxed formulation to obtain the results shown in Fig. 6.7. Then a constraint is added to eliminate the subtour involving cities 1 and 4. Solving the problem again results in a new solution where a subtour involving cities 1 and 2 appears. Using subtour elimination constraints for subtours involving cities 1 & 4 and 1 & 2:

Fig. 6.8 Optimal solution of the TSP example problem

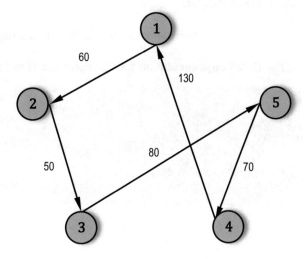

```
$ TITLE TSP Example Problem with 5 cities
SET i cities /1*5/; ALIAS (i,j);

TABLE c(i,j) cost of travelling between i & j
     1    2    3    4    5
1       60   90  130  160
2   60           50  100  100
3   90   50            80   80
4  130  100   80             70
5  160  100   80   70           ;

BINARY VARIABLE x(i,j);
VARIABLE cost;

EQUATIONS Arrive(i), Depart(i), Subtour45, Subtour12, ObjFun;
Arrive(i).. SUM(j,x(j,i)) =E= 1;
Depart(i).. SUM(j,x(i,j)) =E= 1;
ObjFun..   SUM((i,j),c(i,j)*x(i,j)) =E= Cost;
Subtour45.. x('4','5')+x('5','4') =L= 1;
Subtour12.. x('1','2')+x('2','1') =L= 1;
x.fx(i,i)=0;

MODEL TSPexample /ALL/;
OPTIONS MIP=CPLEX;
SOLVE TSPexample USING MIP MINIMIZING Cost;
```

results again in the solution shown in Fig. 6.8.

A second, fundamentally different approach to incorporate the subtour elimination constraints is also commonly presented in the literature. In this alternative approach, dummy variables $0 \leq u_i \leq n - 1$, $\forall\, i$ are intruded that satisfy the constraint:

$$u_i - u_j + nx_{ij} \leq n - 1, \ u_1 = 1, \ 2 \leq i \neq j \leq n$$

The GAMS implementation for this approach is as follows:

```
$ TITLE TSP Example Problem with 5 cities
SET i cities /1*5/; ALIAS (i,j);
SET a(i,j);
a(i,j)=1$(ord(i)>1 AND ord(j)>1 AND ord(i) ne ORD(j));

PARAMETER n; n=card(i);
TABLE c(i,j) cost of travelling between i & j
```

                                                                (continued)

```
      1    2   3    4    5
1          60  90  130  160
2    60        50  100  100
3    90   50        80   80
4   130  100  80         70
5   160  100  80   70        ;
BINARY VARIABLE x(i,j);
VARIABLES u(i), cost;

EQUATIONS Arrive(i), Depart(i), Subtour, ObjFun;
Arrive(i).. SUM(j,x(j,i)) =E= 1;
Depart(i).. SUM(j,x(i,j)) =E= 1;
ObjFun..   SUM((i,j),c(i,j)*x(i,j)) =E= Cost;
Subtour(i,j)$a(i,j).. u(i)-u(j)+n*x(i,j) =L= n-1;
u.lo(i)=1; u.up(i)=n-1;
x.fx(i,i)=0;
u.fx('1')=0;

MODEL TSPexample /ALL/;
OPTIONS MIP=CPLEX;
SOLVE TSPexample USING MIP MINIMIZING Cost;
```

The solution obtained is as follows:

```
---- VAR x
      LOWER    LEVEL    UPPER    MARGINAL
1.4     .      1.000    1.000    130.000
2.1     .      1.000    1.000    60.000
3.2     .      1.000    1.000    50.000
4.5     .      1.000    1.000    70.000
5.3     .      1.000    1.000    80.000
---- VAR u
      LOWER    LEVEL    UPPER    MARGINAL
2    1.000    4.000    4.000      EPS
3    1.000    3.000    4.000       .
4    1.000    1.000    4.000      EPS
5    1.000    2.000    4.000       .
                       LOWER    LEVEL    UPPER    MARGINAL
---- VAR cost          -INF    390.000   +INF       .
```

The solution is the same as the one obtained before. It is interesting to note that the dummy variable $u_i, i \geq 2$ is actually an indication variable that denotes tour ordering, such that $u_i < u_j$ implies that city i is visited before city j.

The TSP problem is the most notorious problem in mathematical programming and combinatorial optimization (problems at which the optimal solution is a subset of a finite set). A characteristic case study of problems that are particularly easy to describe but extremely difficult to solve is known as NP-hard problems. One reason for that is the combinatorial explosion of the alternative solutions or number of equations that are needed to have a complete formulation. For the TSP problem, we have $(n - 1)$ choices when leaving city 1, $(n - 2)$, when leaving city 2, etc. The alternative feasible solutions are therefore $(n - 1)!$ which is a number that increases faster than e^n! If we manage at some point in the future to develop an algorithm for solving the TSP problem efficiently, then the same algorithm can be the basis for solving numerous combinatorial problems and certainly many difficult problems in chemical engineering.

Learning Summary

In this chapter, we introduced GAMS, a powerful modelling environment for optimization studies. We also have used GAMS to solve, among others, two interesting recreational problems that are of direct interest to chemical engineering.

Terms and Concepts

You must be able to discuss the concept of:

GAMS modelling environment and its advantage over MATLAB.
TSP problem as an example of NP-hard problems.

Problems

6.1 Solve Example 5.2 using GAMS.

6.2 Solve Example 5.3 (magic square problem) with $n = 3$, using GAMS.

6.3 Solve the Magic Square problem with $n = 3,4,...,7$ with GAMS. Record the computational time and the number of variables and the number of equations as a function of n. Prepare a short report and comment on the reason(s) why the problem can be more difficult to solve relative to your first impression. Use the minimization of the sum of the diagonal elements as the dummy objective function of the problem.

6.4 Solve the problem presented in Example 5.4 for Orders 1, 2, and 3. Consider the case where our aim is to minimize quantity adjustments. Use GAMS to generate a solution.

6.5 Use GAMS to solve Problem 5.4.

6.6 Use GAMS to solve Problem 5.5.

6.7 Use GAMS to solve Problem 5.6.

6.8 Use GAMS to solve Problem 5.7.

6.9 Use GAMS to solve Problem 5.11.

6.10 In the paper by Pritsker, Watters, and Wolfe (*A zero-one programming Approach to Scheduling with Limited Resources*, The Rand Corporation, RM-5561-PR, 1968), one of the most successful approaches in the scheduling of activities has been presented. Study the paper, and explain the derivation of the following MILP formulation:

$$\min_{y_{jt}} \sum_{j \in J} p_j \left(\sum_{t \in T, \, t > d_j} (t - d_j) \cdot y_{jt} \right)$$

s.t.

$$\sum_{j \in J} \sum_{k=t}^{t+(\tau_j - 1)} y_{jk} \leq 1, \forall t \in T$$

$$\sum_{k=\tau_j}^{\lfloor T \rfloor} y_{jk} = 1, \forall j \in J$$

$$y_{jk} = \{0, 1\}, \forall j \in J, k \in T$$

where $t \in T = \{0, 1, 2, \ldots, H\}$ is the discretized time horizon of interest, $j \in J = \{$Job 1, Job 2,$\ldots$, Job $N\}$ are the activities to be performed, τ_j is the time (duration) necessary to complete Job j, d_j is the day that the job j must have been completed, or otherwise o penalty p_j is imposed for each time interval (until is finally completed and delivered). y_{jt} is a binary variable that denotes that job j is completed at the end of time interval t.

In a particular example, we need to execute six jobs with the following characteristics:

Job	τ_j (h)	d_j (h)	p_j ($/h)
1	5	2	500
2	4	4	400
3	3	8	200
4	5	12	100
5	2	13	700
6	7	17	100
Total	26		

Solve the problem in GAMS to determine the best schedule for completing the jobs.

6.11 An alternative formulation to the Pritsker, Watters, and Wolfe (PWW) formulation for the job scheduling problem, presented in Problem 6.10, is based on a continuous representation of time (and thus avoids the time discretization which is the major drawback of the PWW formulation). In this formulation, we define the binary variables y_{ij} to denote whether job i precedes (executed before) job j ($y_{ij} = 1$). This alternative model introduces the continuous variables x_j which denote that (continuous) time at which the processing of job j starts (the completion time is $x_j + \tau_j$). The complete mathematical formulation is as follows:

$$\min_{y_{ij}, x_j, s_j^+, s_j^-} \sum_{j \in J} p_j s_j^-$$

s.t.

$$\left(x_j + \tau_j\right) + s_j^+ - s_j^- = d_j, \forall j \in J$$

$$\left.\begin{array}{l} My_{ij} + x_i \geq x_j + \tau_j \\ \\ \\ M\left(1 - y_{ij}\right) + x_j \geq x_i + \tau_i \end{array}\right\}, \forall i,j : i < j$$

$$y_{ij} = \{0, 1\}, x_j, s_j^+, s_j^- \geq 0$$

$$y_{ij} = 0, \forall i,j : i \leq j$$

where M is a sufficiently large number, s_j^+ is a non-negative number that denotes the time that job has completed in advance of the due time d_j, and s_j^- is a non-negative number that denotes the delay time in completing job j.

Solve the case study considered in Problem 6.10, and compare the results between the two approaches.

6.12 The thermal isomerization of a-pinene to dipentene and allo-ocimene which in turn yields α- and β-pyronene and a dimer was studied by Fuguitt and Hawkins (1947). The proposed reaction scheme for this homogeneous chemical reaction is:

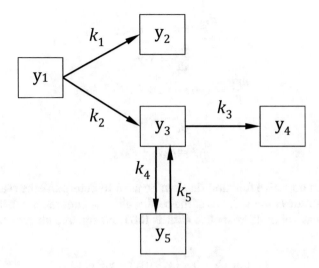

The concentrations of the reactant and the four products were reported by Fuguitt and Hawkins (*J. Am. Chem. Soc.*, 1945, 67 (2), 242–245, & 1947, 69(3), 319–322) at eight time intervals, and these data are reproduced in the following table:

Time (h)	y_1 a-pinene	y_2 dipentene	y_3 allo-ocimene	y_4 pyronene	y_5 dimmer
0	100	0	0	0	0
1230	88.35	7.3	2.3	0.4	1.75
3060	76.4	15.6	4.5	0.7	2.8
4920	65.1	23.1	5.3	1.1	5.8
7800	50.4	32.9	6.0	1.5	9.3
10,680	37.5	42.7	6.0	1.9	12.0
15,030	25.9	49.1	5.9	2.2	17.0
22,620	14.0	57.4	5.1	2.6	21.0
36,420	4.5	63.1	3.8	2.9	25.7

Mathematical models can be derived which give the concentration of the various species as a function of time if the chemical reaction orders are known. Hunter and MacGregor (1967), assuming first-order kinetics throughout, derived the following equations:

$$\frac{dy_1}{dt} = -(k_1 + k_2)y_1$$

$$\frac{dy_2}{dt} = k_1 y_1$$

$$\frac{dy_3}{dt} = k_2 y_1 - (k_3 + k_4)y_3 + k_5 y_5$$

$$\frac{dy_4}{dt} = k_3 y_3$$

$$\frac{dy_5}{dt} = k_4 y_3 - k_5 y_5$$

An objective function that can be used to determine the reaction rate constants k_i, $i = 1, 2, ..., 5$ of the model, although not the best available for the case of multi-response data, is the least squares objective:

$$\min_{k_i} \sum_{j=1}^{5} \sum_{k=1}^{8} \left(y_j^{\exp}(t_k) - y_j(t_k) \right)^2$$

Use GAMS to solve the optimization problem, and estimate the reaction rate constants. Solve the differential equations $dy(t)/dt = f(y(t))$ using the following simple rule of integration:

$$y(t) = y(t - \delta t) + \frac{1}{2}[f(y(t)) + f(y(t - \delta t))]$$

Increase the number of points in the discretization until you have sufficient accuracy.

Hint: as the value of the reaction rate constants is extremely small, it is beneficial to use the transformation $k_i = e^{\kappa_i}$ and determine the $\kappa_i's$.

Chapter 7
Representative Optimization Problems in Chemical Engineering Solved in GAMS®

7.1 Introduction

In this chapter, GAMS® will be used to solve representative optimization problems from the chemical engineering discipline. We will study several examples taken from a wide spectrum of application areas. The aim is for the reader to improve her/his modelling skills, adopt good modelling practices, appreciate the strengths and limitation of currently available optimization software, and use these case studies as the basis for building more detailed optimization models.

7.2 Optimization of a Multiple-Effect Evaporation System

Evaporation systems are common in the food industry primarily as a means of concentrating milk, fruit juices, and sugar solutions prior to crystallization. Evaporation can also be used to raise the solids content of dilute solutions prior to spray- or freeze-drying. The process of evaporation involves the application of heat to vaporize water at low temperatures by heating the product in a vacuum to avoid quality degradation. Energy utilization can be improved by using heat exchangers to extract heat from the vapors to preheat the feed or by using multiple effects where the vapors produced from one effect are used to provide heat in the next effects. The description of the operation of an evaporation unit involves simple material and energy balances. The cost of energy and the cost of equipment must be put in a common, annualized basis to be able to design cost-optimal systems.

I. K. Kookos, *Practical Chemical Process Optimization*, Springer Optimization and Its Applications 197, https://doi.org/10.1007/978-3-031-11298-0_7

Fig. 7.1 A single effect
evaporator

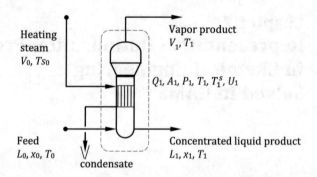

A single-effect evaporator system is shown in Fig. 7.1. A liquid stream with mass flow rate L_0, solids concentration x_0 (mass fraction), and temperature T_0 is fed to the system. Two product streams are obtained: a vapor stream with mass flow rate V_1 and a liquid product stream with mass flow rate L_1 with solids concentration x_1 (mass fraction). Both streams are at the operating temperature of the evaporator T_1. The mathematical model of the evaporator consists of the overall material balance, the material balance of the solids, and the energy balance:

$$L_0 = L_1 + V_1 \tag{7.1}$$

$$L_0 x_0 = L_1 x_1 \tag{7.2}$$

$$L_0 h(x_0, T_0) + Q_1 = L_1 h(x_1, T_1) + V_1 H(T_1) \tag{7.3}$$

where Q_1 is the rate of heat transfer from the heating steam to the solution. The heating steam has a mass flow rate V_0 and is saturated at temperature T_0^s. h is the enthalpy of the liquid stream which is a function of the mass fraction of the solids and the temperature, i.e., $h = h(x, T)$. H is the enthalpy of the steam, which is a function of the steam temperature only, i.e., $H = H(T)$. Q_1 can be expressed through the following equations:

$$Q_1 = V_0 \left(H(T_0^s) - h(T_0^s) \right) \tag{7.4}$$

$$Q_1 = A_1 U_1 \left(T_0^s - T_1 \right) \tag{7.5}$$

where A_1 is the heat transfer area and U_1 is the overall heat transfer coefficient which is assumed to be a decreasing function of the mass fraction of the solids, i.e., $U = U(x)$.

An important element of the modelling of aqueous solution of solids is the rise in the boiling point (BPR) relative to that of pure water at the same pressure:

$$BPR(x) = T(x) - T^s \tag{7.6}$$

where T is the boiling temperature of the solution with solids concentration x and T^s is the boiling point (saturation temperature) of pure water. BPR is usually a linear function of the mass fraction and becomes nonlinear at very high concentrations. It is reminded that the saturation temperature of pure water is related to the saturation pressure (P^s) of water through the well-known Antoine equation:

$$\ln P^s = A - \frac{B}{C + T^s} \tag{7.7}$$

In Table 7.1, we summarize the equations necessary to calculate the boiling point rise, the enthalpy, and the vapor pressure of a specific sugar solution.

The equations that we have presented up to this point are a complete mathematical model for the evaporation process. What is missing is an appropriate cost function. The appropriate cost function to minimize is the total annual operating cost that consists of the annualized cost of the installed evaporator and the cost of utility steam:

$$\text{TAC} = \text{CCF} \cdot \text{FBM} \cdot C_{\text{evap}} + t_y \cdot V_0 \cdot c_{\text{lps}} \tag{7.8}$$

where: TAC is the approximate total annual cost (\$/y); CCF is the capital charge factor (1/3); FBM is the bare module factor (2.25); t_y is the operating time (8322 h/y); c_{lps} is the unit cost of the low-pressure steam (30 \$/t); C_{evap} is the purchase cost of a falling film evaporator given by (Turton et al., *Analysis, Synthesis, and Design of Chemical Processes*, 4th ed., Prentice Hall, 2013):

$$\log_{10} C_{\text{evap}} = 3.9119 + 0.8627 \log_{10} A - 0.008 (\log_{10} A)^2 \tag{7.9}$$

Table 7.1 Properties of a sugar solution (x is the mass fraction of sugars, T is the temperature, and T^s is the saturation temperature in °C)

Property	Equation	Units
Boiling point rise	$\text{BPR} = (1.8 + 6.2x)x$	°C
Overall heat transfer coefficient	$U = \exp(8.14 - 0.078 \ln(x) + 0.26 \ln(x)^2)$	$\frac{\text{kJ}}{\text{h m}^2 \text{°C}}$
Specific enthalpy of liquid water	$h_w = 9.14 + 3.8648T + 3.3337 \cdot 10^{-3} T^2 - 9.8 \cdot 10^{-6} T^3$	$\frac{\text{kJ}}{\text{kg}}$
Specific enthalpy of steam	$H = 2502.04 + 1.8125T + 2.585 \cdot 10^{-4} T^2 - 9.8 \cdot 10^{-6} T^3$	$\frac{\text{kJ}}{\text{kg}}$
Heat capacity of sugars (solids)	$\text{cp}_s = 1.84$	$\frac{\text{kJ}}{\text{kg °C}}$
Specific enthalpy of solution	$h = x \cdot \text{cp}_s T + (1 - x) \cdot h_w(T)$	$\frac{\text{kJ}}{\text{kg}}$
Saturation pressure of pure water	$P^s = \exp\left(11.622 - \frac{3798.3}{T^s + 227.03}\right)$	bar

We consider the following feed: $L_0 = 6.3$ kg/s, $x_0 = 0.1$, and $T_0 = 26.7°C$. We also consider the case where saturated steam at $T_0^s = 160°C$ is available. The aim is to produce a solution that is at least 50% sugars ($x_1 \geq 0.5$). The GAMS model of Table 7.2 is developed. We set the minimum pressure in the evaporator to be 0.1 bar (see Problem 7.1 where the reader is asked to investigate the effect of varying the minimum evaporator pressure on economics). The solution obtained is summarized in the following extract of the *.lst file:

	LOWER	LEVEL	UPPER	MARGINAL
— VAR Ts	26.700	45.747	160.000	.
— VAR Ps	0.100	0.100	2.000	9.7873E+5
— VAR T	26.700	48.197	160.000	.
— VAR L	.	1.260	+INF	.
— VAR V	.	5.040	+INF	.
— VAR x	0.100	0.500	1.000	.
— VAR A	EPS	98.545	500.000	.
— VAR U	.	1.139	+INF	.
— VAR Q	.	12550.939	+INF	.
— VAR hL	.	145.369	+INF	.
— VAR hW	.	202.057	+INF	.
— VAR HV	.	2588.900	+INF	.
— VAR V0	.	6.017	+INF	.
— VAR Cutil	.	5.4082E+6	+INF	.
— VAR Cequip	.	3.9521E+5	+INF	.
— VAR TAC	.	5.7043E+6	+INF	.
— VAR ObjFun	-INF	5.7043E+6	+INF	.

Note that the solution is defined by the active constraints $P^s \geq 0.1$ bar and $x \geq 0.5$ and that the cost-optimal evaporator has a heat transfer area of 98.545 m^2 and consumes 6.017 kg lps steam/s. The steam economy (the mass of water evaporated over the utility steam consumed) is $(5.040/6.017) = 0.838$, which is typical for single evaporator systems (usually in the range 0.75–0.95).

In trying to improve the economics of the process, we may consider the case of a double-effect evaporator shown in Fig. 7.2. We modify the GAMS program of Table 7.2 to that of Table 7.3 (the only important difference being the introduction of the index i to denote the evaporator). By solving the model for two evaporators, we obtain the following solution:

```
---- VAR Ps  saturation pressure in bar
```
	LOWER	LEVEL	UPPER	MARGINAL
evap1	0.100	0.645	2.000	.
evap2	0.100	0.100	2.000	1.2111E+6

(continued)

Table 7.2 GAMS program for the single-effect evaporator optimization

```
$ TITLE Single Effect Evaporator

SCALAR L0   feed mass flowrate in kg per s              /6.3/;
SCALAR x0   mass fraction of sugars in feed             /0.1/;
SCALAR T0   temperature of feed in deg C                /26.7/;
SCALAR cps  heat capacity of sugars in kJ per kg deg C /1.84/;
SCALAR Ts0  utility steam temperature i deg C           /160/;
SCALAR ty   operating hours per year                    /8322/;
SCALAR cs   unit cost of utility steam in $ per kg      /0.03/;
SCALAR FBM  bare module factor                          /2.25/;
SCALAR CCF  capital charge factor                       /0.333/;


PARAMETERS h0   spec. enthalpy of the feed in kJ per kg,
           Dhs utility steam enthalpy of vaporization;
h0 = x0*cps*T0+(1-x0)*(9.14+3.8648*T0+3.3337E-3*T0**2-9.8E-
6*T0**3);
Dhs= 2492.9-2.0523*Ts0-30.752*1E-4*Ts0**2;


POSITIVE VARIABLE Ts saturation temperature in deg C,
                  Ps saturation pressure in bar
                  T  operating temperature in deg C
                  L  liquid product mass flowrate in kg per s
                  V  vapour product mass flowrate in kg per s
                  x  sugars mass fraction
                  A  heat transfer area in m^2
                  U  heat transfer coef kW per m^2 deg C
                  Q  heat transfer rate in kW
                  hL specific enthalpy of liquid in kJ per kg
                  hW spec. enth. of liquid water in kJ per kg
                  HV spec. enth. of vapour water in kJ per kg
                  V0 utility steam consumption in kg per s
                  Cutil utility annual cost in $ per y
                  Cequip purchase cost of evaporator in $
                  TAC approximate total cost in $ per y;

VARIABLE          ObjFun;

EQUATIONS  TotalMassBalance, SugarMassBalance, EnergyBalance,
           UtilitySteam,  OverHeatTrCoef,   HeatTransfArea,
           BoilPointElev, SaturationPress, ObjFunDefinition,
           QualityConstr, CalculatehL,       Calculatehw,
           CalculateHV,    UtilityCost,      EquipCost,
           TAnnualCost;

TotalMassBalance.. L0   =E= L + V;
SugarMassBalance.. L0*x0 =E= L*x;
EnergyBalance..    L0*h0 =E= L*hL + V*HV - Q;
UtilitySteam..     Q    =E= V0*Dhs;
OverHeatTrCoef..   U*3600=E= exp(8.14+(-
0.078+0.26*log(x))*log(x));
HeatTransfArea..   Q    =E= A*U*(Ts0-T);
BoilPointElev..    T    =E= Ts + (1.8+6.2*x)*x;
SaturationPress..  Ps   =E= exp(11.622-3798.3/(Ts+227.03));
CalculatehL..      hL   =E= x*cps*T+(1-x)*hw;
Calculatehw..      hw =E= 9.14+3.8648*T+3.3337E-3*T**2-9.8E-
6*T**3;
```

(continued)

Table 7.2 (continued)

```
CalculateHV..      HV =E= 2502.04+1.8125*T+2.5850E-4*T**2-9.8E-
6*T**3;
QualityConstr..      x    =G= 0.5;
ObjFunDefinition.. ObjFun=E= TAC;
UtilityCost..        Cutil =E= V0*(3600*ty)*cs;
EquipCost.. Cequip=E=10**(3.9119+(0.8627-
0.0088*log10(A))*log10(A));
TAnnualCost..        TAC =E= CCF*FBM*Cequip+Cutil;

x.lo=x0;    x.up=1;    x.l=0.5;
T.lo=T0;    T.up=Ts0;  T.L=100;
Ts.lo=T0;   Ts.up=Ts0; Ts.l=100;
Ps.lo=0.1;  Ps.up=2;   Ps.l=1;
A.lo=EPS;   A.up=500;  A.l=100;

MODEL SingleEfectEvaporator /ALL/;
OPTION NLP=CONOPT4;
SOLVE SingleEfectEvaporator USING NLP MINIMIZING ObjFun;
```

```
---- VAR T operating temperature in deg C
         LOWER    LEVEL      UPPER      MARGINAL
evap1   26.700   88.369    160.000    .
evap2   26.700   48.197    160.000    .


---- VAR V vapour product mass flowrate in kg per s
         LOWER    LEVEL      UPPER      MARGINAL
evap1   .         2.447     +INF       .
evap2   .         2.593     +INF       .


---- VAR x sugars mass fraction
         LOWER    LEVEL      UPPER      MARGINAL
evap1   0.100    0.164     1.000      .
evap2   0.100    0.500     1.000      .


---- VAR A heat transfer area in m^2
         LOWER    LEVEL      UPPER      MARGINAL
evap1   EPS      38.617    500.000    EPS
evap2   EPS     123.810    500.000    .


---- VAR Q heat transfer rate in kW
         LOWER    LEVEL      UPPER      MARGINAL
evap1   .       7117.756   +INF       .
evap2   .       5600.975   +INF       .
```

(continued)

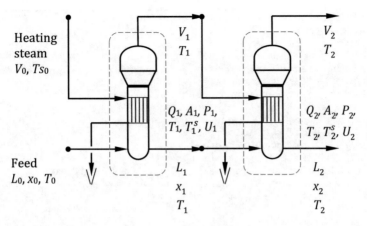

Fig. 7.2 A double-effect evaporator

	LOWER	LEVEL	UPPER	MARGINAL
---- VAR V0	.	3.412	+INF	.
---- VAR Cutil	.	3.0670E+6	+INF	.

---- VAR Cequip purchase cost of evaporator in $

	LOWER	LEVEL	UPPER	MARGINAL
evap1	.	1.8141E+5	+INF	.
evap2	.	4.7729E+5	+INF	.

	LOWER	LEVEL	UPPER	MARGINAL
---- VAR TAC	.	3.5606E+6	+INF	.

It is important to note that the TAC decreases from M$5.7 for a single-effect evaporator to M$3.56 for a double-effect evaporator system. The total heat transfer area increases from 98.5 to 162.43 m^2, but the steam consumption decreases from 6.017 to 3.412 kg/s (and the steam economy increases from 0.838 to 1.477). Further improvements can be obtained by increasing the number of evaporators and/or introducing a heat recovery heat exchanger that preheats the fresh feed.

7.3   Complex Chemical Reaction Equilibrium

We have already analyzed the problem of equilibrium reactions in the gas phase in Chap. 3 and in Examples 3.4 and 3.9. We now present a more realistic case study, and we will compare the theoretical results with experimental data. The mathematical formulation has been presented in Chap. 3 and is repeated here:

Table 7.3 GAMS program for the multi-effect evaporator optimization

```
$ TITLE Multiple Effect Evaporator
SET i evaporator number /evap1*evap2/; ALIAS (i,j);
OPTION optcr=1e-9;
SCALAR L0   feed mass flowrate in kg per s              /6.3/;
SCALAR x0   mass fraction of sugars in feed             /0.1/;
SCALAR T0   temperature of feed in deg C                /26.7/;
SCALAR cps heat capacity of sugars in kJ per kg deg C /1.84/;
SCALAR Ts0 utility steam temperature i deg C            /160/;
SCALAR ty   operating hours per year                    /8322/;
SCALAR cs   unit cost of utility steam in $ per t       /0.03/;
SCALAR FBM bare module factor                           /2.25/;
SCALAR CCF capital charge factor                        /0.333/;
PARAMETERS h0    spec. enthalpy of the feed in kJ per kg,
           Dhs0 utility steam enthalpy of vaporization;
h0  = x0*cps*T0+(1-x0)*(9.14+3.8648*T0+3.3337E-3*T0**2-9.8E-
6*T0**3);
Dhs0= 2492.9-2.0523*Ts0-30.752*1E-4*Ts0**2;

POSITIVE VARIABLE Ts(i) saturation temperature in deg C,
                  Ps(i) saturation pressure in bar
                  T(i)  operating temperature in deg C
                  L(i)  liquid prod. mass flowrate in kg per s
                  V(i) vapor product mass flowrate in kg per s
                  x(i)  sugars mass fraction
                  A(i)  heat transfer area in m^2
                  U(i)  heat transfer coef. in kW per m^2 degC
                  Q(i)  heat transfer rate in kW
                  hL(i) spec. enthalpy of liquid in kJ per kg
                  hW(i) spec. enth. of liq. water in kJ per kg
                  HV(i) spec. enth. of vap water in kJ per kg
                  Dhs(i) heat of condensation in kJ per kg
                  V0 utility steam consumption in kg per s
                  Cutil utility annual cost in $ per y
                  Cequip(i) purchase cost of evaporator in $
                  TAC approximate total annual cost in $ per y;
VARIABLE          ObjFun;

EQUATIONS   TotalMassBal, SugarMassBal, EnergyBal,
            Utility,     OverHeatTrCoef,   HeatArea,
         BoilPointElev,   SaturationPress,  ObjFunDefinition,
         QualityConstr,    CalculatehL,       Calculatehw,
         CalculateHV,       UtilityCost,      EquipCost,
         CalcDhs, TAnnualCost;

TotalMassBal(i).. L0=E=L(i)+sum(j$(ord(j) le ord(i)),V(j));
SugarMassBal(i).. L0*x0 =E= L(i)*x(i);
EnergyBal(i).. L(i-1)*hL(i-1)+(L0*h0)$(ord(i) eq 1) =E=
L(i)*hL(i)+V(i)*HV(i)-Q(i);
Utility(i).. Q(i) =E= V(i-1)*Dhs(i-1)+(V0*Dhs0)$(ord(i) eq 1);
OverHeatTrCoef(i).. U(i)*3600 =E= exp( 8.14+(-
0.078+0.26*log(x(i)))*log(x(i)) );
HeatArea(i) Q(i)=E=A(i)*U(i)*(Ts(i-1)+Ts0$(ord(i) eq 1)-T(i));
BoilPointElev(i).. T(i) =E= Ts(i) + (1.8+6.2*x(i))*x(i);
SaturationPress(i).. Ps(i) =E= exp(11.622-
3798.3/(Ts(i)+227.03));
CalculatehL(i).. hL(i) =E= x(i)*cps*T(i)+(1-x(i))*hw(i);
```

(continued)

Table 7.3 (continued)

```
Calculatehw(i).. hw(i) =E= 9.14+3.8648*T(i)+3.3337E-3*T(i)**2-
9.8E-6*T(i)**3;
CalculateHV(i).. HV(i) =E= 2502.04+1.8125*T(i)+2.5850E-
4*T(i)**2-9.8E-6*T(i)**3;
CalcDhs(i).. Dhs(i) =E= 2492.9-2.0523*Ts(i)-30.752*1E-
4*Ts(i)**2;
QualityConstr.. x('evap2') =G= 0.5;
ObjFunDefinition.. ObjFun=E= TAC;
UtilityCost.. Cutil =E= V0*(3600*ty)*cs;
EquipCost(i).. Cequip(i)=E= 10**(3.9119+(0.8627-
0.0088*log10(A(i)))*log10(A(i))));
TAnnualCost.. TAC =E= CCF*FBM*sum(i,Cequip(i))+Cutil;
V.up(i)=L0; L.up(i)=L0;
x.lo(i)=x0; x.up(i)=1; x.l(i)=0.5;
T.lo(i)=T0; T.up(i)=Ts0; T.L(i)=100;
Ts.lo(i)=T0; Ts.up(i)=Ts0; Ts.l(i)=100;
Ps.lo(i)=0.1; Ps.up(i)=2; Ps.l(i)=1;
A.lo(i)=1; A.up(i)=500; A.l(i)=100;

MODEL SingleEfectEvaporator /ALL/;
OPTION NLP=conopt4;
SOLVE SingleEfectEvaporator USING NLP MINIMIZING ObjFun;
```

$$\min_{n_i \geq 0} \quad G(n_i) = \sum_{i=1}^{NC} n_i \left[\left(\frac{\Delta G_{f,i}^o}{RT} \right) + \ln n_i - \ln \sum_{i=1}^{NC} n_i \right]$$

$$\text{s.t.} \tag{7.10}$$

$$\sum_{i=1}^{NC} n_i \, a_{i,k} = a_k^0, \quad k = 1, 2, \dots, NE$$

where n_i are the mole of chemical species i at the product stream (there are NC chemical species present), $\Delta G_{f,i}^o$ is the standard-state Gibbs free energy of formation of chemical species i, T is the temperature of the reaction, and R is the ideal gas constant. a_{ik} denote the number of atoms of element k in chemical species i, and NE are the elements present in the system. The atoms of element k present in the feed are a_k^0. The main computational problem arising when we need to determine the equilibrium at different temperatures is due to the fact that $\Delta G_{f,i}^o$ is a function of temperature. To automatically determine the equilibrium composition as a function of the temperature, we need to include into the GAMS program the calculation of $\Delta G_{f,i}^o$ at each temperature.

The analytic determination of $\Delta G_{f,i}^o$ is discussed in most chemical engineering thermodynamics textbooks (such as Elliott & Lira, *Introductory Chemical Engineering Thermodynamics*, Pearson, 2012, or Abbot Van Ness, *Thermodynamics with Chemical Applications*, Schaum's Series, 2nd ed., McGraw Hill,

1989). The calculation makes use of the definition of the ideal gas heat capacity of chemical species i in polynomial form:

$$Cp_i = Cp_{1,i} + Cp_{2,i}T + Cp_{3,i}T^2 + Cp_{4,i}T^3 = \sum_{k=1}^{4} Cp_{ik}T^{k-1} \qquad (7.11)$$

The question that we need to answer is how, for any given chemical species, to calculate $\Delta G_{f,T}^\circ$, i.e., the standard Gibbs free energy of formation at temperature T, if $\Delta G_{f,T_r}^\circ$ is known, i.e., the standard Gibbs free energy of formation at the reference temperature $T_r = 298.15$ K. The answer is given by the van't Hoff equation:

$$\frac{\Delta G_{f,T}^\circ}{RT} = \frac{\Delta G_{f,T_r}^\circ}{RT} - \int_{T_r}^{T} \frac{\Delta h_{f,T}^\circ}{RT^2} dT \qquad (7.12)$$

where:

$$\Delta h_{f,T}^\circ = \sum_i v_i \Delta h_{f,T_r,i}^\circ + \int_{T_r}^{T} \sum_{k=1}^{4} \Delta Cp_k T^{k-1} dT \qquad (7.13)$$

where $\Delta Cp_k = \sum_i v_i \Delta Cp_{ki}$, and v_i are the stoichiometric numbers for the formation reaction of species i from the elements in the naturally occurring molecular form at the standard state (298.15 K and 1 bar). By taking these into consideration, Eq. (7.13) can be written as:

$$\Delta h_{f,T}^\circ = \Delta h_{f,T_r}^\circ + \sum_{k=1}^{4} \Delta Cp_k T^{k-1}$$

or:

$$\Delta h_{f,T}^\circ = J + \Delta Cp_1 T + \frac{\Delta Cp_2}{2}T^2 + \frac{\Delta Cp_3}{3}T^3 + \frac{\Delta Cp_4}{4}T^4 \qquad (7.14)$$

where J is a constant that can be calculated from the known pair of data: $T_r, \Delta h_{f,T_r}^\circ$. We finally substitute (7.14) into (7.13) to obtain:

$$\frac{\Delta G_{f,T}^\circ}{RT} = \frac{J}{RT} - \frac{1}{R}\left(\Delta Cp_1 \ln T + \frac{\Delta Cp_2}{2}T + \frac{\Delta Cp_3}{6}T^2 + \frac{\Delta Cp_4}{12}T^3\right) - I \qquad (7.15)$$

The final equation involves two integration constants J and I which are determined by applying Eqs. (7.14) and (7.15) for the reference temperature $T_r = 298.15$ K, for which values of $\Delta h_{f,T_r}^\circ$ and $\Delta G_{f,T_r}^\circ$ are available in tables of thermodynamic data.

Fig. 7.3 Catalytic steam reforming of ethanol problem

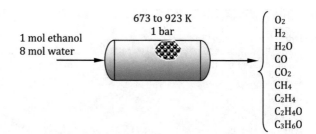

Table 7.4 Experimental data for the catalytic (30% Ni/ZrO2) steam reforming of ethanol

Temperature (°C/K)	mol H_2/mol EtOH	mol CO/mol EtOH	mol CO_2/mol EtOH	mol CH_4/mol EtOH
400/673.15	2.1	0.58	0.54	0.63
450/723.15	3.1	0.38	1.05	0.51
500/773.15	3.8	0.33	1.14	0.45
550/823.15	4.4	0.39	1.25	0.36
600/873.15	5.2	0.49	1.28	0.23
650/923.15	5.8	0.54	1.35	0.12

Data from Biswas & Kunzru, *International Journal of Hydrogen Energy*, 32, pp. 969–980, 2007

We will study the case study of the catalytic steam reforming of ethanol at high temperatures (see Fig. 7.3) for which experimental results are available and are summarized in Table 7.4. The GAMS file that solves this problem is based on Formulation (7.10), and most of its lines implement Eqs. (7.11)–(7.13). The GAMS file is given in Table 7.5. The results of the optimization model are compared to the experimental results in Fig. 7.4. The agreement is considered satisfactory.

7.4   Optimal Design of a Methanol-Water Distillation Column

Distillation is arguably the most important separation process in chemical engineering. It is a separation process that is based on the generation of a new phase, by thermal means, which is enriched in the most volatile components. By stacking many equilibrium trays together (cascade), in which vapor and liquid phases are in equilibrium, high purities can be achieved. To discover the cost-optimal design for a given separation is not a trivial task as it involves, among others, the selection of the optimal number of trays, the optimal location of the feed tray, and the selection of the optimal reflux ratio.

Table 7.5 GAMS program for the catalytic steam ethanol reforming problem

```
$ TITLE Chemical Equilibrium
Options optcr=1e-8;Options optca=1e-8;
SET i components
/O2,C,H2,H2O,CO,CO2,CH4,C2H4,CH3CH2OH,CH3CHO,CH3COCH3/
    k elements    /C, H, O/
    m equation for cp       /1*4/
    TEMP /1*26/;
ALIAS(i,ip);

* units kJ per kmol
PARAMETER Dhfig(i) /     O2        =       0
                         C         =       0
                         H2        =       0
                         H2O       =    -241835
                         CO        =    -110530
                         CO2       =    -393510
                         CH4       =    -74894
                         C2H4      =     52510
                         CH3CH2OH  =   -234950
                         CH3CHO    =   -166200
                         CH3COCH3  =   -215700/;
PARAMETER Dgfig(i) /     O2        =       0
                         C         =       0
                         H2        =       0
                         H2O       =     -228614
                         CO        =     -137160
                         CO2       =     -394380
                         CH4       =      -50450
                         C2H4      =      68430
                         CH3CH2OH  =     -167730
                         CH3CHO    =     -133302
                         CH3COCH3  =     -151200/;
TABLE Cp(i,m) heat capacity polynomial in kJ per kmol K
*               cp1        cp2           cp3            cp4
                1          2             3              4
O2              28.110    -3.6800E-06    1.7460E-05    -1.0650E-08
C              -10.352     9.2635E-02   -1.0577E-04     4.4127E-08
H2              27.140     9.2740E-03   -1.3810E-05     7.6450E-09
H2O             32.240     1.9240E-03    1.0550E-05    -3.5960E-09
CO              30.870    -1.2850E-02    2.7890E-05    -1.2720E-08
CO2             19.800     7.3440E-02   -5.6020E-05     1.7150E-08
CH4             19.250     5.2130E-02    1.1970E-05    -1.1320E-08
C2H4             3.806     1.5660E-01   -8.3480E-05     1.7550E-08
CH3CH2OH         9.014     2.1410E-01   -8.3900E-05     1.3730E-09
CH3CHO          17.531     1.3239E-01   -2.1550E-05    -1.5900E-08
CH3COCH3         6.301     2.6060E-01   -1.2530E-04     2.0380E-08 ;
TABLE A(k,i) elements in species
   O2 C H2 H2O CO CO2 CH4 C2H4 CH3CH2OH CH3CHO  CH3COCH3
C  0  1  0   0   1  1   2    2       2        2       3
H  0  0  2   2   0  0   4    4       6        4       6
O  2  0  0   1   1  2   0    0       1        1       1;
TABLE Stoich_f(i,ip) stoichiometric table of formation reactions
          O2    C  H2 H2O  CO  CO2  CH4  C2H4  CH3CH2OH  CH3CHO
CH3COCH3
O2         0    0  0  0    0   0    0    0     0         0         0
C          0    0  0  0    0   0    0    0     0         0         0
H2         0    0  0  0    0   0    0    0     0         0         0
H2O       -0.5  0 -1 +1    0   0    0    0     0         0         0
CO        -0.5 -1  0  0   +1   0    0    0     0         0         0
CO2       -1   -1  0  0    0  +1    0    0     0         0         0
CH4        0   -1 -2  0    0   0   +1    0     0         0         0
C2H4       0   -2 -2  0    0   0    0   +1     0         0         0
```

(continued)

Table 7.5 (continued)

```
CH3CH2OH -0.5 -2 -3  0    0   0    0    0   +1         0       0
CH3CHO   -0.5 -2 -2  0    0   0    0    0    0        +1       0
CH3COCH3 -0.5 -3 -3  0    0   0    0    0    0         0      +1 ;
SCALAR T  /1000/; SCALAR Tr /298.15/; SCALAR Rg /8.314/;
PARAMETER DHf, DGf, Dcp(i,m), lnKeq, Jc, Ic, DGf_RT;
DHf(i)   = sum(ip, Stoich_f(i,ip)*Dhfig(ip) );  Display DHf;
DGf(i)   = sum(ip, Stoich_f(i,ip)*Dgfig(ip) );  Display DGf;
Dcp(i,m) = sum(ip, Stoich_f(i,ip)*Cp(ip,m) ) ;  Display Dcp;
lnKeq(i) = -DGf(i)/(Rg*Tr); Display lnKeq;
Jc(i)    = DHf(i) - sum(m,Dcp(i,m)*Tr**ord(m)/ord(m)); Display Jc;
Ic(i)=lnKeq(i)+Jc(i)/(Rg*Tr)-
(cp(i,'1')*log(Tr)+Dcp(i,'2')*Tr/2+Dcp(i,'3')*Tr**2/6+Dcp(i,'4')*Tr**3
/ 12)/Rg;
DGf_RT(i)= -(Ic(i)*Rg-
Jc(i)/T+(Dcp(i,'1')*log(T)+Dcp(i,'2')*T/2+Dcp(i,'3')*T**2/6+Dcp(i,'4')
*T**3/ 12))/Rg;
*************8 MOLE H2O per MOLE of CH3CH2OH are fed to the reactor
Parameter Water_2_EtOH_Ratio /8/;
Parameter b0;  b0(k) = A(k,'CH3CH2OH')+A(k,'H2O') *
Water_2_EtOH_Ratio;

POSITIVE VARIABLES n(i), NT;
VARIABLES ObjFun;

EQUATIONS ELEM_BAL, TOTAL_MOLE, Objective;
ELEM_BAL(k).. sum(i,a(k,i)*n(i))=E=b0(k);
TOTAL_MOLE..  sum(i,n(i))=E=NT;
Objective.. ObjFun=E=sum(i,n(i)*DGf_RT(i))+sum(i,n(i)*log(n(i)))-
NT*log(NT);

* INITIAL VALUES AND BOUNDS
n.lo(i)=1e-6; NT.lo=sum(i,n.lo(i));
n.up('O2') = (1+1*Water_2_EtOH_Ratio)/2;
n.up('C')  = 1e-6; n.up('H2')= (6+2*Water_2_EtOH_Ratio)/2;
n.up('H2O') = Water_2_EtOH_Ratio+1;
n.up('CO')=2; n.up('CO2') = 2;
n.up('CH4')       = 2; n.up('C2H4')     = 1;
n.up('CH3CH2OH') = 1; n.up('CH3CHO')   = 1;
n.up('CH3COCH3') =2/3; NT.up=sum(i,n.up(i));
n.l(i)  = n.lo(i); n.l('H2O')=Water_2_EtOH_Ratio;
n.l('CH3CH2OH')=1; NT.l =sum(i,n.l(i));
ObjFun.l = sum(i,n.l(i)*DGf_RT(I))+sum(i,n.l(i)*log(n.l(i)))-
NT.l*log(NT.l);

MODEL CHEM_EQUILIBRIUM /ALL/;
OPTION NLP=BARON;
SOLVE  CHEM_EQUILIBRIUM USING NLP MINIMIZING ObjFun;

file ChReaEq /ChReaEq.m/;
PUT  ChReaEq ;
PUT #1@7'TEMP ',#1@17 'H2',#1@27 'H2O', #1@37 'CO',#1@47 'CO2',#1@57
'CH4', #1@67 'C2H4',#1@77 'CH3CH2OH',#1@87 'CH3CHO',#1@97 'CH3COCH3',
#1@107 'NT' /;
loop(TEMP, T =  273.15+390+10*ord(TEMP);
DGf_RT(i)= -(Ic(i)*Rg-
Jc(i)/T+(Dcp(i,'1')*log(T)+Dcp(i,'2')*T/2+Dcp(i,'3')*T**2/6+Dcp(i,'4')
*T**3/12))/Rg;
SOLVE  CHEM_EQUILIBRIUM USING NLP MINIMIZING ObjFun;
PUT
T:10:4,n.l('H2'):10:4,n.l('H2O'):10:4,n.l('CO'):10:4,n.l('CO2'):10:4,
n.l('CH4'):10:4,n.l('C2H4'):10:4,n.l('CH3CH2OH'):10:4,n.l('CH3CHO'):10
:4,n.l('CH3COCH3'):10:4,Nt.l:10:4/;         );
PUTCLOSE ChReaEq;
```

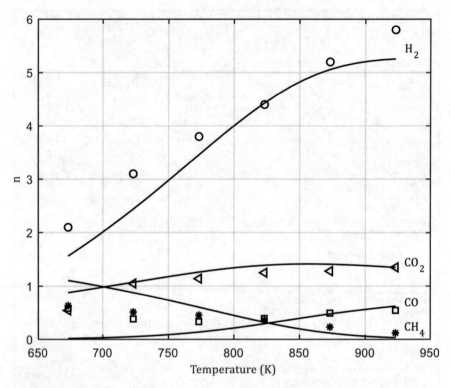

Fig. 7.4 Comparison of experimental data and model predictions (continuous lines) for the catalytic steam reforming of ethanol

To take structural alternatives, which involve different number of trays and different feed locations, into consideration, Viswanathan and Grossmann (*Computers and Chemical Engineering*, 17(9), pp. 949–955, 1993) have proposed the "superstructure" presented in Fig. 7.5a. The column has a feed stream that is fed at a fixed tray (floc). A number of potential trays are postulated for the rectifying section (section above the feed and below condenser), and each tray on this section is a candidate position for returning the reflux.

Reflux is assumed to return to one tray only. The exact tray at which the reflux is returned denotes the end (top) tray of the column. The trays that are above this tray are nonexistent trays in the design selected.

In the same manner, a number of potential trays are postulated for the stripping section (section below the feed and above the reboiler). The boilup is assumed to return to a single tray. The tray that the boilup is fed to the column denotes the first tray inside the column, and all trays below that tray are nonexistent trays in the design selected. In this way, we select the number of existent equilibrium stages in our design and the position of the feed tray.

Fig. 7.5 Structure and
notation used in the
distillation column
optimization where (**a**) the
complete column, and (**b**)
a single tray

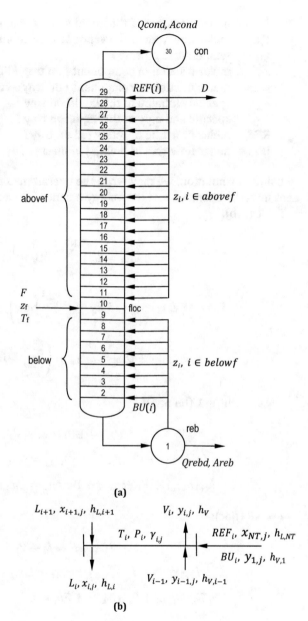

To develop the optimization model use is made of the following standard
notation (index j denotes the component, $j \in \{1,2,\dots,NC\}$, and index i denotes
the tray $i \in \{1,2,\dots,NF,\dots,NT\}$, $1 =$ reboiler, $NT =$ condenser, and $NF =$ feed
tray; $\{2,3,\dots,NF - 1\}$ is the section below the feed, and $\{NF + 1, NF + 2,\dots, NT - 1\}$ is the section above the feed):

L_i: molar flow rate of the liquid stream coming out from tray i
V_i: molar flow rate of the vapor stream coming out from tray i
t_i: temperature on tray i
$x_{i,j}$:   molar fraction of component j on tray i-liquid phase
$y_{i,j}$:   molar fraction of component j on tray i-vapor phase
$h_{L,i}$:   specific enthalpy of the liquid on tray i
$h_{V,i}$:   specific enthalpy of the vapor on tray i
REF_i: molar flow rate of reflux fed on tray i
BU_i: molar flow rate of boilup fed on tray i

Using this notation, we may write the overall material balance, the component material balances, and the energy balance for each tray:

$i = 1$ (reb):

$$L_{i+1} = L_i + \sum_{k=2}^{NF-1} BU_k$$

$$L_{i+1}x_{i+1,j} = L_i x_{i,j} + \left(\sum_{k=2}^{NF-1} BU_k \right) y_{i,j}$$

$$L_{i+1}h_{L,i+1} + Q_{reb} = L_i h_{L,i} + \left(\sum_{k=2}^{NF-1} BU_k \right) h_{V,i}$$

$i = 2,\dots,NF - 1$ (belowf):

$$L_{i+1} + V_{i-1} + BU_i = L_i + V_i$$

$$L_{i+1}x_{i+1,j} + V_{i-1}y_{i-1,j} + BU_i y_{1,j} = L_i x_{i,j} + V_i y_{i,j}$$

$$L_{i+1}h_{L,i+1} + V_{i-1}h_{V,i-1} + BU_i h_{V,1} = L_i h_{L,i} + V_i h_{V,i}$$

$i = NF$ (floc):

$$L_{i+1} + V_{i-1} + F = L_i + V_i$$

$$L_{i+1}x_{i+1,j} + V_{i-1}y_{i-1,j} + Fz_{f,j} = L_i x_{i,j} + V_i y_{i,j}$$

$$L_{i+1}h_{L,i+1} + V_{i-1}h_{V,i-1} + Fh_f = L_i h_{L,i} + V_i h_{V,i}$$

$i = NF + 1, \dots, NT - 1$ (abovef):

$$L_{i+1} + V_{i-1} + REF_i = L_i + V_i$$

$$L_{i+1}x_{i+1,j} + V_{i-1}y_{i-1,j} + REF_i x_{NT,j} = L_i x_{i,j} + V_i y_{i,j}$$

$$L_{i+1}h_{L,i+1} + V_{i-1}h_{V,i-1} + REF_i h_{L,NT} = L_i h_{L,i} + V_i h_{V,i}$$

$i = NT$ (con):

$$V_{i-1} = D + \sum_{k=NF+1}^{NT-1} REF_k$$

$$V_{i-1}y_{i-1,j} = Dx_{i,j} + \left(\sum_{k=NF+1}^{NT-1} REF_k \right) x_{i,j}$$

$$V_{i-1}h_{v,i-1} - Q_{cond} = Dh_{L,i} + \left(\sum_{k=NF+1}^{NT-1} REF_k \right) h_{L,i}$$

The model is completed when we add the vapor-liquid equilibrium model (in our case, an activity model is assumed to be valid):

$$y_{i,j}P_i = \gamma_{i,j}x_{i,j}P_j^{sat}(T_i)$$

$$\sum_{j=1}^{NC} x_{i,j} = \sum_{j=1}^{NC} y_{i,j} = 1$$

where $\gamma_{i,j}$ is the activity coefficient of component j on tray i which is a function of the liquid composition and temperature. Finally, we need also to add equations for the calculation of the specific enthalpy of the vapor and liquid, the activity coefficients, the saturation pressures, and any other thermophysical property that is necessary.

To handle the structural decision relative to the number of trays in the stripping and the rectifying section, we define the binary variables $z_i \in \{0, 1\}$ and the following linear inequalities:

$$0 \leq REF_i \leq REF^U z_i, \forall i = \{NF + 1, NF + 2, \ldots, NT - 1\}$$

$$\sum_{I=NF+1}^{NT-1} z_i = 1$$

$$0 \leq BU_i \leq BU^U z_i, \forall i = \{2, 3, \ldots, NF - 1\}$$

$$\sum_{I=2}^{NF-1} z_i = 1$$

REF^U and BU^U are upper bound on the corresponding variables. These constraints ensure that only one z is equal to 1, and all others are equal to 0 in both the stripping and the rectifying sections. The trays for which $z_i = 1$ define the first and the last tray of the column.

To present an application of the model, we consider the case of an equimolar mixture of methanol (Me) and water (W) with molar flow rate $F = 100$ kmol/h available as liquid close to its saturation temperature $T_f = 72.6$ °C and atmospheric pressure ($P = 101.3$ kPa). Our aim is to separate the mixture into a top product which is at least 98% by mole in methanol and a bottoms product that contains no more than 2% by mol methanol. For this binary mixture, the Van Laar model can be used for calculating the activity coefficient:

$$\ln \gamma_{Me} = A_{Me,W} \left(\frac{A_{W,Me} x_W}{A_{Me,W} x_{Me} + A_{W,Me} x_W} \right)^2$$

$$\ln \gamma_W = A_{W,Me} \left(\frac{A_{Me,W} x_{Me}}{A_{Me,W} x_{Me} + A_{W,Me} x_W} \right)^2$$

where $A_{Me,W} = 0.7989$ and $A_{W,Me} = 0.5625$. The saturation pressure of the two components is calculated using the Antoine equation (temperature in °C and pressure in kPa):

$$\ln P^s_{Me} = 16.5938 - \frac{3644.297}{T + 239.765}$$

$$\ln P^s_W = 16.262 - \frac{3799.887}{T + 226.346}$$

To calculate the cost optimal solution, we minimize the following approximate total annual cost:

$$TAC = CCF \cdot F_{BM} \cdot C_{equip} + t_y \cdot (Q_{reb} c_{lps} + Q_{cond} c_{cw})$$

where: TAC is an approximate total annual cost ($); CCF is the capital charge factor (selected 1/3); F_{BM} is the installed equipment cost factor; t_y is the operating time (8322 h/y); c_{lps} is the unit cost of the low-pressure steam (14.05 $/GJ); c_{cw} is the unit cost of the cooling water (0.354 $/GJ); C_{equip} is the purchase cost of the column, trays, and the heat exchangers.

The purchase cost of the column (c_{col}), trays (c_{tray}), and heat exchangers (c_{he}) is calculated through the Guthrie's correlations (in $@1968):

$$C_{col} = 935.5 \, H_c^{0.81} D_c^{1.05}, \quad H_c, D_c \text{ in m}$$

$$C_{tray} = 125.2 \left(\frac{H_c}{1.1} \right)^{0.97} D_c^{1.45}$$

$$C_{he} = 477 \, A^{0.65}, \quad A \text{ in m}^2$$

To transform the equipment cost into \$@2020, an approximate factor of 5 (=\$@2020/\$@1968) can be used. The installation factors correspond to equipment constructed from stainless steel and are 5.5, 2.7, and 4.07 for the column, trays, and heat exchangers, correspondingly. To determine the height of the column H_c, we assume 0.61 m (24″) tray spacing and 10% extra space for the bottom and top sections of the column. To determine the diameter, we assume a vapor velocity of 0.6 m/s and that the net area of each tray is 90% of the total area of the tray that is based on the tray diameter D_c.

$$H_c = 1.1 \cdot \left(N_{trays} - 1\right) \cdot 0.61$$

$$0.6 \left[0.9 \left(\frac{\pi D_c^2}{4} \right) \right] \rho_{V,i} \geq V_i$$

The number of trays can be inferred from the values of the binary variables z_i:

$$N_{trays} = 1 + \sum_{i=NF+1}^{NT-1} i \cdot z_i - \sum_{i=2}^{NF-1} i \cdot z_i$$

The model is now complete, and based on this model, a GAMS program is developed to solve the methanol-water separation problem. The model is presented in Table 7.6. The solution of the problem gives the optimal design shown on Fig. 7.6. The column has 18 equilibrium stages (including the reboiler), a diameter of 1.282 m, and a height of 10.746 m. The reflux ration is 0.806, the reboiler duty is 3,275.95 MJ/h, and the condenser duty is 3221.19 MJ/h. The overall approximate TAC is \$518,731.8 and is dominated by the utilities cost which is \$392,527.6 (75.6% of TAC).

7.5 A Representative Optimal Control Problem

All problems that we have studied in this book involved an underlying mathematical description that consists of algebraic equalities and inequalities. However, there are important problems in chemical engineering in which the performance of a systems is described by ordinary differential equations (ODEs) or partial differential equations (PDEs) or a combination of differential equations and algebraic equations (DAEs). The fact that differential equations are involved in the description of the behavior of certain systems complicates the solution of the optimization problems. This is due to the fact that the ODEs, PDEs, or DAEs need first to be transformed into algebraic equations in order for the problem to be posed in the general frameworks that have been presented so far. An additional complication that arises is due to the fact that the optimization variables can be a function of the independent variables (time or space) and the size of the corresponding optimization variable vector is infinite.

Table 7.6 GAMS program for the methanol-water distillation column design

```
$Ontext
Methanol-Water Distillation Column
$Offtext
OPTION OPTCR=1E-6;
OPTION OPTCA=1E-6;

SETS  j   components  / Me , W /
      i   stages        / 1*30  /
      reb(i)      reboiler
      con(i)      condenser
      col(i)      stages in the col
      floc(i)     location of the feed stage    / 10 /
      abovef(i)   stages above the feed stage
      belowf(i)   feed stage and those below it ;

      reb(i) = yes$(ord(i) eq 1) ;
      con(i) = yes$(ord(i) eq card(i)) ;
      col(i) = yes - (reb(i)+con(i)) ;
      belowf(i) = yes $ ( (ord(i) le 9 ) and (ord(i) ge 2 ) );
      abovef(i) = yes $ ( (ord(i) ge 11 ) and (ord(i) le 29 ) );
ALIAS (i,k);

* ANTOINE EQ CONSTANTS logPs[kPa] = A-B/(C+T[degC])
PARAMETER  A(j)    / Me   16.5938, W   16.262 /;
PARAMETER  B(j)    / Me 3644.2970, W 3799.887 /;
PARAMETER  C(j)    / Me  239.7650, W  226.346 /;

*   LIQUID  ENTHALPY  POLYNOMIAL  hL[kJ/kmol]=  (HL1+HL2*T[degC]
)*T[degC]
PARAMETER  HL1(j) / Me      72.35, W    73.10/;
PARAMETER  HL2(j) / Me    0.16077, W   0.01507/;

*LIQUIDENTHALPY POLYNOMIAL hV[kJ/kmol]=HV0+(HV1+HV2*T)*T[degC]
PARAMETER  HV0(j) / Me   40528.2, W   45175.6/;
PARAMETER  HV1(j) / Me    -4.737, W    27.507/;
PARAMETER  HV2(j) / Me   0.16077, W   0.01507/;

PARAMETER        zf(j)   mole fractions in feed /Me 0.5, W 0.5/;
PARAMETER        p(i)    pressure in tray i (kPa);
                 p(i) = 101.3;

SCALARS       F       feed molar flowrate in kmol per h
              tf      temperature of the feed (in deg C)
              hf      specific enthalpy of feed in MJ per kmol
              AMW     vanLaar constant
              AWM     vanLaar constant
              Rg      Ideal gas constant (m3 kPa per kmol K)
              pi;
              F   = 100 ;
              tf = 72.6;
        hf = sum(j,zf(j)*( Hl1(j)*tf + Hl2(j)*tf*tf ))/1E3;
              AMW= 0.7989;
              AWM= 0.5625;
              Rg=8.3143;
              pi=3.1416;
```

(continued)

Table 7.6   (continued)

```
DISPLAY reb, con, floc, abovef, belowf, col;
DISPLAY A,B,C,Hl1,Hl2,Hv0,Hv1,Hv2,zF,p,F,tf,hf,AMW,AWM;

POSITIVE VARIABLES
x(i,j)        mole-fr of j-th component in liquid on i-th tray.
y(i,j)        mole-fr of j-th component in vapour on i-th tray
gama(i,j)     activity of j-th component in liquid on i-th tray
L(i)          flowrate of liquid leaving tray i (kmol per h)
V(i)          flowrate of vapour leaving tray i (kmol per h)
t(i)          temperature of tray i (degC)
REF(i)        reflux molar flowrate entering tray i   (kmol per h)
BU(i)         boilup molar flowrate entering tray i   (kmol per h)
D             top product molar flow rate (kmol per h)
RefluxRatio   reflux ratio
Qreb          heat load of the reboiler (MJ per h)
Qcond         heat load of the condenser (MJ per h)
termM(i)      auxiliary (dummy) variable in the van Laar model
termW(i)      auxiliary (dummy) variable in the van Laar model
densV(i)      vapour density (kmol per m3)
Dc            column diameter in m
Hc            column height in m
Areb          Heat transfer area in reboiler
Acond         Heat transfer area in condenser
Ntrays        number of equilibrium stages(reboiler plus trays)
UtlCost       cost of utility steam in $ per y
EquipCost     installed cost of column and trays in $ @ 1968
hl(i)         molar sp.enthalpy of liquid on tray i in MJ per kmol
hv(i)         molar sp.enthalpy of vapour on tray i in MJ per kmol;

VARIABLES     cost
              ObjFun;

BINARY VARIABLES
              z(i)          return tray (z=1) of reflux OR reboil;

EQUATIONS     tmbr(i)       total material balance reboiler
              tmbbf(i)      total material balance below feed
              tmbf(i)       total material balance at feed
              tmbaf(i)      total material balance above feed
              tmbc(i)       total material balance condenser

              cmbr(i,j)     component material balance reboiler
              cmbbf(i,j)    component material balance below feed
              cmbf(i,j)     component material balance at feed
              cmbaf(i,j)    component material balance above feed
              cmbc(i,j)     component material balance condenser

              ebr(i)        enthalpy balance reboiler
              ebbf(i)       enthalpy balance below feed
              ebf(i)        enthalpy balance at feed
              ebaf(i)       enthalpy balance above feed
              ebc(i)        enthalpy balance condenser

              defrr         reflux ration definition
              xsum(i)       sum of mole fractions equals to one
              ysum(i)       sum of mole fractions equals to one
```

(continued)

Table 7.6 (continued)

```
          defhl(i)      definition of hl(i)
          defhv(i)      definition of hv(i)

          phe(i,j)      phase equilibrium relation

          DefTermW(i)   van Laar model
          DefTermE(i)   van Laar model
          DefgamaW(i)   van Laar model
          DefgamaE(i)   van Laar model

          DefdensV(i)   vapor density
          DefDiam(i)    column diameter
          DefNtray      number of equilibrium stages
          DefAcond      calculate the condenser heat transfer
area
          DefAreb       calculate the reboiler heat transfer
area
          DefColH       calculate the height of the column
          DefCost       calculate the approximate TAC
          DefCost1      calculate installed equipment cost
          DefCost2      calculate utility cost

          const1a       reflux returns in only one tray only
          const2a(i)

          const1b       boilup returns in one tray only
          const2b(i)

          purity1       top product purity
          purity2       bottom's product purity;

tmbr(i)$reb(i)..   L(i) + sum( k, BU(k)$belowf(k) ) =E= L(i+1);
tmbbf(i)$belowf(i).. L(i)+V(i)-L(i+1)-V(i-1) - BU(i)=E=0;
tmbf(i)$floc(i)..    L(i)+V(i)-L(i+1)-V(i-1)-F=E= 0        ;
tmbaf(i)$abovef(i).. L(i)+V(i)-L(i+1)-V(i-1)-REF(i)=E=0;
tmbC(i)$con(i)..     D+sum(k,REF(k)$abovef(k))=e=V(i-1);

cmbr(i,'Me')$reb(i)..      L(i+1)*x(i+1,'Me') =E=
L(i)*x(i,'Me')+sum(k,BU(k)$belowf(k))*y(i,'Me');

cmbbf(i,'Me')$belowf(i)..  L(i)*x(i,'Me') + V(i)*y(i,'Me') =E=
L(i+1)*x(i+1,'Me')+V(i-1)*y(i-1,'Me')+BU(i)*y('1','Me');

cmbf(i,'Me')$floc(i)..     L(i)*x(i,'Me') + V(i)*y(i,'Me') =E=
L(i+1)*x(i+1,'Me') + V(i-1)*y(i-1,'Me') + F*zf('Me');

cmbaf(i,'Me')$abovef(i)..  L(i)*x(i,'Me') + V(i)*y(i,'Me') =E=
L(i+1)*x(i+1,'Me') + V(i-1)*y(i-1,'Me') + REF(i)*x('30','Me');
cmbC(i,'Me')$con(i)..                              (      D
+sum(k,REF(k)$abovef(k)))*x(i,'Me')=E= V(i-1)*y(i-1,'Me');

ebr(i)$reb(i)..            L(i)*hl(i)  +sum(k, BU(k)$belowf(k)
)*hv(i)=E=L(i+1)*hl(i+1)+ Qreb  ;

ebbf(i)$belowf(i)..  L(i)*hl(i) + V(i)*hv(i)     =E=
L(i+1)*hl(i+1)+ V(i-1)*hv(i-1)+ BU(i)*hv('1');
```

(continued)

Table 7.6 (continued)

```
ebf(i)$floc(i)..        L(i)*hl(i)    + V(i)*hv(i)  =E=
                        L(i+1)*hl(i+1) + V(i-1)*hv(i-1) + F*hf ;

ebaf(i)$abovef(i)..  L(i)*hl(i) + V(i)*hv(i)       =E=
              L(i+1)*hl(i+1)+V(i-1)*hv(i-1)+REF(i)*hl('30');

ebC(i)$con(i)..         (D + sum(k,REF(k)$abovef(k)) )*hl(i) =E=
                        V(i-1)*hv(i-1) - Qcond ;

defrr.. RefluxRatio * D =E= sum( k,REF(k)$abovef(k) );

xsum(i)..               sum(j,x(i,j))  =e= 1;
ysum(i)..               sum(j,y(i,j))  =e= 1;

defhl(i)..              1E3 * hl(i) =E=
            sum(j, (        (HL1(j)+HL2(j)*t(i))*t(i))*x(i,j));

defhv(i)..              1E3 * hv(i) =E=
sum(j,(Hv0(j)+(HV1(j)+HV2(j)*t(i))*t(i))*y(i,j));

phe(i,j)..              y(i,j)* p(i) =E=
          x(i,j)*gama(i,j) * exp(A(j)-B(j)/(C(j)+t(i)));

DefTermW(i)..
termM(i)*(AMW*x(i,'Me')+AWM*x(i,'W'))=E=AWM*x(i,'W');
DefTermE(i)..
termW(i)*(AMW*x(i,'Me')+AWM*x(i,'W'))=E=AWM*x(i,'Me');
DefgamaW(i).. gama(i,'Me') =E= exp( AMW * termM(i) * termM(i) );
DefgamaE(i).. gama(i,'W')  =E= exp( AWM * termW(i) * termW(i) );

DefdensV(i)$col(i).. densV(i) * (t(i)+273.15) =E= p(i)/Rg;
DefDiam(i)$col(i)..          (3600*0.6)*densV(i)*(0.9   *   pi   *
Dc*Dc/4)=g=V(i);
DefNtray..             Ntrays =e= sum(i,ord(i)*z(i)$abovef(i))
 -sum(i,ord(i)*z(i)$belowf(i))+1;
DefColH..              Hc =g= 0.61*(Ntrays-1)*1.1;

DefAcond..             (Qcond/3.6) =E= Acond*1*(t('30')-25);
DefAreb..              (Qreb/3.6)  =E= Areb *1*(160-t('1'));

DefCost..          cost=E=(1/3) *(600/120)* EquipCost + UtlCost;
DefCost1..       EquipCost =E= 5.50*935.6*Hc**0.81      *Dc**1.05
                        +
2.70*125.2*(Hc/1.1)**0.97*Dc**1.45
                        + 4.07*477*Areb**0.65
                        + 4.07*477*Acond**0.65;

DefCost2..             UtlCost =E=( 14.050 * Qreb /1e+3
             +   0.354 * Qcond/1e+3  )*365*24*0.95;

const1a..              sum(i$abovef(i),z(i)) =e= 1    ;

const2a(i)$abovef(i).. REF(i) =l= z(i)*F;

const1b..              sum(i$belowf(i),z(i)) =e= 1    ;

const2b(i)$belowf(i).. BU(i)  =l= z(i)*(2*F)    ;
```

(continued)

Table 7.6 (continued)

```
purity1..                x('30','Me') =g= 0.98;

purity2..                x('1' ,'Me' )=l= 0.02;

* BOUNDS AND INITIAL VALUES
t.lo(i)=-C('Me')+B('Me')/(A('Me')-log(p(i)));
t.up(i)=-C('W')+B('W')/(A('W')-log(p(i))) ;
t.l(i)=t.up(i)-(t.up(i)-t.lo(i))*(ord(i)-1)/(card(i)-1);

x.l(i,'W') = 0.98-0.96*(ord(i)-1)/(card(i)-1);
x.l(i,'Me')=1-x.l(i,'W'); x.up(i,j)=1; y.up(i,j)=1;

gama.l(i,'Me')=exp(AMW*(AWM*x.l(i,'W')
/(AMW*x.l(i,'Me')+AWM*x.l(i,'W')))**2);
gama.l(i,'W')
=exp(AWM*(AMW*x.l(i,'Me')/(AMW*x.l(i,'Me')+AWM*x.l(i,'W')))**2
);
gama.lo(i,J)=1; gama.up(i,J)=3;
termM.up(i)=1;termW.up(i)=1;
y.l(i,j) = x.l(i,j) * gama.l(i,j) * exp(A(j)-B(j)/(C(j)+t.l(i)))
/p(i);

hv.l(i)=40; hl.l(i)=10;
D.l=F/zf('Me'); L.l('1')=F-D.l;
RefluxRatio.l   =0.8; RefluxRatio.lo=0.25; RefluxRatio.up=2;

L.l(i)$abovef(i)=RefluxRatio.l*D.l;
L.l(i)$belowf(i)=RefluxRatio.l*D.l+F;
L.l(i)$floc(i)  =RefluxRatio.l*D.l+F;

V.l(i)$abovef(i)=(1+RefluxRatio.l)*D.l;
V.l(i)$belowf(i)=(1+RefluxRatio.l)*D.l;
V.l(i)$floc(i)  =(1+RefluxRatio.l)*D.l;

Qcond.lo=1E2; Qcond.up=1E5; Qcond.l=3.2E3;
Qreb.lo =1E2; Qreb.up=1E5;  Qreb.l=3.2E3;

Acond.lo=1;Acond.up=1000;Acond.l=10;
Areb.lo=1 ;Areb.up=1000; Areb.l=10;

z.fx('10')=0; z.fx('1')=0; z.fx('30')=0;
z.l(i) =0;
L.fx('30')=0; V.fx('1')=0;

densV.L(i)=0.04;
Dc.lo=0.5; Dc.up=5; Dc.l=1;
Ntrays.lo=3; Ntrays.up=30; Ntrays.l=15;
Hc.lo=1; Hc.up=Ntrays.up*0.6*1.2; Hc.l=Ntrays.l*0.6*1.2;
EquipCost.l=100000; UtlCost.l=300000;

MODEL MeWDist /ALL/;

OPTIONS NLP=CONOPT4;
OPTIONS miNLP=SBB;
SOLVE MeWDist USING miNLP MinIMIZING cost;
```

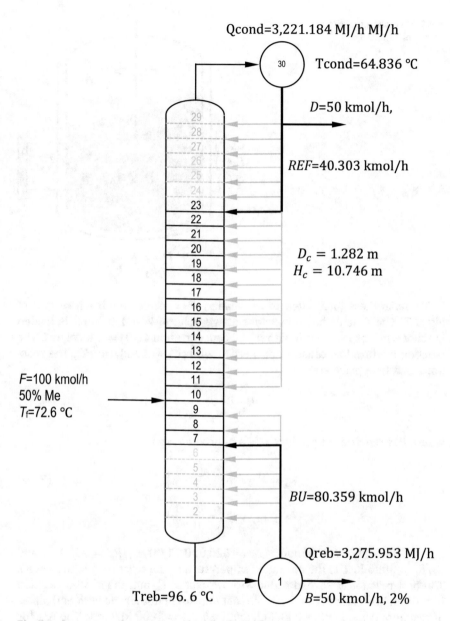

Qcond=3,221.184 MJ/h MJ/h

Tcond=64.836 °C

D=50 kmol/h,

REF=40.303 kmol/h

$D_c = 1.282$ m
$H_c = 10.746$ m

F=100 kmol/h
50% Me
T_f=72.6 °C

BU=80.359 kmol/h

Qreb=3,275.953 MJ/h

Treb=96. 6 °C   B=50 kmol/h, 2%

Fig. 7.6 Structure and notation used in the distillation column optimization

Fig. 7.7 A batch reactor
for the production of
product P

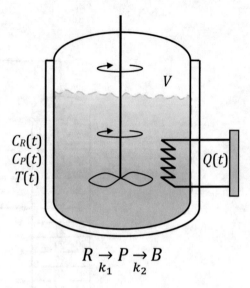

$$R \to P \to B$$
$$\quad k_1 \quad k_2$$

To make these ideas clear to the reader, we consider the batch reactor of
Fig. 7.7. The reactor has a constant active volume $V = 1$ m^3 and is loaded
initially with the reactant R with initial concentration $C_R(0) = 1$ kmol/m^3. The
reacting mixture has constant properties with $\rho C_p = 10$ kJ/(m^3 K). The reactions that take place are:

$$R \to P \to B$$
$$\;\; k_1 \;\; k_2$$

where P is the desired product, B is a byproduct, and:

$$k_1 = k_{1,0} \exp\left(-\frac{E_1}{R} \cdot \frac{1}{T}\right) \tag{7.16}$$

$$k_2 = k_{2,0} \exp\left(-\frac{E_2}{R} \cdot \frac{1}{T}\right) \tag{7.17}$$

$K_{1,0} = 4000$ (1/h)(m^3/kmol), $k_{2,0} = 620{,}000$ (1/h), $E_1/R_g = 2500$ K, and
$E_2/R_g = 5000$ K. T is the absolute temperature of the reacting mixture (in K).
The first reaction is second order with respect to C_R and the second reaction
first order with respect to C_P. Both reactions are endothermic with enthalpies
of reaction: $\Delta h_{rxn,1} = 3600$ kJ/kmol and $\Delta h_{rxn,2} = 1800$ kJ/kmol. The reactor
can be heated by an immersed 1 kW electrical resistance driven by a silicon
controlled rectifier (SCR) and can deliver heat at a rate up to 3600 kJ/h.

The mathematical model of the batch reactor consists of the material balances for components R and P and the energy balance of the reactor (it is
reminded that the reactor volume and all properties are constant):

$$\frac{dC_R(t)}{dt} = -r_1(t) \tag{7.18}$$

$$\frac{dC_P(t)}{dt} = r_1(t) - r_2(t) \tag{7.19}$$

$$\frac{dT(t)}{dt} = \frac{(-\Delta H_{\text{rxn},1})}{\rho C_p} \cdot r_1(t) + \frac{(-\Delta H_{\text{rxn},2})}{\rho C_p} \cdot r_2(t) + \frac{Q(t)}{\rho C_p V} \tag{7.20}$$

$$r_1(t) = k_{1,0} \exp\left(-\frac{E_1}{R} \cdot \frac{1}{T(t)}\right) C_R^2(t) \tag{7.21}$$

$$r_2(t) = k_{2,0} \exp\left(-\frac{E_2}{R} \cdot \frac{1}{T(t)}\right) C_P(t) \tag{7.22}$$

There are several constraints that the reacting mixture needs to satisfy in order to ensure a safe operation:

$$298.15 \leq T(t) \leq 398.15 \text{ K} \tag{7.23}$$

$$0 \leq Q(T) \leq 3600 \text{ kJ/h} \tag{7.24}$$

If we fix the duration of the operation of the batch reactor to $t_f = 1$ h, then our objective is to determine the heat transfer rate $Q(t)$ that maximizes the final concentration of the product:

$$\max_{Q(t), T(t), C_R(t), C_P(t)} C_P(t_f) \tag{7.25}$$

The problem that we have just presented has two main differences compared with the problems that we have seen in all other applications. The first one is due to the fact that all variables are defined for $t \in [0, t_f]$ which is an infinite dimensional set. The second one is that all variables are given implicitly as the solution of the set of ODEs describing the problem. One method for solving ODEs is to use finite differences (other methods, such as polynomial approximations, are available, and although they are more accurate, they are also much more complex to present). We consider a discretization of the time domain, which is the independent variable, using $n + 1$ equidistance nodes (see Fig. 7.8). Node 1 corresponds to $t = 0$ and node $n + 1$ to $t = t_f$. Node i corresponds to $t = (i - 1)\Delta t$, where $\Delta t = \frac{t_f}{n}$ is the distance between the nodes. For the general ODE:

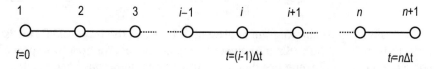

Fig. 7.8 Discretization of the solution domain using $n + 1$ equidistance nodes

$$\frac{dx}{dt} = f(x) \tag{7.26}$$

the solution can be approximated by the following algebraic equation:

$$\frac{1}{\beta \Delta t} \sum_j \alpha_j x_j = f(x_i) \tag{7.27}$$

Characteristic examples are the *forward finite differences*:

$$\frac{x_{i+1} - x_i}{\Delta t} = f(x_i) \tag{7.28a}$$

central finite differences:

$$\frac{x_{i+1} - x_{i-1}}{2\Delta t} = f(x_i) \tag{7.28b}$$

backward finite differences:

$$\frac{x_i - x_{i-1}}{\Delta t} = f(x_i) \tag{7.28c}$$

The accuracy of the forward and backward approximations is first order with respect to Δt, while the accuracy of the central approximation is second order. There are more complex but more accurate approximations such as the following (fourth-order accuracy):

$$\frac{x_{i-2} - 8x_{i-1} + 8x_{i+1} - x_{i+2}}{12\Delta t} = f(x_i) \tag{7.28d}$$

An alternative way to solve (7.26) is through numerical integration:

$$x_{i+1} = x_i + \int_{t_i}^{t_{i+1}} f(x)dt \tag{7.29}$$

A characteristic example which combines acceptable accuracy and simplicity in programming is the trapezoidal rule:

$$x_{i+1} = x_i + \frac{1}{2}\Delta t(f(x_i) + f(x_{i+1})) \tag{7.30}$$

In Table 7.7, a GAMS program is presented in which the trapezoidal rule is used to transform the ODEs into algebraic equations. In addition, to improve convergence, the following profile for the rate of heat input has been used:

Table 7.7 GAMS program for the dynamic optimization of the batch reactor

```
$ TITLE BATCH REACTOR DYNAMIC OPTIMIZATION PROBLEM

SETS j time points /1*201/
     i states       /R,P,T/
     r reactions    /RtoP,PtoB/;

SCALAR   rhoCp density times heat capacity in kJ per m^3 K /10/;
SCALAR   V       reactor volume in m^3 /1/;

PARAMETER NP number of elements; NP=card(j)-1;
PARAMETER dt discetization ; dt=1/NP;
PARAMETER time(j); time(j)=(ord(j)-1)*dt;
PARAMETER Dhrxn(r) heats of reaction kJ per kmol
          / RtoP 3600
            PtoB 1800/;

TABLE n(i,r) stoichiometric numbers
    RtoP PtoB
R  -1    0
P  +1   -1
T   0    0;

VARIABLES            C0           constant in profile of u
                     C1           constant in profile of u;
POSITIVE VARIABLES x(i,j)         state variables
                   rate(j,r)      reaction rates in kmol per h
                   u(j)           scaled manipulated variable
                   Q(j)           dimensionless heat rate profile;
VARIABLE             f(i,j)       rhs of differential equations
                     Obj          objective function;

EQUATIONS eq1, eq2, eq3,eq4, eq5, eq6, eq7, eq8, ObjFun;

eq1(i,j)$(ord(j) ge 2).. x(i,j) =E= x(i,j-
1)+.5*dt*(f(i,j)+f(i,j-1));
eq2(j).. f('R',j)       =E= -rate(j,'RtoP') ;
eq3(j).. f('P',j)       =E= +rate(j,'RtoP') - rate(j,'PtoB');
eq4(j).. f('T',j)*rhoCp =E=
sum(r,(-Dhrxn(r))*rate(j,r))+Q(j)/V;
eq5(j).. rate(j,'RtoP') =E=
4000*exp(-2500/x('T',j))*x('R',j)**2;
eq6(j).. rate(j,'PtoB') =E=
620000*exp(-5000/x('T',j))*x('P',j);
eq7(j).. u(j)           =E= C0*exp(-C1*time(j)) ;
eq8(j).. Q(j)           =E= 3600*u(j);
ObjFun.. Obj            =E= x('P','201');


* -----------------------------------------------------
* I N I T I A L    G U E S S E S
* -----------------------------------------------------
* Known bounds
x.lo('R',j)=0;        x.up('R',j)=1;
x.lo('P',j)=0;        x.up('P',j)=1;
x.lo('T',j)=298.15;   x.up('T',j)=398.15;
u.lo(j) = 0;          u.up(j) = 1;
Q.lo(j) = 0;          Q.up(j) = 3600;
```

(continued)

Table 7.7 (continued)

```
rate.lo(j,r)=0; rate.up(j,r)=10;
* Assume constant temperature profile
x.l('T',j)=350;
* Calculate concentrations using Eulers method
x.l('R',j+1)=x.l('R',j)-dt*4000*exp(-
2500/x.l('T',j))*x.l('R',j)**2;
x.l('P',j+1)=x.l('P',j)+dt*4000*exp(-
2500/x.l('T',j))*x.l('R',j)**2-
620000*exp(-5000/x.l('T',j))*x.l('P',j);
* Calculate rates and objective function
rate.l(j,'RtoP') =   4000*exp(-2500/x.l('T',j))*x.l('R',j)**2;
rate.l(j,'PtoB') = 620000*exp(-5000/x.l('T',j))*x.l('P',j);
Obj.l=x.l('P','201');
* -INITIAL CONDITIONS (temperature is free)
x.fx('R','1')=1;
x.fx('P','1')=0;

MODEL Batch_Reactor /ALL/ ;
options nlp=conopt4;
SOLVE Batch_Reactor USING NLP MAXIMIZING Obj;
```

$$\frac{Q(t)}{Q_{max}} = u(t) = C_0 \exp(-C_1 t) \tag{7.31}$$

This turns out to be a reasonable profile for the manipulated variable that is easy to implement in practice. Other functional form, such as polynomial functions of time:

$$\frac{Q(t)}{Q_{max}} = C_0 + C_1 t + C_2 t^2 + \ldots \tag{7.32}$$

are equally valid, but for this particular application, (7.31) is sufficient from the practical point of view. It is interesting to note that we can also optimize the value obtained by the manipulated variable at each node. This will always result in an improved objective function, but numerical problems are expected, while the practical application of such a sequence of numbers can be challenging in a real application.

When 201 nodes are used ($\Delta t = 1/200 = 0.005$), then the maximum concentration of P that can be achieved is 0.60848860 (slightly smaller than 0.610071345 which is obtained when no specific profile is assumed). The optimal manipulated input profile is given by (7.31) with $C_0 = 1$ and $C_1 = 1.32404288$. The profiles of the concentrations, temperature, and heat transfer rate are presented in Figs. 7.9, 7.10, and 7.11.

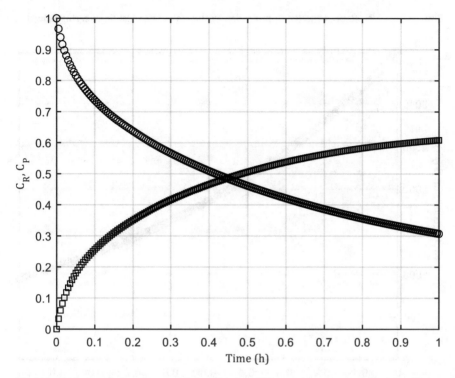

Fig. 7.9 Variation of the reactant (○) and product (□) concentration in the batch reactor

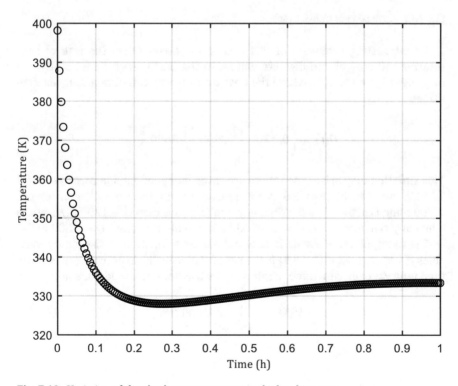

Fig. 7.10 Variation of the absolute temperature in the batch reactor

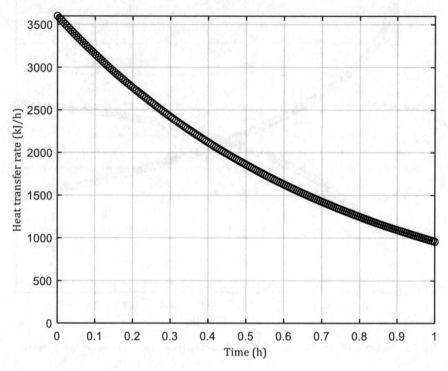

Fig. 7.11 Optimal heat transfer rate

An interesting extension of this case study arises when the rate of heat transfer can only obtain discrete values. In the batch reactor case study, we may investigate the case where the heat transfer rate can obtain four discrete values:

$$Q(t) \in \left\{ 0, 1200 \, \frac{kJ}{h}, 2400 \, \frac{kJ}{h}, 3600 \, \frac{kJ}{h} \right\}$$

This situation may arise when there are three independent, identical electrical 1/3 kW resistances installed in the reactor whose state is either on or off. When none is active, $Q = 0$. When any one is active, then $Q = 1200$ kJ/h, while when any two are active, $Q = 2400$ kJ/h. When all are active, $Q = 3600$ kJ/h.

The manipulated variable is now defined as a function of the binary variables $y(k, s)$ where $s \in \{1,2,3,4\}$ is the set of states for the heat input and k is the time interval over which the manipulated variable is assumed constant:

$$Q(k) = \sum_{s \in \{1, 2, 3, 4\}} Q_s \cdot y(k, s) \tag{7.33a}$$

$$\sum_{s\in\{1,2,3,4\}} y(k,s) = 1, \forall k \tag{7.33b}$$

In this case study, we consider that the heat input can be changed every 12 min or 0.2 h, and it changes at the beginning of each interval and then remains constant until the end of the time interval.

Given the fact that we have transformed the ODEs into nonlinear algebraic equality constraints, the introduction of binary variables transforms our model from an NLP model to a MINLP model. The GAMS model in Table 7.8 presents a potential implementation of a representative mixed-integer, dynamic optimization problem. The reader must study the model carefully and to realize that the model is developed by defining time intervals over which the input remains constant and that the intervals are connected together by equating the values of all differential variables at the end of each interval with their values at the beginning of the next interval. The optimal profiles are presented in Figs. 7.12, 7.13, and 7.14. The objective function obtains the value of 0.60883088 (slightly better than the one obtained with the control profile given by (7.31)).

7.6   Optimal Design of a Renewable Energy Production System

The ever-increasing energy demand, our dependence on depleting fossil resources, and their effects on the environment have become areas of major concern. A strategy that can be used to tackle both problems is the use of renewable energy sources (RES). Within this framework, the concept of a "smart grid" has been re-invented. A smart grid system enables interaction and inter-change of information between the power generation units and the consumers which allow the optimal overall operation to be achieved by increasing reliability, facilitating energy security, decreasing overall and individuals' costs, and reducing environmental burden.

In this section, we will study the optimal design of a hybrid, smart RES system consisting of (see Fig. 7.15):

- A wind turbine generator system
- A photovoltaic energy generator system
- A diesel generator system
- A battery bank system
- A connection to the national electricity grid

to satisfy in a cost-optimal way the electricity demand of a public building.

The electricity demand of the building $P_D(t)$ is shown in Fig. 7.16. Note that the demand is given for 72 h, which corresponds to 3 representative days of a calendar year. The first day (0–24 h) corresponds to a winter day and the third

Table 7.8 GAMS program for the dynamic optimization of the batch reactor—discrete inputs

```
$ TITLE BATCH REACTOR DYNAMIC OPTIMIZATION PROBLEM WITH DISCRETE
INPUTS
option optcr=1e-8;
option optca=1e-8;
SETS k finite elements        /1*5/
     j nodes in each element /1*41/
     i states                /CR,CP,T/
     r reactions             /RtoP,PtoB/
     s discrete Q states     /1*4/;

SCALAR    rhoCp density times heat capacity in kJ per m^3 K /10/;
SCALAR    V       reactor volume in m^3 /1/;

PARAMETER NP, dt; NP=(card(j)-1)*card(k); dt=1/NP;
PARAMETER time(j,k); time(j,k)=((ord(k)-1)*(card(j)-1)+ord(j)-
1)*dt;
PARAMETER Dhrxn(r) heats of reaction kJ per kmol
          / RtoP 3600
            PtoB 1800/;

PARAMETER QS(s)/ 1 0, 2 1200, 3 2400, 4 3600/;

TABLE n(i,r)
    RtoP   PtoB
CR  -1      0
CP  +1     -1
T    0      0;

VARIABLES f(i,j,k),  x(i,j,k),  rate(j,r,k),  Q(k),  Obj;
BINARY VARIABLE y(k,s);

EQUATIONS
eq1(i,j,k),eq2(i,k),eq3(i,j,k),eq4(i,j,k),eq5(j,k),eq6(j,k),eq
7(k),eq8(k),
ObjFun;
eq1(i,j,k)$(ord(j) gt 1).. x(i,j,k)=E=
x(i,j-1,k)+.5*dt*(f(i,j,k)+f(i,j-1,k));
eq2(i,k)$(ord(k) gt 1)..    x(i,'1',k) =E= x(i,'41',k-1);
eq3(i,j,k)$( ord(i) ne 3).. f(i,j,k) =E=
sum( r, n(i,r)*rate(j,r,k) );
eq4(i,j,k)$(ord(i) eq 3).. f(i,j,k)=E=
(( -DHrxn('RtoP')*rate(j,'RtoP',k)-
DHrxn('PtoB')*rate(j,'PtoB',k) )*V+Q(k))/(rhoCp*V);
eq5(j,k).. rate(j,'RtoP',k)=E=
4000*exp(-2500/x('T',j,k))*x('CR',j,k)**2;
eq6(j,k).. rate(j,'PtoB',k) =E=
620000*exp(-5000/x('T',j,k))*x('CP',j,k);
eq7(k).. Q(k) =e= sum(s, QS(s)*y(k,s) );
eq8(k).. sum(s,y(k,s)) =E= 1;
ObjFun.. Obj=e= x('CP','41','5');

* Known bounds
x.lo(i,j,k)=0;x.lo('T',j,k)=298;x.up(i,j,k)=1;
x.up('T',j,k)=398.15;
```

(continued)

Table 7.8 (continued)

```
rate.lo(j,r,k)=0;  rate.up(j,r,k)=4000*exp(-2500/400);
Q.lo(k)=0;  Q.up(k)=3600; Q.l(k)=0;

* Initial Conditions on the concentrations
x.fx('CR','1','1')=1; x.fx('CP','1','1')=0;

MODEL Ray_Reactor /ALL/ ;

options nlp=CONOPT4;
options mip=CPLEX;
OPTIONS minlp=sbb;
SOLVE Ray_Reactor USING MINLP MAXIMIZING Obj;

file                     BatchRea                    /BatchRea.m/
;
PUT  BatchRea ;
PUT #1@5'Time ',#1@15 'CR',#1@27 'CP', #1@37 'T',#1@51 'Q' /;
loop((k,j),
PUT   time(j,k):10:4,
      x.l('CR',j,k):12:6,
      x.l('CP',j,k):12:6,
      x.l('T',j,k):12:6,
      Q.l(k):8:0    /;
      );
PUTCLOSE BatchRea;
```

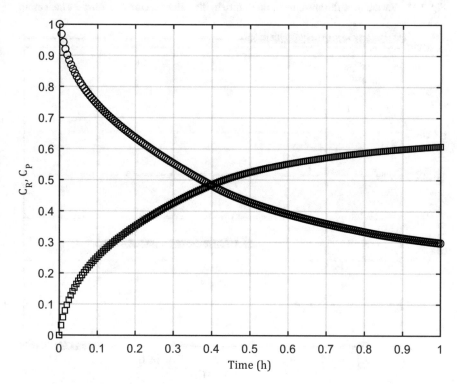

Fig. 7.12 Variation of the reactant (○) and product (□) concentration in the batch reactor for optimal operation with discrete input

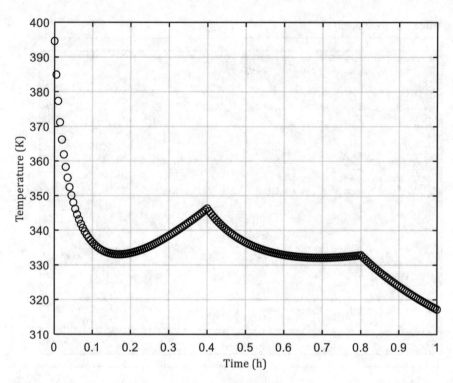

Fig. 7.13 Variation of the absolute temperature in the batch reactor with discrete heat input

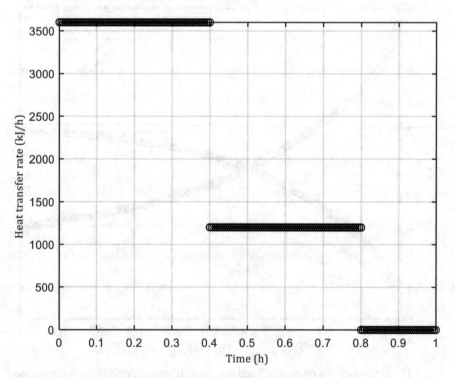

Fig. 7.14 Optimal heat transfer rate with discrete states

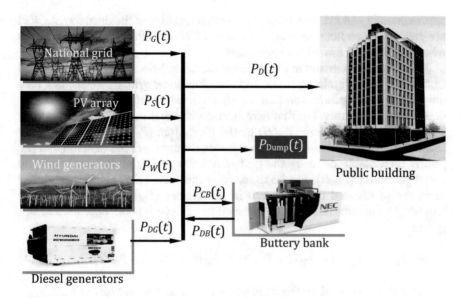

Fig. 7.15 Case study on renewable energy production system

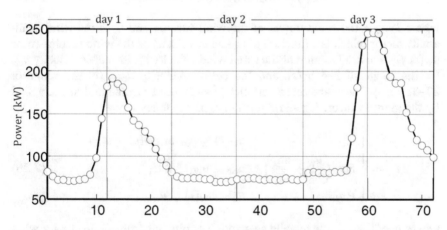

Fig. 7.16 Data on electricity demand for the public building (day 1 is a representative working day in January, day 2 is a public holiday, and day is a working day in July)

day (48–72 h) to a summer day. Both these days are working days. A public holiday (or weekend day) is also included (24–48 h) as weekends, or public holidays account for almost 30% of a year. In a realistic application, data for a whole year (8760 h) would have been used, but here we restrict attention to 3 representative days for clarity of presentation. As can be observed from Fig. 7.16, electricity demand is increasing during the morning hours and reaches a peak around midday. There is also a background electricity demand which is around 70 kW in our case. The maximum power demand is approximately 245 kW and takes place around midday of the third day. As data are given for each hour, we can use units of kW or kWh (this holds true only when the data are available every hour).

The electricity demand of the building can be satisfied using the grid $(P_G(t))$ as this is a public building connected to the national grid. However, this may not be the cost optimal solution, or its environmental performance can be deemed unsatisfactory. For that reason, electric power can be supplied directly from the wind generators $(P_W(t))$ or the PV system $(P_S(t))$ or by the battery $(P_{BD}(t))$. If the renewable energy generators cannot meet the demand and the grid is unavailable or using the grid is not the preferred solution, then the diesel generator $(P_{DG}(t))$ can be used. Any renewable energy that is produced in excess of demand can be stored in the battery $(P_{BC}(t))$. At any time, the following constraint needs to be satisfied to achieve feasible operation of our system:

$$P_G(t) + P_W(t) + P_S(t) + P_{BD}(t) + P_{DG}(t) \geq P_D(t) + P_{BC}(t) \qquad (7.34)$$

The power supplied by the wind generator is usually expressed as:

$$P_W(t) = P_{W,R} \cdot f_w(u_w(t), u_R, u_{cin}, u_{cout}) \qquad (7.35)$$

where $P_{W,R}$ is the rated power of the wind turbine and f_w is the wind availability factor which is a function of the wind velocity at the wind turbine rotor $u_w(t)$. f_w is also a function of the rated wind velocity u_R, the cut-in velocity u_{cin} (in the range of 2–4 m/s), and the cut-out velocity u_{cout} (in the range of 25–35 m/s). These are characteristics of each wind turbine and are supplied by the manufacturer. A general form of f_w is the following:

$$f_w = \begin{cases} 0, & u_w(t) \leq u_{cin} \text{ or } u_w(t) \geq u_{cout} \\ P_{W,R} \left(\dfrac{u_w^n - u_{cin}^n}{u_R^n - u_{cin}^n} \right), & u_{cin} < u_w(t) < u_R \\ P_{W,R}, & u_R < u_w(t) < u_{cout} \end{cases} \qquad (7.36)$$

where $n = 1, 2,$ or 3 is usually selected. It is obvious from (7.36) that a wind turbine operates only when the wind velocity is larger than u_{cin} and smaller than u_{cout}. The power increases monotonically as the wind speed increases above u_{cin} up until it exceeds u_R. It then delivers the rated power when $u_w > u_R$, and finally, for safety reasons, the turbine is shut down for wind velocities above u_{cout}. For a given set of hourly data for the local wind velocity and a given wind turbine, Eq. (7.36) is applied to determine f_w.

The power supplied by the PV array generator is usually expressed as:

$$P_S(t) = P_{S,R} \cdot \frac{I(t)}{I_S} \qquad (7.37)$$

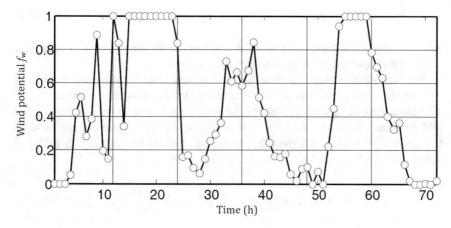

Fig. 7.17   Measured wind potential for the public building RES model optimization

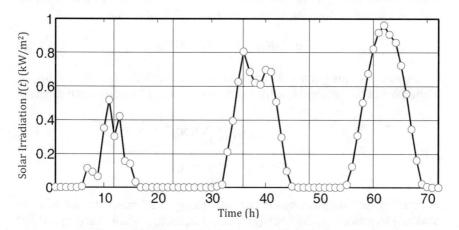

Fig. 7.18   Measured solar radiation for the public building RES model optimization

where $P_{S,R}$ is the rated power of the solar PV array, I_s is the incident radiation at standard test conditions (1 kW/m^2), and $I(t)$ is the solar radiation incident on the PV array (which is measured using specialized equipment). The measured $f_W(t)$ and $I(t)$ at the specific location of the public building under study are presented in Figs. 7.17 and 7.18. The wind potential is not significant, while the solar irradiation is significant during day 3 (summer day) and less significant during day 1 (winter day).

For the diesel generator, the important characteristic is the dependence of the diesel consumption D_{cons} on the diesel generator rated power $P_{DG, R}$ and the electrical power output of the generator ($P_{DG}(t)$):

$$D_{cons}(t) = d_0 P_{DG,R} + d_1 P_{DG}(t), \text{ in } L/h \qquad (7.38)$$

where $P_{DG,R}$ is the rated power output of the diesel generator. d_0 and d_1 are constants characteristic for each diesel generator.

The final element of the grid is the battery bank. Energy storage is vital to all renewable energy systems as the wind potential and solar radiation are available for a limited time only, and energy storage is necessary to balance supply and demand. The state of a battery is described by its State of Charge (SOC(t), the level of charge of an electric battery relative to its capacity in kWh). A simple power "balance" gives the following equation:

$$SOC(t+1) = SOC(t) + \left(\eta_C P_{BC}(t) - \frac{P_{BD}(t)}{\eta_D}\right)\Delta t \qquad (7.39)$$

where η_C is the efficiency of charging and η_D the efficiency of discharging the battery. To maintain good performance of the battery for its entire lifetime, the Depth of Discharge must be maintained above 20% of its rated capacity:

$$0.2 SOC_R \leq SOC(t) \leq SOC_R \qquad (7.40)$$

For the same reason (to avoid damage of the battery), the rate of charging or discharging of the battery is restricted to be less than 10% of its rated capacity:

$$0 \leq P_{BC}(t) \leq \frac{1}{10} SOC_R \qquad (7.41a)$$

$$0 \leq P_{BD}(t) \leq \frac{1}{10} SOC_R \qquad (7.41b)$$

As charging and discharging cannot take place simultaneously, we need to introduce two binary variables: $\varphi(t)$ which is equal to 1 when charging and $\varepsilon(t)$ which is equal to 1 when discharging. We then exclude the possibility of simultaneous charging and discharging using the following linear constraints:

$$0 \leq P_{BC}(t) \leq \frac{1}{10} SOC_R \qquad (7.42a)$$

$$0 \leq P_{BC}(t) \leq \varphi(t) SOC_{R,\,max} \qquad (7.42b)$$

$$0 \leq P_{BD}(t) \leq \frac{1}{10} SOC_R \qquad (7.42c)$$

$$0 \leq P_{BD}(t) \leq \varepsilon(t) SOC_{R,\,max} \qquad (7.42d)$$

$$\varphi(t) + \varepsilon(t) \leq 1 \qquad (7.42e)$$

$$\varphi(t), \varepsilon(t) \in \{0, 1\} \qquad (7.42f)$$

We have up to this point presented all equations necessary to describe the system dynamics for the time horizon of interest $t = 1,2,\ldots,H$ (in our case $H = 72$ h). What remains to be done is the development of an appropriate cost model. An appropriate approximate model is based on the calculation of the annualized capital cost and the operating and maintenance cost. The overall objective is:

$$
\begin{aligned}
\text{Cost} = \text{CCF} \sum_{i \in \{W, S, \text{DG}, B\}} \text{UC}_i \cdot P_{i,\text{R}} \left(\lambda_i + \frac{\text{Inst}_i}{100} \right) \\
+ \sum_{i \in \{W, S, \text{DG}, B\}} \text{UC}_i \cdot P_{i,\text{R}} \cdot \frac{M_i}{100} \\
+ \sum_t (\text{UC}_{\text{el}} \cdot P_\text{G}(t) + \text{UC}_{\text{Dsl}} \cdot D_{\text{cons}}(t)) \frac{365 \cdot 24}{H}
\end{aligned}
\tag{7.43}
$$

The first term in the rhs accounts for the annualized cost of installed equipment (including equipment replacement). CCF is the capital recovery factor. $P_{i,\text{R}}$ is the rated power for the wind turbines $(i = W)$, the PV cells $(i = S)$, the diesel generator $(i = \text{DG})$, or the rated capacity of the battery bank $(i = B)$. When the rated power or capacity is multiplied by the unit equipment cost (UC_i, in \$/kW rated or \$/kWh rated) gives the purchase cost of the particular equipment. To account for all costs incurring when installing the equipment, an installation factor Inst_i is introduced for each type of equipment. This factor gives the additional cost incurring when installing the equipment as percentage of the purchase cost of the equipment. When the equipment lifetime exceeds the lifetime of the project, then no replacement is necessary, and $\lambda_i = 1$. However, if the lifetime of the equipment LT_i is smaller than the lifetime of the project LT_p, then replacement is necessary every LT_i years and:

$$
\lambda_i = \frac{1}{(1+r)^{0 \cdot \text{LT}_i}} + \frac{1}{(1+r)^{1 \cdot \text{LT}_i}} + \cdots + \frac{1}{(1+r)^{k \cdot \text{LT}_i}}
$$

or:

$$
\lambda_i = \sum_{k=0,1,2\ldots:k \cdot \text{LT}_i \leq \text{LT}_p} \frac{1}{(1+r)^{k \cdot \text{LT}_i}}
\tag{7.44}
$$

The second term in the rhs accounts for the maintenance cost which we simply assume that it is proportional to the purchase cost of the equipment (many other options are available, but this is the simplest one).

The third term in the rhs accounts for the cost of electricity supplied by the grid (UC_{el} is the unit cost of electricity in \$/kWh) and the cost of diesel consumed by the diesel generator (UC_{Dsl} is the unit cost of diesel in \$/L). This term is multiplied by the ratio of the hours in a complete calendar year over the time horizon of the model (in h). When data for a complete calendar year is available, then $H = 365 \cdot 24$, and the ratio is equal to 1. We summarize the complete model:

$$\min_{\substack{P_G(t), P_W(t), P_S(t), P_{BD}(t), P_{DG}(t), P_{BC}(t), SOC(t), \varphi(t), \varepsilon(t) \\ P_{DG,R}, P_{W,R}, P_{S,R}, SOC_R}}$$

$$\text{Cost} = \text{CCF} \sum_{i \in \{W, S, DG, B\}} \text{UC}_i \cdot P_{i,R}\left(\lambda_i + \frac{\text{Inst}_i}{100}\right)$$

$$+ \sum_{i \in \{W, S, DG, B\}} \text{UC}_i \cdot P_{i,R} \cdot \frac{M_i}{100}$$

$$+ \sum_t (\text{UC}_\text{el} \cdot P_G(t) + \text{UC}_\text{Dsl} \cdot D_\text{cons}(t)) \frac{365 \cdot 24}{H}$$

$$\text{s.t.}$$

$$\left.\begin{aligned}
P_G(t) + P_W(t) + P_S(t) + P_{BD}(t) + P_{DG}(t) &\geq P_D(t) + P_{BC}(t) \\
P_W(t) &= P_{W,R} \cdot f_w(t) \\
P_S(t) &= P_{S,R} \cdot \frac{I(t)}{I_S} \\
D_\text{cons}(t) &= d_0 P_{DG,R} + d_1 P_{DG}(t) \\
SOC(t+1) &= SOC(t) + \left(\eta_C P_{BC}(t) - \frac{P_{BD}(t)}{\eta_D}\right)\Delta t \\
P_{DG}(t) &\leq P_{DG,R} \\
0.2 SOC_R \leq SOC(t) &\leq SOC_R \\
0 \leq P_{BC}(t) &\leq \frac{1}{10} SOC_R \\
0 \leq P_{BD}(t) &\leq \frac{1}{10} SOC_R \\
P_{BC}(t) &\leq \varphi(t) SOC_{R,\max} \\
P_{BD}(t) &\leq \varepsilon(t) SOC_{R,\max} \\
\varphi(t) + \varepsilon(t) &\leq 1
\end{aligned}\right\} \forall t$$

$$P_G(t), P_W(t), P_S(t), P_{BD}(t), P_{DG}(t), P_{BC}(t), SOC(t) \geq 0$$

$$P_{DG,R}, P_{W,R}, P_{S,R}, SOC_R \geq 0$$

$$\varphi(t), \varepsilon(t)\{0, 1\}$$

An implementation of this mixed integer, linear programming model is given in Table 7.9. All constants appearing in the model can be obtained from the GAMS program to save space. The cost optimal design features a total annual cost of \$135,308.2 that corresponds to a unitary cost of 0.141 \$/kWh. It only uses a wind turbine with a rated capacity of 141.57 kW, and no PV cell array,

Table 7.9 GAMS program for the optimal energy management system case study

```
$ TITLE Renewable Energy Resource Design for Public Building
SET       t time in h /1*72/ ;
PARAMETER PD(t) power demand in kW
          I(t)  solar irradiation in kW per square m
          fw(t) wind power availability (fr of rated power);

TABLE data(t,*) problem data
          D         S         W
    1    81.7280    0         0
    2    76.4160    0         0
    3    72.4480    0         0
    4    71.6160    0         0.0485
    5    70.9760    0         0.4206
    6    70.5920    0.0047    0.5128
    7    71.8080    0.1082    0.2805
    8    72.8320    0.0906    0.3863
    9    78.7200    0.0601    0.8838
   10    97.7920    0.3485    0.1948
   11   144.5100    0.5152    0.1492
   12   181.5700    0.2991    1.0000
   13   191.1700    0.4206    0.8354
   14   184.8300    0.1530    0.3398
   15   181.1200    0.1390    1.0000
   16   157.3800    0.0291    1.0000
   17   141.5700    0         1.0000
   18   137.0900    0         1.0000
   19   128.7700    0         1.0000
   20   119.6800    0         1.0000
   21   108.1000    0         1.0000
   22    96.3840    0         1.0000
   23    89.2160    0         1.0000
   24    81.2800    0         0.8359
   25    75.9040    0         0.1595
   26    74.3040    0         0.1687
   27    74.3680    0         0.0924
   28    74.3040    0         0.0637
   29    73.8560    0         0.1463
   30    72.9600    0         0.2530
   31    73.1520    0.0001    0.2954
   32    70.2720    0.0139    0.3615
   33    69.5040    0.2095    0.7292
   34    69.8880    0.3934    0.6090
   35    71.0400    0.6255    0.6653
   36    73.5360    0.8040    0.5809
   37    73.2800    0.6844    0.6768
   38    74.1760    0.6194    0.8433
   39    74.6880    0.6078    0.5113
   40    73.1520    0.6959    0.4230
   41    72.5120    0.6846    0.2448
   42    73.1520    0.5075    0.1651
   43    71.9360    0.2925    0.1572
   44    71.7440    0.0951    0.1800
   45    73.8560    0.0025    0.0551
   46    73.6000    0         0.0194
   47    72.4480    0         0.0877
```

(continued)

Table 7.9 (continued)

```
     48   72.6400        0      0.0992
     49   81.0240        0        0
     50   81.6640        0      0.0749
     51   81.1520        0        0
     52   80.7680        0      0.2248
     53   80.7680        0      0.4478
     54   81.6640        0      0.9401
     55   82.0480     0.0115    1.0000
     56   83.8400     0.1195    1.0000
     57  121.2200     0.3048    1.0000
     58  180.8600     0.4980    1.0000
     59  230.2700     0.6755    1.0000
     60  244.6100     0.8201    0.7806
     61  244.8600     0.9152    0.6968
     62  244.2200     0.9578    0.6321
     63  224.0000     0.9074    0.4017
     64  193.5400     0.8576    0.3238
     65  185.6000     0.7238    0.3627
     66  186.3700     0.5535    0.1168
     67  152.0600     0.3447    0.0181
     68  132.9900     0.1577      0
     69  119.3000     0.0207      0
     70  112.7700     0.0000    0.0036
     71  107.5200        0        0
     72   98.9440        0      0.0212 ;
PD(t) = data(t,'D'); I(t) = data(t,'S'); fw(t) = data(t,'W');


SCALARS
interest interest rate                                     /0.1/
years time horizon of the project                          /25/
cS   unit cost of solar panels per rated kW                /2000/
cW   unit cost of wind turbine per rated kW                /2500/
cDG  unit cost of diesel generator per rated kW            /150/
cB   unit cost of baterry per rated kWh                    /200/
MS   maintenance cost of PV panels in % of initial invest  /1/
MW   maintenance cost of wind turbines in % of init invest /1/
MDG  maintenance cost of diesel gen in % of init invest.   /1/
MB   maintenance cost of battery in % of initial invest    /1/
cel  unit cost of grid electricity in $ per kWh            /0.18/
cD   unit cost of diesel fuel in $ per L                   /0.6/
hCh  efficiency of battery charging                        /0.95/
hDCh efficiency of battery discharging                     /0.85/
FinstS instal expens as %of initial inv in solar panels    /30/
FinstW instal expens as %of initial inv in wind turbines   /20/
FinstDG instal expens as %of initial inv. diesel gen.      /5/
FinstB instal expens as %of initial inv in batteries       /5/
CEF_El CO2 emission for grid elec. in kgCO2 per kWh        /0.750/
CEF_Dsl CO2 emission for diesel fuel in kgCO2 per L        /2.900/
CEF_S  CO2 emission for PV electr. in kgCO2 per kWh        /0.050/
CEF_W  CO2 emission for wind turb. in kgCO2 per kWh        /0.035/;


SCALARS lamdaS ,lamdaW,lamdaDG,lamdaB, TotalCO2_grid;
lamdaS = 1; lamdaW = 1;
lamdaDG= 1+1/(1+interest)**10 + 1/(1+interest)**20;
lamdaB = 1 + 1/(1+interest)**5  + 1/(1+interest)**10
           + 1/(1+interest)**15 + 1/(1+interest)**20;
TotalCO2_Grid = CEF_El*SUM(t,PD(t))*365/3;
```

(continued)

Table 7.9 (continued)

```
DISPLAY TotalCO2_grid;
PARAMETER CCF capital charge factor;
CCF = interest*(1+interest)**years/((1+interest)**years-1);

POSITIVE VARIABLES
PG(t)   electrical energy from the grid              in kW
PS(t)   electrical energy from the solar panels      in kW
PW(t)   electrical energy from the wind turbines     in kW
PDG(t)  electrical energy from the diesel generator  in kW
ConsD(t)  diesel consumption                         in L per h
PCB(t)  electrical energy charging battery           in kW
PDB(t)  electrical energy discharging battery        in kW
Pdump(t) electrical energy dumped                    in kW
SOC(t)  battery state of charge                      in kWh
PSR     rated power of solar panels                  in kW
PWR     rated power of wind turbine                  in kW
PDGR    rated power of diesel generator              in kW
SOCR    rated capacity of battery                    in kWh
TotalCO2eq kg CO2eq emitted per year                 in kg;

BINARY VARIABLES     phi(t), epsilon(t);
*AnnualCost          total annual cost in $ per year
*Unit AnnualCost     unit total annual cost in $ per kWh
*CO2PerCentInitial CO2 emitted relative to grid      in %;
VARIABLE  AnnualCost, UnitAnnualCost, CO2PerCentInitial;

EQUATIONS EnergyBalance(t), SolarPanels(t), PerCentCO2,
          WindTurbine(t), DieselGener(t), StateB(t)
          InitialStateBat, UpperDG(t), UpperB1(t), UpperB2(t),
          UpperB3(t), UpperB4(t),UpperB5(t), Lower1(t),
          LogicalB(t), CostCalc, UCostCalc,CO2emissions;

EnergyBalance(t)..
PG(t)+PS(t)+PW(t)+PDG(t)+PDB(t)=G=PD(t)+PCB(t);
SolarPanels(t)..   PS(t)     =E=I(t) *PSR;
WindTurbine(t)..   PW(t)     =E=fw(t)*PWR;
DieselGener(t)..   ConsD(t)  =E=0.025*PDGR+0.275*PDG(t);
StateB(t)$(ord(t)>1).. SOC(t)=E=SOC(t-1)+hCh*PCB(t-1)-PDB(t-
1)/hDCh;
InitialStateBat..  SOC('1')  =E=SOC('72');
UpperDG(t)..       PDG(t)    =L=PDGR;
UpperB1(t)..       SOC(t)    =L=SOCR;
Lower1(t)..        SOC(t)    =G=0.2*SOCR;
UpperB2(t)..       PCB(t)    =L=(SOCR/10);
UpperB3(t)..       PDB(t)    =L=(SOCR/10);
UpperB4(t)..       PCB(t)    =L=1000*phi(t);
UpperB5(t)..       PDB(t)    =L=1000*epsilon(t);
LogicalB(t)..      phi(t) + epsilon(t) =L= 1;
CostCalc..  AnnualCost=E=CCF *(
 cS*PSR *(lamdaS + FinstS/100) +
 cW*PWR *(lamdaW + FinstW/100) +
cDG*PDGR*(lamdaDG+FinstDG/100)+
 cB*SOCR*(lamdaB + FinstB/100))
+(MS /100)*cS  *PSR +(MW /100)*cW  *PWR
+(MDG/100)*cDG *PDGR               +(MB /100)*cB  *SOCR
+ SUM(t,PG(t)*cel+ConsD(t)*cD)*365/3;
```

(continued)

Table 7.9   (continued)

```
UCostCalc..UnitAnnualCost=E=AnnualCost/((365/3)*SUM(t,PD(t)));
CO2emissions.. TotalCO2eq =E= SUM( t,
PG(t)*CEF_El+PS(t)*CEF_S+PW(t)*CEF_W +ConsD(t)*CEF_Dsl)*365/3;
PerCentCO2.. CO2PerCentInitial=E=100*TotalCO2eq/TotalCO2_Grid;

MODEL EnergyGrid /ALL/;
OPTIONS MIP=CPLEX;
* USE THIS OBJECTIVE TO DETERMINE BEST ECONOMIC PERFORMANCE
SOLVE EnergyGrid USING MIP MINIMIZING AnnualCost;

$ontext
* USE THIS OBJECTIVE TO DETERMINE BEST ENVIRONMENTAL
PERFORMANCE
SOLVE EnergyGrid USING MIP MINIMIZING CO2PerCentInitial;
$offtext
```

diesel generator, or energy storage facility is required. The CO_{2eq} emitted is 373.7 t CO_{2eq}/y (see the GAMS listing for the emissions estimation details). The optimal operation is presented in Fig. 7.19.

The GAMS model of Table 7.9 also determines the total CO_{2eq} (equivalent CO_2) emitted for the satisfaction of the electricity demand of the public building under study. This is calculated at the current state where electricity from a national grid is used exclusively to satisfy demand. The carbon emission factor for the electricity supplied by the national grid has been assumed to be $CEF_{El} = 0.75$ kg CO_{2eq}/kWh. The emission factors for the renewable energy resources reported in the literature vary over two orders of magnitude, and average (and possibly optimistic) values have been used ($CEF_W = 0.035$ kg CO_{2eq}/kWh, $CEF_S = 0.050$ kg CO_{2eq}/kWh). In the cost optimal solution, the CO_{2eq} emissions are reduced from 719.96 t CO_{2eq}/y to 373.7 t CO_{2eq}/y. Although this corresponds to an impressive 48.1% reduction, further reduction is possible. Using the emitted CO_{2eq} as the objective function to minimize results in the solution shown in Fig. 7.20 which is using a wind turbine with rated capacity of 258.9 kW and a battery bank with rated storage capacity of 2314.5 kWh. The idea is to overdesign the wind turbine and store any surplus electrical energy into the battery. The overall annual cost is now $221,320 that corresponds to a unitary cost of 0.231 $/kWh (60% increase). However, the reduction in the CO_2 emissions is impressive as in the final solution, only 35.8 t CO_{2eq}/y are emitted achieving an almost 95% emissions reduction. Note that solar PV system is not selected in any of the two alternative solutions.

Needless to say, all parameters that have been used in this linear model are only approximations of the real world. However, the reader can adapt these numbers to any case study and thus obtain results that are closer to reality for any particular case study. It is nonetheless important for the reader to develop

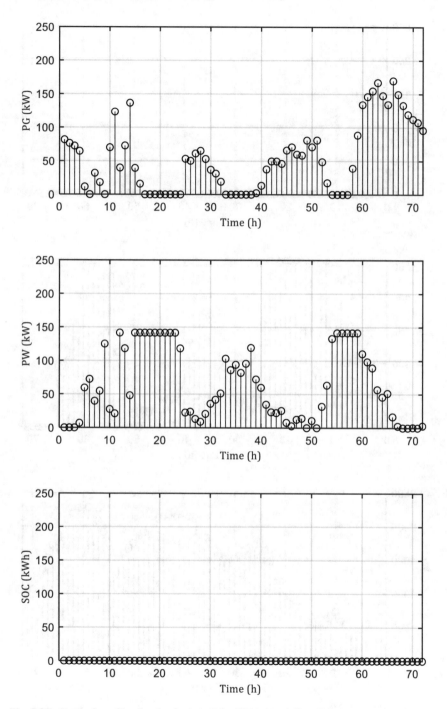

Fig. 7.19 Optimal profiles for the design of the RES system for minimum cost

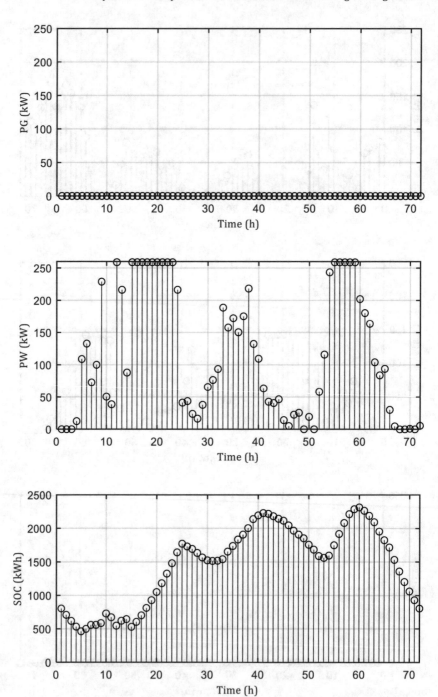

Fig. 7.20 Optimal profiles for the design of the RES system for minimum CO_2 emissions

a strategy to study the sensitivity of the results obtained to the parameters of the model. This can be done, in the simplest approach, by examining any one of the parameters in isolation from the others, i.e., study the variation of the optimal solution while varying one parameter at the time (and keeping all others constant at their nominal values). Alternatively, if sufficient information is available, we may postulate statistical distributions for the parameters and then perform Monte Carlo simulations to generate the statistics of the optimal solution. This will give us invaluable information on the main characteristics of the optimal solutions and on the parameters that are most important.

7.7   Metabolic Flux Analysis

Metabolic flux analysis (MFA) is a powerful technique for, among others, the determination of the fluxes through various biochemical reaction networks. MFA can, therefore, offer assistance to studies of metabolite production. The basis of MFA is a stoichiometric model that describes the biochemistry of a microorganism. The stoichiometric model is developed based on a postulated set of biochemical reactions relating consumed substrates to metabolic products. As this reaction network can be extremely complex, a subset of the most relevant reactions is usually considered. Using the assumption of the pseudo-steady state for the concentrations of the intracellular metabolites, mass balancing around each intracellular metabolite is performed resulting in a set of linear equality constraints:

$$\mathbf{S} \cdot \begin{bmatrix} \mathbf{v} \\ \mathbf{r} \end{bmatrix} = \mathbf{0} \tag{7.45}$$

where $\mathbf{v} \in R^n$ is the vector of the intracellular reaction rates (fluxes) and $\mathbf{r} \in R^p$ is the vector of the (extracellular) metabolite consumption or production rates. $\mathbf{S} \in R^{m \cdot (n+p)}$ is the stoichiometric matrix (which has as many rows as there are intracellular metabolites and as many columns as there are fluxes).

  To present a simple example that will clarify the idea of the stoichiometric matrix and the fluxes in a reaction network, we consider the example shown in Fig. 7.21. In this example, three extracellular metabolites (A, P_1, P_2) and four intracellular metabolites (B, C, D & E) are considered. If we assume pseudo-steady state for the intracellular metabolites, then we may write one material balance for each intracellular metabolite:

$$\text{B:} \quad r_1 - v_1 = 0 \tag{7.46a}$$

$$\text{C:} \quad 0.5\, v_1 - v_2 - v_3 = 0 \tag{7.46b}$$

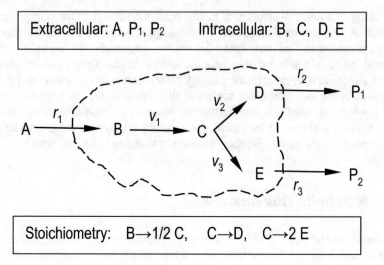

Fig. 7.21 A simple metabolic network

$$D: \quad v_2 - r_2 = 0 \tag{7.46c}$$
$$E: \quad 2v_3 - r_3 = 0 \tag{7.46d}$$

To make the presentation compact, we may define the intracellular fluxes v and the "extracellular" fluxes r:

$$v = \begin{bmatrix} v_1 \\ v_2 \\ v_3 \end{bmatrix}, \quad r = \begin{bmatrix} r_1 \\ r_2 \\ r_3 \end{bmatrix}$$

and Eq. (7.46) can be written as:

$$\begin{bmatrix} -1 & 0 & 0 & +1 & 0 & 0 \\ 0.5 & -1 & -1 & 0 & 0 & 0 \\ 0 & +1 & 0 & 0 & -1 & 0 \\ 0 & 0 & +2 & 0 & 0 & -1 \end{bmatrix} \cdot \begin{bmatrix} v \\ r \end{bmatrix} = 0 \tag{7.47}$$

which is of the general form given by (7.45) where $m = 4$, $n = 3$, and $p = 3$. As there are six unknown fluxes and four equations, in order to be able to calculate all fluxes, we only need to measure two fluxes. If, for instance, we can measure r_1 and r_2 and assume that $r_1 = 3$, $r_2 = 1$, then from (7.46a) $v_1 = 3$; from (7.46c) $v_2 = 1$; from (7.46b) $v_3 = 0.5$; and finally from (7.46d) $r_3 = 1$.

As in biochemical reaction networks the number of metabolites is significantly less than the number of fluxes, it is unlikely that sufficient measurements will be available to elucidate all internal fluxes. In these cases, optimization can be a useful alternative to obtain an estimate of the fluxes. What is needed is an appropriate and plausible objective function which must be a function of the fluxes and as many as possible additional equality or inequality constraints to reduce the space of feasible solutions. If a linear objective is selected, then the overall optimization problem is a LP problem of the following general form:

$$\min_{v,r} c^T v + d^T r$$

$$\text{s.t.}$$

$$S \cdot \begin{bmatrix} v \\ r \end{bmatrix} = 0 \tag{7.48}$$

$$\begin{bmatrix} v^L \\ r^L \end{bmatrix} \leq \begin{bmatrix} v \\ r \end{bmatrix} \leq \begin{bmatrix} v^U \\ r^U \end{bmatrix}$$

We will consider a relatively complex metabolic network. We will study the metabolic network of *Basfia succiniciproducens*, a well-known microorganism with high succinate production capacities (bio-succinate is among the most promising platform chemicals). In Tables 7.10 and 7.11, we present the basic metabolic reactions and metabolites considered (adapted from Becker et al., *Biotechnology and Bioengineering*, 110(11), p. 3013, 2013 & Kim et al., *Biotechnology and Bioengineering*, 97(4), p. 657, 2007). The network is also represented graphically in Fig. 7.22. The metabolic network consists of 47 metabolites, and the quasi-steady-state assumption is applicable for 35 of them. The fluxes involved are 46.

To demonstrate the potential use of this metabolic network model for *B. succiniciproducens*, we solve the optimization model (7.48) by setting a basis for the flux of glucose and with the following objective function:

$$\max_{v,r} r_{\text{biomass}}$$

The implementation in GAMS is presented in Table 7.12.

It is important to note that in the GAMS implementation, we have (based on information provided in the references given above and in Nelson & Cox, *Lehninger Principles of Biochemistry*, 4th ed., Freeman, 2005):

- Categorize all reactions as reversible (unbounded) or irreversible (flux ≥ 0)
- Categorize metabolites as intracellular (zero accumulation or at pseudo-steady state) and extracellular (consumed with negative flux or produced with positive flux)
- Implemented Formulation (7.48) with (7.45) written in the following convenient form:

Table 7.10 The reduced metabolic network of *Basfia succiniciproducens*

Reaction#	Reactants						Products				
Assimilation											
1.	GLC(e)	+	2 ATP			⇑	G6P	+	2ADP		
2.	SO4(e)	+	3NADPH	+	4ATP	⇑	H2S	+	3NADP	+	4ADP
Pentose phosphate T (p. 552 Lehninger)											
3.	G6P	+	NADP			⇑	6PG	+	NADPH		
4.	6PG	+	NADP			⇑	RIB5P	+	NADPH	+	CO₂
5.	RIB5P					⇕	XYL5P				
6.	RIB5P					⇕	RIBO5P				
7.	S7P	+	GAP			⇕	RIBO5P	+	XYL5P		
8.	S7P	+	GAP			⇕	E4P	+	F6P		
9.	F6P	+	GAP			⇕	E4P	+	XYL5P		
Glycolysis/gluconeogenesis (p. 524 Lehninger)											
10	G6P					⇕	F6P				
11	F6P	+	ATP			⇑	FBP	+	ADP		
12	FBP					⇑	F6P				
13	FBP					⇕	GAP	+	DAHP		
14	DAHP					⇕	GAP				
15	GAP	+	NAD			⇕	PG13	+	NADH		
16	PG13	+	ADP			⇑	PG3	+	ATP		
17	PG3					⇕	PG2				
18	PG2					⇕	PEP				
19	PEP	+	ADP			⇑	PYR	+	ATP		
20	PYR	+	NAD	+	CoA	⇑	AcCoA	+	NADH	+	CO₂

TCA cycle (p. 607 Lehninger)											
21	AcCoA	+	OAA			⇑	CIT	+	CoA		
22	CIT					⇕	ICIT				
23	ICIT	+	NADP			⇑	AKG	+	NADPH	+	CO$_2$
24	AKG	+	NAD	+	CoA	⇑	SUCC-CoA	+	NADH	+	CO$_2$
25	SUCC-CoA	+	ADP			⇕	SUCC	+	ATP	+	CoA
26	SUCC	+	MK			⇕	FUM	+	MKH		
27	FUM					⇕	MAL				
28	MAL	+	NAD			⇕	OAA	+	NADH		
Pyruvate metabolism											
29	PEP	+	CO$_2$			⇑	OAA				
30	PEP	+	CO$_2$	+	ADP	⇑	OAA	+	ATP		
31	MAL	+	NADP			⇑	PYR	+	NADPH	+	CO$_2$
32	OAA					⇑	PYR	+	CO$_2$		
Fermentation pathways											
33	PYR	+	NADH			⇕	LAC	+	NAD		
34	PYR	+	CoA			⇕	FORM	+	AcCoA		
35	FORM	+	NAD			⇑	NADH	+	CO$_2$		
36	AcCoA					⇕	AcP	+	CoA		
37	AcP	+	ADP			⇕	ACE	+	ATP		
38	AcP					⇑	ACE				
39	AcCoA	+	NADH			⇑	AcAld	+	NAD	+	CoA
40	AcAld	+	NADH			⇑	EtOH	+	NAD		
41	AcAld	+	NAD			⇑	ACE	+	NADH		
Energy metabolism											
43	3 NADH	+	3 NADP	+	ATP	⇑	3 NAD	+	3NADPH	+	ADP
44	NADPH	+	NAD			⇑	NADP	+	NADH		
45	NADH	+	MK			⇑	NAD	+	MKH		
46	ATP					⇑	ADP	+	ATPmain		

(continued)

Table 7.10 (continued)

Reaction#	Reactants			Products
Biomass formation (reaction # 42)		1.724	+	OAA
		1.221	+	PG3
		0.853	+	RIBO5P
		36.655	+	ATP
		15.754	+	NADPH
		9.284	+	NH3
		3.094	+	NAD
		2.887	+	AcCoA
		2.745	+	PYR
		1.186	+	AKG
		0.223	+	E4P
		0.495	+	PEP
		0.091	+	H2S
		0.071	+	F6P
		0.205	+	G6P
		0.129	⇑	GAP
		1.000		Biomass
		2.887	+	CoA
		15.754	+	NADP
		3.094	+	NADH
		36.655	+	ADP
		1.381	+	CO2

Adapted from Becker et al., *Biotechnology and Bioengineering*, 110(11), p. 3013, 2013

Table   7.11   The   reduced   set   of   metabolites   for   the   metabolic   network   of
B. succiniciproducens

Metabolites for which pseudo-steady state is assumed			
ATP	Adenosine triphosphate	$C_{10}H_{16}O_{13}P_3N_5$	1
ADP	Adenosine diphosphate	$C_{10}H_{15}O_{10}P_2N_5$	2
NAD	Nicotinamide adenine dinucleotide	$C_{21}H_{27}N_7O_{14}P_2$	3
NADH			4
NADP	Nicotinamide adenine dinucleotide phosphate	$C_{21}H_{29}N_7O_{17}P_3$	5
NADPH			6
MK	Menaquinone (coenzyme Q)	$C_{59}H_{90}O_4$	7
MKH	Menaquinol (reduced form of coenzyme Q)		8
G6P	Glucose-6-phosphate	$C_6H_{13}O_9P$	9
6PG	6-phosphogluconate	$C_6H_{13}O_{10}P$	10
RIB5P	Ribulose 5-phosphate	$C_5H_{11}O_8P$	11
XYL5P	D-Xylulose 5-phosphate	$C_5H_{11}O_8P$	12
RIBO5P	D-ribose 5-phosphate	$C_5H_{11}O_8P$	13
GAP	Glyceraldehyde 3-phosphate	$C_3H_7O_6P$	14
S7P	Sedoheptulose 7-phosphate	$C_7H_{15}O_{10}P$	15
E4P	Erythrose 4-phosphate	$C_4H_9O_7P$	16
F6P	Fructose 6-phosphate	$C_6H_{13}O_9P$	17
FBP	Fructose 1,6-bisphosphate	$C_6H_{14}O_{12}P_2$	18
DAHP	Dihydroxyacetone phosphate	$C_3H_7O_6P$	19
PG13	1,3-Bisphosphoglycerate	$C_3H_8O_{10}P_2$	20
PG3	3-Phosphoglycerate	$C_3H_7O_7P$	21
PG2	2-Phosphoglycerate	$C_3H_7O_7P$	22
PEP	Phosphoenolpyruvate	$C_3H_5O_6P$	23
PYR	Pyruvate	$C_3H_4O_3$	24
AcCoA	Acetyl-CoA	$C_{23}H_{38}N_7O_{17}P_3S$	25
OAA	Oxaloacetate	$C_4H_4O_5$	26
CIT	Citrate	$C_6H_8O_7$	27
ICIT	Isocitrate	$C_6H_8O_7$	28
AKG	a-Ketoglutarate	$C_5H_6O_5$	29
SUCC-CoA	Succinyl-CoA	$C_{25}H_{40}N_7O_{19}P_3S$	30
FUM	Fumarate	$C_4H_4O_4$	31
MAL	Malate	$C_4H_6O_5$	32
AcP			33
AcAld			34
H2S			35
Metabolites with positive fluxes (products)			
BIOMASS			36
ACE	Acetate	$C_2H_4O_2$	37
LAC	Lactate	$C_3H_6O_3$	38
FORM	Formate	CH_2O_2	39
EtOH	Ethanol	C_2H_6O	40
SUCC	Succinate	$C_4H_6O_4$	41

(continued)

Table 7.11 (continued)

ATPmain	ATP cons for maintenance		42
Metabolites with negative fluxes (consumed substrates)			
GLC	Glucose	$C_6H_{12}O_6$	43
CO2	Carbon dioxide	CO_2	44
NH3	Ammonia	NH_3	45
SO4		SO_4	46
CoA	CoA	$C_{21}H_{36}N_7O_{16}P_3S$	47

Adapted from Becker et al., *Biotechnology and Bioengineering*, 110(11), p. 3013, 2013

$$\begin{bmatrix} S_i & 0 \\ S_e & -I \end{bmatrix} \cdot \begin{bmatrix} v \\ r \end{bmatrix} = 0$$

i.e., the stoichiometric matrix has been partitioned in a part involving only internal fluxes and a part involving both internal and external fluxes. This is not the only option available but is handy for this case study.

The results obtained using the GAMS model of Table 7.12 are presented in Fig. 7.23. The continuous line represents the maximum theoretical yield of the microorganism. All experimental points are expected to lie inside the feasible region. Maximum theoretical yields can also be obtained from the figure.

7.8   Optimal Design of Proportional-Integral-Derivative (PID) Controllers

Classical PID controllers have been used for almost one century in the chemical industry and remain the most widely used type of controllers. Their design is normally performed through rules of thumb that are the result of accumulated experience and innovative research work. More recently, there has been a lot of interest in designing optimal PID controllers through classical optimization formulations. These methodologies are usually restricted by the type of the objective function that can be used. Classical, least squares objectives are normally implemented as they facilitate the analytic or numerical solution of the resulting optimization problem. In addition, apart from the requirement for the closed loop system to be stable, there is an additional requirement that the system possesses robustness to variability in its parameters. In other words, the parameters of the model obtain nominal values which are obtained through classical process identification experiments (such as step response or frequency response tests). However, the parameters of the model may vary due to changes in the environment or operating point or through ageing. The requirement is for the closed loop system to be stable and also have acceptable

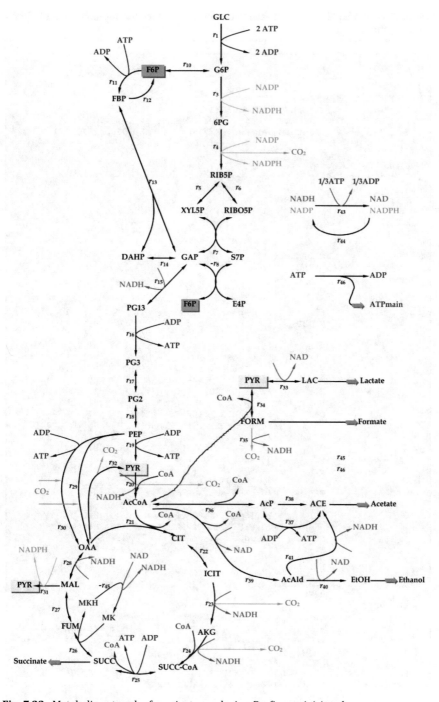

Fig. 7.22 Metabolic network of succinate producing *Basfia succiniciproducens*

Table 7.12 GAMS program for the analysis of the metabolic capabilities of *Basfia succiniciproducens*

```
$ TITLE METABOLIC FLUX ANALYSIS OF BASFIA SUCCINICIPRODUCENS
$ONTEXT
DATA TAKEN FROM
1. Becker et al., BIOTECHNOLOGY&BIOENGINEERING,110(11),p.3013,2013
2. Kim et al,     BIOTECHNOLOGY & BIOENGINEERING,97( 4),p.657, 2007
3. Ladakis et al.,BIOCHEMICAL ENGINEERING JOURNAL,137,p.262, 2018
$OFFTEXT

SET iter /1*100/;
SET  m "Metabolites"
/
* NOT CROSSING THE MEMBRANE
                ATP      # 1 Adenosine triphosphate C10 H16 O13 P3 N5
                ADP      # 2 Adenosine diphosphate
                NAD      # 3
                NADH     # 4
                NADP     # 5
                NADPH    # 6
                MK       # 7
                MKH      # 8
                G6P      # 9 Glucose-6-phosphate    C6  H13 O9  P
                6PG      #10 6-phosphogluconate     C6  H13 O10 P
                RIB5P    #11 Ribulose 5-phosphate   C5  H11 O8  P
                XYL5P    #12 D-XYLULOSE 5-PHOSPHATE  C5  H11 O8  P
                RIBO5P   #13 D-RIBOSE 5-PHOSPHATE    C5  H11 O8  P
                GAP      #14 GLYCERALDEHYDE 3-PHOSP  C3  H7  O6  P
                S7P      #15 SEDOHEPTULOSE 7-PHOSPH  C7  H15 O10 P
                E4P      #16 ERYTHROSE 4-PHOSPHATE   C4  H9  O7  P
                F6P      #17 FRUCTOSE 6-PHOSPHATE    C6  H13 O9  P
                FBP      #18 FRUCTOSE 1 6 BISPHOSPH  C6  H14 O12 P2
                DAHP     #19 DIHYDROXYACETONE PHOSP  C3  H7  O6  P
                PG13     #20 1 3 BISPHOSPHOGLYCERAT  C3  H8  O10 P2
                PG3      #21 3 PHOSPHOGLYCERATE      C3  H7  O7  P
                PG2      #22 2 PHOSPHOGLYCERATE      C3  H7  O7  P
                PEP      #23 PHOSPHOENOLPYRUVATE     C3  H5  O6  P
                PYR      #24 PYRUVATE                C3  H4  O3
                AcCoA    #25 Acetyl-CoA              C23 H38 O17 P3 S N7
                OAA      #26 OXALOACETATE            C4  H4  O5
                CIT      #27 CITRATE                 C6  H8  O7
                ICIT     #28 ISOCITRATE              C6  H8  O7
                AKG      #29 a-KETOGLUTARATE         C5  H6  O5
                SUCCCoA  #30 SUCCINYL-CoA            C25 H40 O19 P3 S N7
                FUM      #31 FUMARATE                C4  H4  O4
                MAL      #32 MALATE                  C4  H6  O5
                AcP      #33
                AcAld    #34
                H2S      #35
*PRODUCED
                Biomass  #36 Biomass
                ACE      #37 ACETATE                 C2  H4  O2
                LAC      #38 LACTATE                 C3  H6  O3
                FORM     #39 FORMATE                 C   H2  O2
                EtOH     #40 ETHANOL                 C2  H6  O
                SUCC     #41 SUCCINATE               C4  H6  O4
                ATPmain  #42 ATP FOR MAINTENANCE
*CONSUMED
                GLC      #43 Glucose                 C6  H12 O6
                CO2      #44 CARBON DIOXIDE          C       O2
                NH3      #45
```

(continued)

Table 7.12   (continued)

```
                    SO4     #46
                    CoA     #47 CoA                              C21 H36 O16 P3 S
N7/;

* METABOLITES NOT CROSSING THE MEMBRANE - PSEUDO STEADY STATE
SET i(m) /          ATP, ADP, NAD, NADH, NADP, NADPH, MK, MKH, G6P, 6PG,
                    RIB5P, XYL5P, RIBO5P, GAP, S7P, E4P, F6P, FBP, DAHP,
                    PG13, PG3, PG2, PEP, PYR, AcCoA, OAA, CIT, ICIT, AKG,
                    SUCCCoA, FUM, MAL, AcP, AcAld, H2S /;
* METABOLITES CROSSING THE MEMBRANE
* p r o d u c e d
SET j(m) /          Biomass, ACE, LAC, FORM, EtOH, SUCC, ATPmain,
* c o n s u m e d
                    GLC, CO2, NH3, SO4, CoA /;
DISPLAY m,i,j;
SET Flux /v1*v46/; ALIAS(Flux,n);
SET reversible_Flux(n)
/v5*v10,v13*v15,v17,v18,v22,v25*v28,v33,v34,v36,v37/;
SET irreversible_Flux(n)
/v1*v4,v11*v12,v16,v19*v21,v23*v24,v29*v32,v35,v38*v46/;
ALIAS(  reversible_Flux,nr);
ALIAS(irreversible_Flux,nir);


* STOICHIOMETRY
TABLE S(m,n)
```

	v1	v2	v3	v4	v5	v6	v7	v8	v9	v10	v11	v12	v13	v14	v15	v16	v17
ATP	-2	-4	0	0	0	0	0	0	0	0	-1	0	0	0	0	+1	0
ADP	+2	+4	0	0	0	0	0	0	0	0	+1	0	0	0	0	-1	0
NAD	0	0	0	0	0	0	0	0	0	0	0	0	0	0	-1	0	0
NADH	0	0	0	0	0	0	0	0	0	0	0	0	0	0	+1	0	0
NADP	0	+3	-1	-1	0	0	0	0	0	0	0	0	0	0	0	0	0
NADPH	0	-3	+1	+1	0	0	0	0	0	0	0	0	0	0	0	0	0
MK	0	0	0	0	0	0	0	0	0	0	0	0	0	0	0	0	0
MKH	0	0	0	0	0	0	0	0	0	0	0	0	0	0	0	0	0
G6P	+1	0	-1	0	0	0	0	0	0	-1	0	0	0	0	0	0	0
6PG	0	0	+1	-1	0	0	0	0	0	0	0	0	0	0	0	0	0
RIB5P	0	0	0	+1	-1	-1	0	0	0	0	0	0	0	0	0	0	0
XYL5P	0	0	0	0	+1	0	+1	0	+1	0	0	0	0	0	0	0	0
RIBO5P	0	0	0	0	0	+1	+1	0	0	0	0	0	0	0	0	0	0
GAP	0	0	0	0	0	0	-1	-1	-1	0	0	0	+1	+1	-1	0	0
S7P	0	0	0	0	0	0	-1	-1	0	0	0	0	0	0	0	0	0
E4P	0	0	0	0	0	0	0	+1	+1	0	0	0	0	0	0	0	0
F6P	0	0	0	0	0	0	0	+1	-1	+1	-1	+1	0	0	0	0	0
FBP	0	0	0	0	0	0	0	0	0	0	+1	-1	-1	0	0	0	0
DAHP	0	0	0	0	0	0	0	0	0	0	0	0	+1	-1	0	0	0
PG13	0	0	0	0	0	0	0	0	0	0	0	0	0	0	+1	-1	0
PG3	0	0	0	0	0	0	0	0	0	0	0	0	0	0	0	+1	-1
PG2	0	0	0	0	0	0	0	0	0	0	0	0	0	0	0	0	+1
PEP	0	0	0	0	0	0	0	0	0	0	0	0	0	0	0	0	0
PYR	0	0	0	0	0	0	0	0	0	0	0	0	0	0	0	0	0
AcCoA	0	0	0	0	0	0	0	0	0	0	0	0	0	0	0	0	0
OAA	0	0	0	0	0	0	0	0	0	0	0	0	0	0	0	0	0
CIT	0	0	0	0	0	0	0	0	0	0	0	0	0	0	0	0	0
ICIT	0	0	0	0	0	0	0	0	0	0	0	0	0	0	0	0	0
AKG	0	0	0	0	0	0	0	0	0	0	0	0	0	0	0	0	0
SUCCCoA	0	0	0	0	0	0	0	0	0	0	0	0	0	0	0	0	0
FUM	0	0	0	0	0	0	0	0	0	0	0	0	0	0	0	0	0
MAL	0	0	0	0	0	0	0	0	0	0	0	0	0	0	0	0	0
AcP	0	0	0	0	0	0	0	0	0	0	0	0	0	0	0	0	0
AcAld	0	0	0	0	0	0	0	0	0	0	0	0	0	0	0	0	0
H2S	0	+1	0	0	0	0	0	0	0	0	0	0	0	0	0	0	0
Biomass	0	0	0	0	0	0	0	0	0	0	0	0	0	0	0	0	0
ACE	0	0	0	0	0	0	0	0	0	0	0	0	0	0	0	0	0
LAC	0	0	0	0	0	0	0	0	0	0	0	0	0	0	0	0	0

(continued)

Table 7.12 (continued)

FORM	0	0	0	0	0	0	0	0	0	0	0	0	0	0	0	0
EtOH	0	0	0	0	0	0	0	0	0	0	0	0	0	0	0	0
SUCC	0	0	0	0	0	0	0	0	0	0	0	0	0	0	0	0
ATPmain	0	0	0	0	0	0	0	0	0	0	0	0	0	0	0	0
GLC	−1	0	0	0	0	0	0	0	0	0	0	0	0	0	0	0
CO2	0	0	0	+1	0	0	0	0	0	0	0	0	0	0	0	0
NH3	0	0	0	0	0	0	0	0	0	0	0	0	0	0	0	0
SO4	0	−1	0	0	0	0	0	0	0	0	0	0	0	0	0	0
CoA	0	0	0	0	0	0	0	0	0	0	0	0	0	0	0	0

+	v18	v19	v20	v21	v22	v23	v24	v25	v26	v27	v28	v29	v30	v31	v32	v33	v34
ATP	0	+1	0	0	0	0	0	+1	0	0	0	0	+1	0	0	0	0
ADP	0	−1	0	0	0	0	0	−1	0	0	0	0	−1	0	0	0	0
NAD	0	0	−1	0	0	0	−1	0	0	0	−1	0	0	0	0	+1	0
NADH	0	0	+1	0	0	0	+1	0	0	0	+1	0	0	0	0	−1	0
NADP	0	0	0	0	0	−1	0	0	0	0	0	0	0	−1	0	0	0
NADPH	0	0	0	0	0	+1	0	0	0	0	0	0	0	+1	0	0	0
MK	0	0	0	0	0	0	0	0	−1	0	0	0	0	0	0	0	0
MKH	0	0	0	0	0	0	0	0	+1	0	0	0	0	0	0	0	0
G6P	0	0	0	0	0	0	0	0	0	0	0	0	0	0	0	0	0
6PG	0	0	0	0	0	0	0	0	0	0	0	0	0	0	0	0	0
RIB5P	0	0	0	0	0	0	0	0	0	0	0	0	0	0	0	0	0
XYL5P	0	0	0	0	0	0	0	0	0	0	0	0	0	0	0	0	0
RIBO5P	0	0	0	0	0	0	0	0	0	0	0	0	0	0	0	0	0
GAP	0	0	0	0	0	0	0	0	0	0	0	0	0	0	0	0	0
S7P	0	0	0	0	0	0	0	0	0	0	0	0	0	0	0	0	0
E4P	0	0	0	0	0	0	0	0	0	0	0	0	0	0	0	0	0
F6P	0	0	0	0	0	0	0	0	0	0	0	0	0	0	0	0	0
FBP	0	0	0	0	0	0	0	0	0	0	0	0	0	0	0	0	0
DAHP	0	0	0	0	0	0	0	0	0	0	0	0	0	0	0	0	0
PG13	0	0	0	0	0	0	0	0	0	0	0	0	0	0	0	0	0
PG3	0	0	0	0	0	0	0	0	0	0	0	0	0	0	0	0	0
PG2	−1	0	0	0	0	0	0	0	0	0	0	0	0	0	0	0	0
PEP	+1	−1	0	0	0	0	0	0	0	0	0	−1	−1	0	0	0	0
PYR	0	+1	−1	0	0	0	0	0	0	0	0	0	0	+1	+1	−1	−1
AcCoA	0	0	+1	−1	0	0	0	0	0	0	0	0	0	0	0	0	+1
OAA	0	0	0	−1	0	0	0	0	0	0	+1	+1	+1	0	−1	0	0
CIT	0	0	0	+1	−1	0	0	0	0	0	0	0	0	0	0	0	0
ICIT	0	0	0	0	+1	−1	0	0	0	0	0	0	0	0	0	0	0
AKG	0	0	0	0	0	+1	−1	0	0	0	0	0	0	0	0	0	0
SUCCCoA	0	0	0	0	0	0	+1	−1	0	0	0	0	0	0	0	0	0
FUM	0	0	0	0	0	0	0	0	+1	−1	0	0	0	0	0	0	0
MAL	0	0	0	0	0	0	0	0	0	+1	−1	0	0	−1	0	0	0
AcP	0	0	0	0	0	0	0	0	0	0	0	0	0	0	0	0	0
AcAld	0	0	0	0	0	0	0	0	0	0	0	0	0	0	0	0	0
H2S	0	0	0	0	0	0	0	0	0	0	0	0	0	0	0	0	0
Biomass	0	0	0	0	0	0	0	0	0	0	0	0	0	0	0	0	0
ACE	0	0	0	0	0	0	0	0	0	0	0	0	0	0	0	0	0
LAC	0	0	0	0	0	0	0	0	0	0	0	0	0	0	0	+1	0
FORM	0	0	0	0	0	0	0	0	0	0	0	0	0	0	0	0	+1
EtOH	0	0	0	0	0	0	0	0	0	0	0	0	0	0	0	0	0
SUCC	0	0	0	0	0	0	0	+1	−1	0	0	0	0	0	0	0	0
ATPmain	0	0	0	0	0	0	0	0	0	0	0	0	0	0	0	0	0
GLC	0	0	0	0	0	0	0	0	0	0	0	0	0	0	0	0	0
CO2	0	0	+1	0	0	+1	+1	0	0	0	0	−1	−1	+1	+1	0	0
NH3	0	0	0	0	0	0	0	0	0	0	0	0	0	0	0	0	0
SO4	0	0	0	0	0	0	0	0	0	0	0	0	0	0	0	0	0
CoA	0	0	−1	+1	0	0	−1	+1	0	0	0	0	0	0	0	0	−1

+	v35	v36	v37	v38	v39	v40	v41	v42	v43	v44	v45	v46
ATP	0	0	+1	0	0	0	0	−36.655	−1	0	0	−1
ADP	0	0	−1	0	0	0	0	+36.655	+1	0	0	+1
NAD	−1	0	0	0	+1	+1	−1	−3.094	+3	−1	+1	0

(continued)

Table 7.12   (continued)

NADH	+1	0	0	0	-1	-1	+1	+3.094	-3	+1	-1	0
NADP	0	0	0	0	0	0	0	+15.754	-3	+1	0	0
NADPH	0	0	0	0	0	0	0	-15.754	+3	-1	0	0
MK	0	0	0	0	0	0	0	0	0	0	-1	0
MKH	0	0	0	0	0	0	0	0	0	0	+1	0
G6P	0	0	0	0	0	0	0	-0.205	0	0	0	0
6PG	0	0	0	0	0	0	0	0	0	0	0	0
RIB5P	0	0	0	0	0	0	0	0	0	0	0	0
XYL5P	0	0	0	0	0	0	0	0	0	0	0	0
RIBO5P	0	0	0	0	0	0	0	-0.853	0	0	0	0
GAP	0	0	0	0	0	0	0	-0.129	0	0	0	0
S7P	0	0	0	0	0	0	0	0	0	0	0	0
E4P	0	0	0	0	0	0	0	-0.223	0	0	0	0
F6P	0	0	0	0	0	0	0	-0.071	0	0	0	0
FBP	0	0	0	0	0	0	0	0	0	0	0	0
DAHP	0	0	0	0	0	0	0	0	0	0	0	0
PG13	0	0	0	0	0	0	0	0	0	0	0	0
PG3	0	0	0	0	0	0	0	-1.221	0	0	0	0
PG2	0	0	0	0	0	0	0	0	0	0	0	0
PEP	0	0	0	0	0	0	0	-0.495	0	0	0	0
PYR	0	0	0	0	0	0	0	-2.745	0	0	0	0
AcCoA	0	-1	0	0	-1	0	0	-2.887	0	0	0	0
OAA	0	0	0	0	0	0	0	-1.724	0	0	0	0
CIT	0	0	0	0	0	0	0	0	0	0	0	0
ICIT	0	0	0	0	0	0	0	0	0	0	0	0
AKG	0	0	0	0	0	0	0	-1.186	0	0	0	0
SUCCCoA	0	0	0	0	0	0	0	0	0	0	0	0
FUM	0	0	0	0	0	0	0	0	0	0	0	0
MAL	0	0	0	0	0	0	0	0	0	0	0	0
AcP	0	+1	-1	-1	0	0	0	0	0	0	0	0
AcAld	0	0	0	0	+1	-1	-1	0	0	0	0	0
H2S	0	0	0	0	0	0	0	-0.091	0	0	0	0
Biomass	0	0	0	0	0	0	0	+1	0	0	0	0
ACE	0	0	+1	+1	0	0	+1	0	0	0	0	0
LAC	0	0	0	0	0	0	0	0	0	0	0	0
FORM	-1	0	0	0	0	0	0	0	0	0	0	0
EtOH	0	0	0	0	0	+1	0	0	0	0	0	0
SUCC	0	0	0	0	0	0	0	0	0	0	0	0
ATPmain	0	0	0	0	0	0	0	0	0	0	0	+1
GLC	0	0	0	0	0	0	0	0	0	0	0	0
CO2	+1	0	0	0	0	0	0	+1.381	0	0	0	0
NH3	0	0	0	0	0	0	0	-9.284	0	0	0	0
SO4	0	0	0	0	0	0	0	0	0	0	0	0
CoA	0	+1	0	0	+1	0	0	+2.887	0	0	0	0 ;

```
$ONTEXT
  v(42) is the biomass production and corresponds to g of biomass
        per mmol => biomass production per mol of glucose
        is given by v42 * 1000/|v(Glucose)|
$OFFTEXT

VARIABLES v(n) internal fluxes
          r(m) external fluxes
          obj;

EQUATIONS MAT_BAL_INTRA(i), MAT_BAL_EXTRA(j), OBJFUN;
MAT_BAL_INTRA(i).. sum(n,S(i,n)*v(n)) =e= 0;
MAT_BAL_EXTRA(j).. sum(n,S(j,n)*v(n)) =E= r(j);
OBJFUN.. obj =E= r('Biomass');

v.up(n)  = +200;  v.lo(n)  = -200; v.lo(nir)=    0;

* positive fluxes
r.lo('Biomass')=0; r.lo('ACE')=0;  r.lo('LAC')=0;  r.lo('FORM')=0;
```

(continued)

Table 7.12 (continued)

```
r.lo('EtOH')=0;      r.lo('SUCC')=0; r.lo('ATPmain')=0;
* negative fluxes
r.up('GLC')=0; r.up('CO2')=0; r.up('NH3')=0; r.up('SO4')=0;
r.up('CoA')=0;
* set biomass growth rate
r.fx('SUCC')=100 ;
* set basis for glucose consumption in mol glucose
r.fx('GLC')=-100 ;

MODEL BASFIA /ALL/;
SOLVE BASFIA USING LP MAXIMIZING obj;

FILE out /result.m/;

PUT out;

LOOP(iter, r.fx('SUCC') = 171*(ord(iter)-1)/(card(iter)-1);
           SOLVE BASFIA USING LP MAXIMIZING obj;
           PUT ord(iter):3:0,
           r.l('SUCC'):16:8,
           obj.l:16:8    /;
    );

PUTCLOSE out;
```

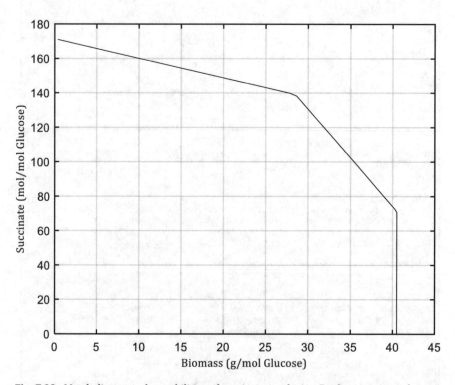

Fig. 7.23 Metabolic network capabilities of succinate producing *Basfia succiniciproducens*

performance not only for the nominal values of its parameters but for all potential values of its uncertain parameters.

In this section, we will consider the case of low-order mathematical models, commonly used to describe chemical process dynamics, such as:

First-order plus deadtime (FOPDT) model:

$$G(s; k, \tau, d) = \frac{Y(s)}{U(s)} = \frac{k}{\tau s + 1} e^{-ds} \qquad (7.49)$$

Integrating plus deadtime (IPDT) model:

$$G(s; k, d) = \frac{Y(s)}{U(s)} = \frac{k}{s} e^{-ds} \qquad (7.50)$$

Second-order plus deadtime (SOPDT) model:

$$G(s; k, \tau, \zeta, d) = \frac{Y(s)}{U(s)} = \frac{k}{\tau s^2 + 2\zeta \tau s + 1} e^{-ds} \qquad (7.51)$$

where k is the gain of the process, τ is the dominant time constant, ζ is the damping coefficient of the second-order system, and d is the deadtime (or time delay) of the process (see Kravaris & Kookos, *Understanding Chemical Process Dynamics and Control*, CUP, 2021, for the physical meaning of these parameters).

All these low-order models can be used to capture the dynamics of the majority of chemical process systems and can be stated in the time domain using the following state-space form with input delay:

$$\frac{d\mathbf{x}(t)}{dt} = \mathbf{A}\mathbf{x}(t) + \mathbf{b}u(t - d) \qquad (7.52a)$$

$$y(t) = \mathbf{c}\mathbf{x}(t) \qquad (7.52b)$$

where $\mathbf{A}$ is a constant matrix and $\mathbf{b}$ and $\mathbf{c}$ are constant vectors. For a FOPDT system, for instance, the state-space model obtains the following form:

$$\frac{dx(t)}{dt} = -\frac{1}{\tau}x(t) + \frac{k}{\tau}u(t - d) \qquad (7.53a)$$

$$y(t) = x(t) \qquad (7.53b)$$

Alternatively, the process systems dynamics can be stated in time domain using the following state-space form with output delay:

$$\frac{dx(t)}{dt} = \mathbf{A}x(t) + \mathbf{b}u(t) \qquad (7.54a)$$

$$y(t) = \mathbf{c}x(t - d) \qquad (7.54b)$$

For a FOPDT system:

$$\frac{dx(t)}{dt} = -\frac{1}{\tau}x(t) + \frac{k}{\tau}u(t) \qquad (7.55a)$$

$$y(t) = x(t - d) \qquad (7.55b)$$

The solution of the linear differential equation (7.54a) for given initial conditions $x(0)$ and constant input u can be obtained through several alternative methods with the simplest to be through the use of the integrating factor $e^{-\mathbf{A}t}$. We first multiply both sides of (7.54a) by the integrating factor:

$$e^{-\mathbf{A}t}\frac{dx(t)}{dt} - e^{-\mathbf{A}t}\mathbf{A}x(t) + e^{-\mathbf{A}t}\mathbf{b}u(t)$$

The last equation can be written as:

$$\frac{d(e^{-\mathbf{A}t}x(t))}{dt} = e^{-\mathbf{A}t}\mathbf{b}u(t)$$

We integrate the last equation between 0 and an arbitrary time t with constant input $u = u(0)$, to obtain:

$$\left[e^{-\mathbf{A}\theta}x(\theta)\right]_{\theta=0}^{\theta=t} = \left(\int_0^t e^{-\mathbf{A}\theta}d\theta \cdot \mathbf{b}\right) \cdot u$$

or:

$$x(t) = e^{\mathbf{A}t}x(0) + \left(\int_0^t e^{-\mathbf{A}(t-\theta)}d\theta \cdot \mathbf{b}\right) \cdot u$$

We may follow exactly the same approach to integrate the differential equation between time t and $t + T_s$ to obtain:

$$x(t + T_s) = e^{\mathbf{A}T_s}x(t) + \left(\int_t^{t+T_s} e^{-\mathbf{A}(t-\theta)}d\theta \cdot \mathbf{b}\right) \cdot u(t) \qquad (7.56)$$

where $u(t)$ is the input at time t which remains constant in the interval $[t, t + T_s)$. We will use the index n to denote the time point with the understanding that $t = n \cdot T_s$ and $t + T_s = (n + 1)T_s$. We may therefore express the analytic solution of the differential equation (7.54a) in the following discrete form:

$$x(n + 1) = A_d x(n) + b_d \cdot u(n) \tag{7.57a}$$

$$y(n) = c x(n - \delta) \tag{7.57b}$$

where $\delta = d/T_s$ and:

$$A_d = e^{A T_s}, \quad b_d = b \int_0^{T_s} e^{A(T_s - \theta)} d\theta \tag{7.58}$$

For the FOPDT system, for instance, $A = -1/\tau$, and $b = k/\tau$ (scalars), and performing the integration, we obtain:

$$A_d = e^{-\frac{T_s}{\tau}}, \quad b_d = k\left(1 - e^{-\frac{T_s}{\tau}}\right) \tag{7.59}$$

i.e., the discrete solution of the FOPDT system to a constant input is given by:

$$x(n + 1) = e^{-\frac{T_s}{\tau}} \cdot x(n) + k\left(1 - e^{-\frac{T_s}{\tau}}\right) \cdot u(n) \tag{7.60a}$$

$$y(n) = x(n - \delta) \tag{7.60b}$$

For the case of input delay, the equivalent discrete form of the dynamics is:

$$x(n + 1) = e^{-\frac{T_s}{\tau}} \cdot x(n) + k\left(1 - e^{-\frac{T_s}{\tau}}\right) \cdot u(n - \delta) \tag{7.61a}$$

$$y(n) = x(n\delta) \tag{7.61b}$$

We may use this description of the FOPDT system dynamics to develop optimal input sequences that minimize a specific objective function or measure of performance. If the objective function is linear, then the final problem is a LP problem. A linear objective function that can be used is the sum of absolute deviations of the output $y(t)$ from a reference output profile or set point y_{sp}:

$$\min_{x(n), y(n), u(n)} \sum_{n=1}^{N} |y_{sp} - y(n)| \tag{7.62}$$

where $t_f = N \cdot T_s$ is the final time of the simulation. The absolute values of the deviations are a more useful measure of system performance when compared to the squared deviations as the latter overemphasize large deviations and almost ignore small deviations. The main drawback of the objective function (7.62) is that it ignores the input variation or any bound or rate constraints on the input. We therefore can use an augmented objective function that penalizes large input variations:

$$\min_{x(n), y(n), u(n)} \sum_{n=1}^{N} |y_{sp} - y(n)| + \rho \cdot \sum_{n=1}^{N} |u(n) - u(n-1)| \qquad (7.63)$$

where ρ is a weighting factor.

The complete formulation of the analysis that has been presented up to this point is the following (note the way we calculate the absolute values of the deviations and the control variations):

$$\min_{x(n), y(n), u(n)} J(u(n)) = \left[\sum_{n=1}^{N} (e^+(n) + e^-(n)) + \rho \cdot \sum_{n=1}^{N} (\Delta u^+(n) + \Delta u^-(n)) \right] \cdot T_s$$

s.t.

$$\left.\begin{array}{l} x(n) = A_d \cdot x(n-1) + b_d \cdot u((n-1)) \\ y(n) = x(n - \delta) \end{array}\right\} \quad n = 1, 2, \ldots, N$$

$$(7.64)$$

$$\left.\begin{array}{l} y_{sp} - y(n) = e(n) = e^+(n) - e^-(n) \\ u(n) - u(n-1) = \Delta u(n) = \Delta u^+(n) - \Delta u^-(n) \end{array}\right\} \quad n = 1, 2, \ldots, N$$

$$\left.\begin{array}{l} \dfrac{\Delta u^+(n)}{T_s} \leq \Delta \dot{u}_{max} \\[2mm] \dfrac{\Delta u^-(n)}{T_s} \leq \Delta \dot{u}_{max} \end{array}\right\} \quad n = 1, 2, \ldots, N$$

$$x(0) = x(t = 0) = x_0$$

$$0 \leq e^+(n), e^-(n), \Delta u^+(n), \Delta u^-(n)$$

Note that the objective function has been multiplied by the sampling time T_s to approximate the continuous time integration.

Formulation (7.64) is an optimistic approach as no special structure has been assumed for the control law. Classical feedback control laws have the drawback that they react after detecting the deviation of the output from the reference signal or set point. This feedback structure is vital to the practical implementation but results in a suboptimal control variation when compared to the results of Formulation (7.64).

The GAMS implementation of Formulation (7.64) is presented in Table 7.13 for a FOPDT process with $\tau = d = 1$ s and $k = 1$. The results obtained are shown in Fig. 7.24. Note that the controller acts as a predictive controller and rumps

Table 7.13 GAMS program for the open-loop optimal control of a FOPDT system

```
$ TITLE FOPDT SYSTEM OPEN LOOP OPTIMAL CONTROL

SET        n time points /0*100/ ;

PARAMETER tau      process time constant            /1/
          d        delay time                       /1/
          k        process gain                     /1/
          ysp      set point value                  /1/
          Dumax    maximum variation of the control /1/
          umin     minimum value of control         /-2/
          umax     maximum value of control         /+2/
          rho      weighting factor                 /1/
          tf       final time                       /10/
          Ts       sampling time
          A_d      A matrix of the state space description
          b_d      b matrix of the state space description
          time(n)  time;
Ts       = tf/(card(n)-1);
A_d      = exp(-Ts/tau);
b_d      = (1-A_d)*k;
time(n)  = Ts*(ord(n)-1);

POSITIVE VARIABLES eplus(n), eminus(n), Duplus(n), Duminus(n);
VARIABLES x(n), u(n), y(n), ObjFun;

EQUATIONS Dynamics(n)     system state dynamics
          Output(n)       system output dynamics
          ABSerror(n)     absolute error
          ABSuvar(n)      absolute control variation
          Criterion       objective function;

Dynamics(n).. x(n) =E= A_d*x(n-1)+b_d*u(n-1);
Output(n)..    y(n) =E= x(n-ceil(d/Ts));
ABSerror(n).. ysp - y(n)    =E= eplus(n)  - eminus(n);
ABSuvar(n)..  u(n)- u(n-1) =E= Duplus(n) - Duminus(n);
Criterion..   ObjFun =E=
Ts*sum(n, (eplus(n)+eminus(n))+rho*(Duplus(n)+Duminus(n)));

Duplus.up(n) =+Dumax*Ts;
Duminus.up(n)=+Dumax*Ts;
u.up(n) = umax;
u.lo(n) = umin;
x.fx('0')=0;

MODEL FOPDT_OPEN_LOOP /ALL/;
OPTION LP=CPLEX;
SOLVE FOPDT_OPEN_LOOP USING LP MINIMIZING ObjFun;

FILE FOPDT_OL /FOPDT_OL.m/;
PUT  FOPDT_OL ;
LOOP(n,
PUT  time(n):10:4,x.l(n):12:8,y.l(n):12:8,u.l(n):12:6 /;
     );
PUTCLOSE FOPDT_OL;
```

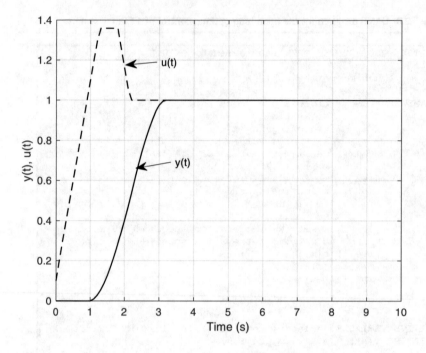

Fig. 7.24 Open-loop optimal control of a FOPDT system with $\tau = 1$, $k = 1$, and $d = 1$ with rate and bound constraints on the control (sampling time $T_s = 0.1$)

up starting from $t = 0$. The objective function achieved is the minimum possible for the problem specifications and is equal to $J = 2.3498$ for $T_s = 0.1$.

We now turn our attention to the closed loop formed when a Proportional-Integral (or PI) controller is used. The time domain description of the action performed by a PI controller is the following:

$$u(t) = k_c \left(e(t) + \frac{1}{\tau_I} \int_0^t e(\theta)d\theta \right) \tag{7.65}$$

where k_c (the controller gain) and τ_I (the integral time) are the two adjustable parameters of the PI controller. For $t = t + T_s$ we have that:

$$u(t + T_s) = k_c \left(e(t + T_s) + \frac{1}{\tau_I} \int_0^{t+T_s} e(\theta)d\theta \right)$$

$$= k_c \left(e(t + T_s) + \frac{1}{\tau_I} \int_0^t e(\theta)d\theta + \frac{1}{\tau_I} \int_t^{t+T_s} e(\theta)d\theta \right)$$

and by subtracting Eq. (7.65), we have that:

$$u(t + T_s) - u(t) = k_c \left(e(t + T_s) - e(t) + \frac{1}{\tau_I} \int\limits_{t}^{t+T_s} e(\theta) d\theta \right)$$

The input remains constant between t and $t + T_s$, and if assumed equal to its value at $t + T_s$, we obtain the discrete form of a PI controller (usually called the velocity form of a PI controller):

$$u(n + 1) = u(n) + k_c \left(e(n + 1) - e(n) + \frac{T_s}{\tau_I} e(n + 1) \right)$$

which can also be written as:

$$u(n) = u(n - 1) + \Delta u(n) \tag{7.66}$$

$$\Delta u(n) = k_c \left[\left(\frac{T_s}{\tau_I} + 1 \right) e(n) - e(n - 1) \right] \tag{7.67}$$

We modify Formulation (7.64) to postulate the mathematical programming formulation of the optimal PI controller design problem:

$$\min_{k_c, \tau_I, x(n), y(n), u(n)} J(u(n)) = \left[\sum_{n=1}^{N} (e^+(n) + e^-(n)) + \rho \cdot \sum_{n=1}^{N} (\Delta u^+(n) + \Delta u^-(n)) \right] \cdot T_s$$

s.t.

$$\left. \begin{array}{l} x(n) = A_d \cdot x(n - 1) + b_d \cdot u((n - 1)) \\ y(n) = x(n - \delta) \end{array} \right\} \quad n = 1, 2, \ldots, N$$

$$\tag{7.68}$$

$$\left. \begin{array}{l} y_{sp} - y(n) = e(n) = e^+(n) - e^-(n) \\ u(n) - u(n - 1) = \Delta u(n) = \Delta u^+(n) - \Delta u^-(n) \\ \Delta u(n) = k_c \left[\left(\frac{T_s}{\tau_I} + 1 \right) e(n) - e(n - 1) \right] \end{array} \right\} \quad n = 1, 2, \ldots, N$$

$$\left. \begin{array}{l} \dfrac{\Delta u^+(n)}{T_s} \leq \Delta \dot{u}_{max} \\ \dfrac{\Delta u^-(n)}{T_s} \leq \Delta \dot{u}_{max} \end{array} \right\} \quad n = 1, 2, \ldots, N$$

$$x(0) = x(t = 0) = x_0$$

$$0 \leq e^+(n), e^-(n), \Delta u^+(n), \Delta u^-(n), \tau_I$$

The GAMS implementation of Formulation (7.68) is presented in Table 7.14. The closed loop results obtained are shown in Fig. 7.25. The optimal controller

Table 7.14 GAMS program for the closed loop optimal PI control of a FOPDT system

```
$ TITLE FOPDT SYSTEM CLOSED LOOP OPTIMAL PI CONTROL
SET        n time points /0*300/ ;

PARAMETER tau       process time constant            /1/
          d         delay time                       /1/
          k         process gain                     /1/
          ysp       set point value                  /1/
          Dumax     maximum variation of the control /1/
          umin      minimum value of control         /-2/
          umax      maximum value of control         /+2/
          rho       weighting factor                 /1/
          tf        final time                       /30/
          Ts        sampling time
          A_d       A matrix of the state space description
          b_d       b matrix of the state space description
          time(n)   time;
Ts       = tf/((card(n)-1);
A_d      = exp(-Ts/tau);
b_d      = (1-A_d)*k;
time(n)  = Ts*(ord(n)-1);

POSITIVE VARIABLES eplus(n), eminus(n), Duplus(n), Duminus(n);
VARIABLES x(n), u(n), y(n), ObjFun;

POSITIVE VARIABLE invtau_I;
VARIABLES             k_c, e(n);

EQUATIONS Dynamics(n)     system state dynamics
          Output(n)       system output dynamics
          ABSerror(n)     absolute error
          ABSuvar(n)      absolute control variation
          Criterion       objective function
          Error(n)        error dynamics
          Control(n)      PI controller;

Dynamics(n).. x(n)          =E= A_d*x(n-1)+b_d*u(n-1);
Output(n)..   y(n)          =E= x( n-ceil(d/Ts) );
ABSerror(n).. ysp - y(n)    =E= eplus(n)   - eminus(n);
ABSuvar(n)..  u(n)- u(n-1)  =E= Duplus(n) - Duminus(n);
Criterion..   ObjFun        =E=
Ts*sum(n, (eplus(n)+eminus(n))+rho*(Duplus(n)+Duminus(n)));

Error(n)..    e(n)          =E= ysp - y(n);
Control(n)..  u(n)-u(n-1)   =E=k_c*((invtau_I*Ts+1)*e(n)-e(n-1));

Duplus.up(n) =+Dumax*Ts;
Duminus.up(n)=+Dumax*Ts;
u.up(n)  = umax;
u.lo(n)  = umin;
x.fx('0')=0;
k_c.l = k;
invtau_I.l =1/tau;
MODEL FOPDT_CLOSED_LOOP_PI /ALL/;
OPTION NLP=CONOPT4;
SOLVE FOPDT_CLOSED_LOOP_PI USING NLP MINIMIZING ObjFun;
```

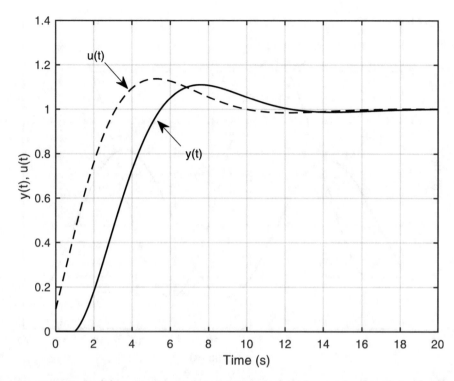

Fig. 7.25 Closed loop optimal control of a FOPDT system with $\tau = 1$, $k = 1$, and $d = 1$ with rate and bound constraints on the control (sampling time $T_s = 0.1$, optimal controller parameters $k_c = 0.06493670$ and $\tau_I = 0.18519845$)

parameters are $k_c = 0.06493670$ and $\tau_I = 0.18519845$. The objective function achieved is $J_{PI} = 3.857$ for $T_s = 0.1$. The objective function of the optimal PI controller is 64% larger than the optimal open-loop control. This performance degradation is due to the restriction that the controller obtains a specific structure (that of a PI controller). The performance degradation is also obvious by comparing Figs. 7.24 and 7.25. The optimal open-loop controller drives the output at the desired value in only 3 s, while the PI controller needs more than 10 s. The open-loop controller has, however, a more aggressive input variation.

What we have achieved is important: we have developed a mathematical programming formulation that can be used to select the optimal parameters of classical PI controllers for any FOPDT process. However, FOPDT models are only (crude) approximations of real processes that exhibit complex, high-order, and time-varying dynamics. The meaning is that although we have developed the optimal PI controller for the nominal values of the FOPDT model, there is no guarantee that the performance of the closed loop system will be acceptable or, at least stable, when the parameters k, τ, and d vary.

To demonstrate how vulnerable our "optimal" design can be to parameter variation, we simulate in MATLAB® the closed loop system when the

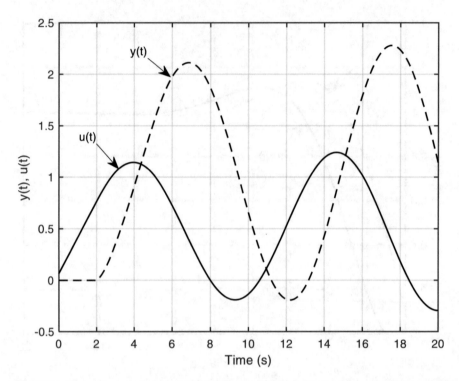

Fig. 7.26 Closed loop control of a FOPDT system with $\tau = 1$, $k = 2$, and $d = 2$ with rate and bound constraints on the control (sampling time $T_s = 0.1$, controller parameters $k_c = 0.06493670$ and $\tau_I = 0.18519845$ which are optimal for $\tau = 1$, $k = 1$ and $d = 1$)

parameters of the FOPDT system become $\tau = 1$ (unchanged), $k = 2$ (100% increase), and $d = 2$ (100% increase). The results are shown in Fig. 7.26. The output of the system exhibits oscillations of increased amplitude which is a clear indication of unstable closed loop behavior. Therefore, we need to devise a method to make our "optimal" design insensitive to parameter variations or robust, as it is usually referred to in Control Theory.

Incorporating robustness into the design of optimal control systems has a long history in Control Systems Theory, and many alternative methodologies for achieving that are currently available. The classical idea of achieving pre-specified gain and phase margins at the design of the controller for the nominal system is a methodology that is particularly appropriate for low-order process systems models.

For the definition and the practical implications of achieving specific values for the phase margin (PM) and the gain margin (GM), the reader is referred to the relevant literature (see, for instance, Kravaris & Kookos, *Understanding Process Dynamics and Control*, CUP, 2021). For our optimization approach, it is sufficient to remind to the reader that the PM and GM can be calculated through the solution of a set of four nonlinear equations in four unknowns.

For a FOPDT system and a PI controller, we have that:

$$L(s) = k_c \frac{(1 + \tau_I s)}{\tau_I s} \cdot \frac{k}{(\tau s + 1)} e^{-ds} \tag{7.69}$$

The magnitude of $L(s)$ is:

$$|L(\omega; k_c, \tau_I, k, \tau, d)| = \frac{kk_c}{\omega \tau_I} \frac{\sqrt{1 + \omega^2 \tau_I^2}}{\sqrt{1 + \omega^2 \tau^2}} \tag{7.70}$$

and the phase:

$$\arg\{L(\omega; k_c, \tau_I, k, \tau, d)\} = -\frac{\pi}{2} + \tan^{-1}(\omega \tau_I) - \omega d - \tan^{-1}(\omega \tau) \tag{7.71}$$

The phase crossover frequency ω_p is the frequency at which the phase of the loop transfer function $L(s) = G_c(s)G(s)$ ($G_c(s)$ is the transfer function of the controller and $G(s)$ is the transfer function of the process) becomes equal to $-180°$.

$$\frac{\pi}{2} + \tan^{-1}(\omega_p \tau_I) - \omega_p d - \tan^{-1}(\omega_p \tau) = 0 \tag{7.72a}$$

The gain crossover frequency ω_g is defined as the frequency at which the magnitude of the loop transfer function becomes equal to 1:

$$\frac{kk_c}{\omega_g \tau_I} \frac{\sqrt{1 + \omega_g^2 \tau_I^2}}{\sqrt{1 + \omega_g^2 \tau^2}} = 1 \tag{7.72b}$$

We finally have the equations for the GM and the PM:

$$\frac{1}{GM} = \frac{kk_c}{\omega_p \tau_I} \frac{\sqrt{1 + \omega_p^2 \tau_I^2}}{\sqrt{1 + \omega_p^2 \tau^2}} \tag{7.72c}$$

$$PM = \frac{\pi}{2} + \tan^{-1}(\omega_g \tau_I) - \omega_g d - \tan^{-1}(\omega_g \tau) \tag{7.72d}$$

It is important to note that in Eqs. (7.72a)–(7.72d) appear four unknowns: PM, GM, ω_g, and ω_p. These unknowns are functions of the FOPDT model parameters (k, τ, d) and the PI controller parameters (k_c, τ_I). To ensure that the closed loop system remains stable, it is sufficient to ensure that, for all potential values of the FOPDT model parameters $k_s, \tau_s, d_s, s = 1,2,3,\ldots,N_s$, PM (k_s, τ_s, d_s) and GM(k_s, τ_s, d_s) satisfy the constraints:

$$0 \leq \mathrm{PM}(k_s, \tau_s, d_s), \quad s = 1, 2, \ldots, N_s \qquad (7.73\mathrm{a})$$

$$1 \leq \mathrm{GM}(k_s, \tau_s, d_s), \quad s = 1, 2, \ldots, N_s \qquad (7.73\mathrm{b})$$

Satisfaction of constraints (7.73a) and (7.73b) ensures closed loop stability. However, apart from ensuring stability, PM and GM can be used to ensure indirectly acceptable performance. This is achieved by setting lower bounds for the PM and the GM:

$$\mathrm{PM}_{\min} \leq \mathrm{PM}(k_s, \tau_s, d_s), \quad s = 1, 2, \ldots, N_s \qquad (7.74\mathrm{a})$$

$$\mathrm{GM}_{\min} \leq \mathrm{GM}(k_s, \tau_s, d_s), \quad s = 1, 2, \ldots, N_s \qquad (7.74\mathrm{b})$$

The minimum values are normally selected to be $\mathrm{GM}_{\min} = 1.7$ and $\mathrm{PM}_{\min} = 30°$ or $\pi/6$.

By adding constraints (7.73) and (7.74) to the optimal PI parameter selection formulation (7.68), we obtain the mathematical formulation for optimizing the nominal performance of the closed loop system while ensuring stability for all potential values of the FOPDT model parameters. There is a more demanding and challenging approach in which Formulation (7.68) is applied for all potential model parameters. A new performance criterion is defined which is a weighted sum of the values of the performance criterion for all potential realizations of the model parameters. This is the most comprehensive approach but can be computationally intensive. Using Eq. (7.74), acceptable performance is expected even for the worst case combination of the parameters of the FOPDT model.

The GAMS model that implements Formulation (7.68) with the additional stability and robustness constraints (7.72) and (7.74) is presented in Table 7.15. The solution of the problem for the FOPDT system:

$$G(s) = \frac{Y(s)}{U(s)} = \frac{k}{\tau s + 1} e^{-ds} \qquad (7.75)$$

$$0.5 \leq k, \tau, d \leq 1.5$$

with nominal parameters $k_N = \tau_N = d_N = 1$ and $\mathrm{PG}_{\min} = 10°$, and $\mathrm{GM}_{\min} = 1.2$ is $k_c = 0.06493670$ and $\tau_I = 0.18519845$. The minimum phase margin is $23.4°$ and the minimum gain margin is 1.610 and correspond to the realization $k_N = \tau_N = d_N = 1.5$.

To demonstrate the effectiveness of the formulation for generating not only acceptable nominal performance but also impressive robustness and overall performance, we perform a crude Monte Carlo simulation of the closed loop system. 10^4 realizations are randomly selected in MATLAB, and the closed loop system simulation is performed. The results are presented in Figs. 7.27 and 7.28. In Fig. 7.27, the parameters $k_s, \tau_s,$ and $d_s, s = 1, 2, \ldots, 10^4$, are shown which are randomly and uniformly selected in $[0.5, 1.5]^3$. The closed loop system is

Table 7.15 GAMS program for the closed loop optimal PI control of a FOPDT system with minimum PM and GM specification

```
$ TITLE FOPDT SYSTEM CLOSED LOOP OPTIMAL PI CONTROL
SET       n time points                      /0*300/
          s schenarios for parameters values /1*8/;

PARAMETER tauN     process time constant             /1/
          dN       delay time                        /1/
          kN       process gain                      /1/
          ysp      set point value                   /1/
          Dumax    maximum variation of the control  /1/
          umin     minimum value of control          /-2/
          umax     maximum value of control          /+2/
          rho      weighting factor                  /1/
          tf       final time                        /30/
          Ts       sampling time
          A_d      A matrix of the state space description
          b_d      b matrix of the state space description
          time(n) time;
Ts        = tf/(card(n)-1);
A_d       = exp(-Ts/tauN); b_d      = (1-A_d)*kN;
time(n)   = Ts*(ord(n)-1);
*                       +    -    +    +    +    -    -    -
*                       +    +    -    +    -    +    -    -
*                       +    +    +    -    -    -    +    -
PARAMETER k(s)    /1 1.5,2 0.5,3 1.5,4 1.5,5 1.5,6 0.5,7 0.5,8 0.5/;
PARAMETER d(s)    /1 1.5,2 1.5,3 0.5,4 1.5,5 0.5,6 1.5,7 0.5,8 0.5/;
PARAMETER tau(s)  /1 1.5,2 1.5,3 1.5,4 0.5,5 0.5,6 0.5,7 1.5,8 0.5/;
SCALAR    pi /3.14159265/;

POSITIVE VARIABLES eplus(n), eminus(n), Duplus(n), Duminus(n);
VARIABLES x(n), u(n), y(n), ObjFun;
POSITIVE VARIABLE invtau_I, wg(s), wp(s), PM(s), GM(s);
VARIABLES          k_c, e(n);

EQUATIONS Dynamics(n)      system state dynamics
          Output(n)       system output dynamics
          ABSerror(n)     absolute error
          ABSuvar(n)      absolute control variation
          Criterion       objective function
          Error(n)        error dynamics
          Control(n)      PI controller
          GainCo(s), PhaseCo(s), GainM(s), PhaseM(s);

Dynamics(n).. x(n)        =E= A_d*x(n-1)+b_d*u(n-1);
Output(n)..   y(n)        =E= x( n-ceil(dN/Ts) );
ABSerror(n).. ysp - y(n)  =E= eplus(n)  - eminus(n);
ABSuvar(n)..  u(n)- u(n-1) =E= Duplus(n) - Duminus(n);
Criterion.. ObjFun =E=
Ts*sum(n, (eplus(n)+eminus(n))+rho*(Duplus(n)+Duminus(n)));
Error(n)..    e(n)        =E= ysp - y(n);
Control(n)..  u(n)- u(n-1) =E= k_c*( (invtau_I*Ts+1)*e(n)-e(n-1) );

GainCo(s)..    wg(s)*sqrt(1+(wg(s)*tau(s))**2) =E=
invtau_I*k(s)*k_c*sqrt(1+(wg(s)/invtau_I)**2);
PhaseCo(s)..   pi/2+arctan(wp(s)/invtau_I)-wp(s)*d(s)-
arctan(wp(s)*tau(s)) =E=0;
GainM(s)..     GM(s)*k_c*k(s)*sqrt(1+(wp(s)/invtau_I)**2) =E=
wp(s)*sqrt(1+(wp(s)*tau(s))**2)/invtau_I;
PhaseM(s)..    PM(s) =E= pi/2+arctan(wg(s)/invtau_I)-wg(s)*d(s)-
arctan(wg(s)*tau(s));

Duplus.up(n) =+Dumax*Ts; Duminus.up(n)=+Dumax*Ts;
u.up(n) = umax; u.lo(n) = umin;
k_c.l = 1/kN; invtau_I.l =1/tauN;
x.fx('0')=0;
GM.lo(s) = 1.2;  PM.lo(s) = 3.14153/12;

MODEL FOPDT_CLOSED_LOOP_robustPI/ALL/;
OPTION NLP=CONOPT4;
SOLVE FOPDT_CLOSED_LOOP_robustPI USING NLP MINIMIZING ObjFun;
```

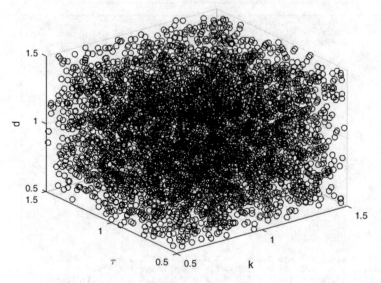

Fig. 7.27 Parameters of the FOPDT system for the closed loop simulations shown in Fig. 7.28

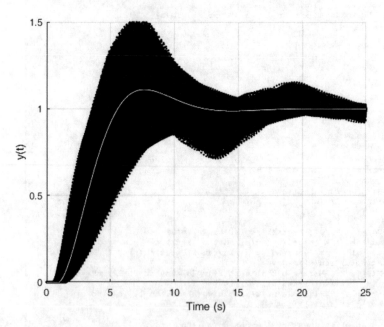

Fig. 7.28 Closed loop dynamics of the FOPDT system with model parameters shown in Fig. 7.27

then formed using the optimal PI controller parameters, and the closed loop system is simulated. In this way, Fig. 7.28 is created (the white line shows the response of the nominal system). The overall performance of the FOPDT system and the optimal PI controller is more than satisfactory.

It is important to stress at this point that many researchers have proposed using mathematical programming formulations for designing classical and modern control laws using computational methods, but the acceptance from the community is limited. Despite the fact that the mathematical programming approach is theoretically consistent and very powerful, the control community insists on accepting only methodologies that can offer results in closed form. This is unfortunate as the methodologies that do offer results in closed form are based on very limiting assumptions and their practical usefulness is questionable.

Before concluding this interesting application, we will present the solution to a closely related problem. Assume that we have a low-order model for a process and we have designed a controller for the nominal system. An interesting question of great practical value is the following: What is the minimum combined variation of the model parameters that drives the system to the verge of instability? Note that the problem now is to determine the parameters of the model for a fixed controller that results in an unstable closed loop system. We are, more specifically, looking for the parameters that cause instability, but their scaled distance from the nominal values is minimal:

$$\min w_k^2 (k - k_N)^2 + w_\tau^2 (\tau - \tau_N)^2 + w_d^2 (d - d_N)^2 \tag{7.76}$$

where w_k, w_τ, and w_d are the scaling parameters (which can be selected to be $w_k = 1/k_N$, $w_\tau = 1/\tau_N$, and $w_d = 1/d_N$). When the systems is at the verge of instability, its GM is one when the phase is $-\pi$ (i.e., $|L(\omega_c)| = 1$, and arg $\{L(\omega_c\} = -\pi$, or GM $= 1$ and PM $= 0$ must hold simultaneously).

$$\min_{k,\tau,d} w_k^2 (k - k_N)^2 + w_\tau^2 (\tau - \tau_N)^2 + w_d^2 (d - d_N)^2$$

$$\text{s.t.} \tag{7.77}$$

$$\frac{\pi}{2} + \tan^{-1}(\omega_c \tau_I) - \omega_c d - \tan^{-1}(\omega_c \tau) = 0$$

$$kk_c \sqrt{1 + \omega_c^2 \tau_I^2} - \omega_c \tau_I \sqrt{1 + \omega_c^2 \tau^2} = 0$$

where ω_c is the critical or crossover frequency.

The GAMS program given in Table 7.16 implements Formulation (7.77). The solution for the case study under investigation is:

Table 7.16 GAMS program for the closed loop stability analysis of FOPDT system and PI control

```
$ TITLE FOPDT SYSTEM CLOSED LOOP STABILITY BOUNDARY
PARAMETERS kN    nominal value of process gain          /1/
           tauN  nominal value of process time constant /1/
           dN    nominal value of dead time             /1/
           kc    controller gain                /0.06493670/
           tauI  integral time                  /0.18519845/;

SCALAR     pi                                   /3.14159265/;

POSITIVE VARIABLES wc, k, tau, d;

VARIABLES L2normSq;

EQUATIONS Magn, Phase, Obj;

Magn..   k*kc*sqrt(1+(wc*tauI)**2)-wc*tauI*sqrt(1+(wc*tau)**2)
=E=0;
Phase..  pi/2+arctan(wc*tauI)-wc*d-arctan(wc*tau) =E=0;
Obj..    L2normSq =E= power(k/kN-1,2)+power(tau/tauN-1,2)+
power(d/dN-1,2);

k.lo   = 0.001; k.up   = 3; k.l=kN;
tau.lo = 0.001; tau.up = 3; tau.l=tauI;
d.lo   = 0.001; d.up   = 3; d.l=dN;
wc.l   = 0;     wc.up  = pi; wc.l=pi/(2*d.l);

MODEL StabilityBountary /ALL/;
OPTION NLP=SNOPT;
SOLVE StabilityBountary USING NLP MINIMIZING L2normSq;
```

	LOWER	LEVEL	UPPER	MARGINAL
---- VAR wc	.	0.568	3.142	6.5595E-6
---- VAR k	0.001	1.934	3.000	.
---- VAR tau	0.001	1.165	3.000	.
---- VAR d	0.001	1.918	3.000	-1.548E-6
---- VAR L2normSq	-INF	1.743	+INF	.

The designer needs to decide at this point whether the distance from the instability is sufficient for the system under study. The decision is usually based on the experience, how critical the process unit is (if this is an unstable reactor with the potential of a run-away reaction we may need to go back to the controller design step and sacrifice performance for improving the robustness), and the uncertainty in the low-order model relative to the actual process.

Fig 7.29 The mixing tank
process

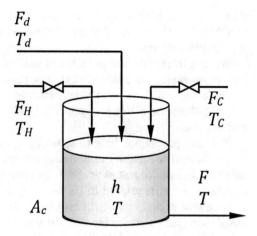

7.9   The Control Structure Selection Problem

In Fig. 7.29, a mixing tank process is presented. A hot stream with temperature T_H and volumetric flow rate F_H and a cold stream with temperature T_C and volumetric flow rate F_C are fed to a mixing tank together with a process stream with temperature T_d and volumetric flow rate F_d. The volumetric flow rate of the product stream is given by $F = k\sqrt{h}$, where k is a constant. We can manipulate the hot and the cold stream with the aim to achieve given specification (or set point value) on the temperature of the product stream T. The liquid level in the tank (h) needs also to remain within certain bounds for avoiding overflow of the tank. The tank has constant cross-sectional area (A_c), while the density and heat capacity of all streams are constant. The objective is to design an optimal decentralized proportional controller (optimal in terms of structure and parameters) that uses the two manipulated variables $(F_H$ and $F_C)$ to control the two controlled variables $(T$ and $h)$.

The problem of selecting the best decentralized multivariable controller is both a structural and parametric optimization problem. There are six potential control structures (two 2×2 control structures and four 1×1 structures) in addition to the open-loop (no control) structure:

Structure 1: $F_H \rightarrow T$ & $F_C \rightarrow h$
Structure 2: $F_H \rightarrow h$ & $F_C \rightarrow T$
Structure 3: $F_H \rightarrow T$
Structure 4: $F_H \rightarrow h$
Structure 5: $F_C \rightarrow T$
Structure 6: $F_C \rightarrow h$

There is also the need to determine, for each alternative control structure, the optimal parameters of the controllers. We may follow a complete enumeration approach in which each structure is evaluated, and then they are compared to determine the best overall structure. This strategy will work for this small problem, but the alternative structures increase very fast with the number of inputs and outputs.

We can formulate the problem of control structure selection as an optimization problem. To achieve that, we need a performance criterion and a methodology to describe mathematically the alternative control structures. We can then solve the problem using classical mathematical programming techniques.

The first problem is how to select an appropriate performance criterion so as to efficiently evaluate the alternative control structures. This is clearly a very complex and multifaceted problem and has no easy solution. The ideal performance criterion is related to (in one way or another) the process economics and the cost of building a process unit or the cost of operating the unit. For the small case study that we investigate, we will use a more humble criterion that is based on the squared deviation of the two controlled variables from their set points h_{sp} and T_{sp}:

$$\min_{\substack{\text{controller structure} \\ \text{and parameters}}} \int_{t=0}^{t=H} \left[\left(\frac{h(t) - h_{sp}}{h_{sp}} \right)^2 + \left(\frac{T(t) - T_{sp}}{T_{sp}} \right)^2 \right] dt \qquad (7.78)$$

The second problem that we face is the definition of the scenario under which we will determine what is the optimal structure. In other words, we need to define the time variation of the uncertain parameters and disturbances that will cause the controlled variables to deviate from their desired values. This is also a very important and complex issue as the best control structure can depend on the assumed scenario. For demonstration purposes and for our small case study, we will consider a relatively simple scenario: step change in the set point of one of the controlled variables.

We now present the mathematical formulation of the problem. We first develop the mathematical model of the process that consists of the material and energy balance:

$$A_c \frac{dh}{dt} = F_H + F_C + F_d - k\sqrt{h} \qquad (7.79)$$

$$A_c h \frac{dT}{dt} = F_H(T_H - T) + F_C(T_C - T) + F_d(T_d - T) \qquad (7.80)$$

The equations of the multivariable proportional controllers are the following:

$$F_H = F_{H,s} + k_{FH,h}(h_{sp} - h) + k_{FH,T}(T_{sp} - T) \qquad (7.81)$$

$$F_C = F_{C,s} + k_{FC,h}(h_{sp} - h) + k_{FC,T}(T_{sp} - T) \qquad (7.82)$$

where subscript s denotes steady-state values and $k_{FH,h}$, $k_{FH,T}$, $k_{FC,h}$, and $k_{FC,T}$ are the controller gains. The controller description is not complete as we need to express the logical condition that the controller needs to be a decentralized controller. To this end, we define the binary variables y_{mn} to denote that manipulated variable m ($m \in \{F_H, F_C\}$) is used to control the controlled variable n ($n \in \{h, T\}$). The following inequality constraints enforce a decentralized control structure:

$$-B_M y_{FH,h} \leq k_{FH,h} \leq B_M y_{FH,h} \tag{7.83}$$

$$-B_M y_{FH,T} \leq k_{FH,T} \leq B_M y_{FH,T} \tag{7.84}$$

$$-B_M y_{FC,h} \leq k_{FC,h} \leq B_M y_{FC,h} \tag{7.85}$$

$$-B_M y_{FC,T} \leq k_{FC,T} \leq B_M y_{FC,T} \tag{7.86}$$

$$0 \leq y_{FH,h} + y_{FH,T} \leq 1 \tag{7.87}$$

$$0 \leq y_{FC,h} + y_{FC,T} \leq 1 \tag{7.88}$$

$$0 \leq y_{FH,h} + y_{FC,h} \leq 1 \tag{7.89}$$

$$0 \leq y_{FH,T} + y_{FC,T} \leq 1 \tag{7.90}$$

$$y_{FH,T}, y_{FC,T}, y_{FC,T}, y_{FC,h} = \{0, 1\} \tag{7.91}$$

Equations (7.83)–(7.86) ensure that when $y_{mn} = 0$, then $k_{mn} = 0$, while k_{mn} can obtain any value in $[-B_M, B_M]$ when $y_{mn} = 1$, where B_M is a sufficient large positive constant. Equations (7.87) and (7.88) ensure that each manipulated variable is assigned to at most one controlled variable. Equations (7.89) and (7.90) ensure that at most one manipulated variable is selected for each controlled variable.

To have a complete mathematical description, we need to define the initial conditions of the system which is assumed to be initially at steady state:

$$F_{H,s} + F_{C,s} + F_{d,s} - k\sqrt{h_s} = 0 \tag{7.92}$$

$$F_{H,s}(T_{H,s} - T_s) + F_{C,s}(T_{C,s} - T_s) + F_{d,s}(T_{d,s} - T_s) = 0 \tag{7.93}$$

where h_s, T_s, $T_{H,s}$, $T_{C,s}$, $F_{d,s}$, and $T_{d,s}$ are known at the initial steady state. h_{sp} and T_{sp} are also known functions of time. The unknowns are the variables y_{mn}, k_{mn}, $h(t)$, and $T(t)$. y_{mn} and k_{mn} are obtained from the solution of the resulting optimization formulation and define the best control structure and its parameters. $h(t)$ and $T(t)$ are determined from the solution of the differential equation for a given control structure and parameters.

To demonstrate the application of the control structure selection formulation, we consider that the mixing tank is initially at steady state with $h_s = h_{sp} = 6.25$ m and $T_s = T_{sp} = 25\,°C$ and that the set point temperatures changes for $t > 0$ to $T_{sp} = 30$ °C. The cross-sectional area of the tank ($A_c = 1$), the constant of the output flow rate ($k = 3$ m$^{5/2}$/min), the temperature of the hot and cold streams ($T_H = 65$ °C, $T_C = 15$ °C), and the flow rate and temperature of the process stream ($F_d = 1$ m^3/min, $T_d = 40$ °C) remain constant (i.e., only the temperature set point varies). The GAMS model presented in Table 7.17 is created, and the control structure selection problem is solved. The best integer solution is the following:

   Optimal Structure : $F_H \rightarrow T \& F_C \rightarrow h$ Objective function : 33.248323

   Integer cuts are then added to find the remaining promising structures:

   2nd Optimal Structure: $F_H \rightarrow h$ & $F_C \rightarrow T$ Objective function: 49.581339
   3rd Optimal Structure: $F_H \rightarrow T$ Objective function: 91.784266
   4th Optimal Structure: $F_C \rightarrow T$ Objective function: 133.564341
   5th Optimal Structure: no control   Objective function: 138.657407

   We note that the two structures that use both manipulated variables to control both controlled variables are significantly better than the two structures that use either F_H or F_C to control only the temperature T. The fifth best structure is the open-loop (or no control) structure. Needless to say that any solution that does not control the level of the liquid in the tank is not acceptable from the practical point of view as overflow is highly possible.

Learning Summary

In the previous chapter, we introduced GAMS, a powerful modelling environment for optimization studies. In this chapter, we solve a number of challenging case studies in GAMS with the aim to build confidence and also realize the limitations of the currently available optimization software.

Terms and Concepts

You must be able to discuss the concept of:

Complex Chemical Equilibrium model and solution
Dynamic optimization and Mixed-Integer Dynamic
GAMS modelling environment and its advantage over MATLAB
Optimal design of Distillation columns
Optimal design of multi-effect Evaporators
Optimal Design of Renewable Energy Systems
Controller design and control structure selection

Table 7.17 GAMS model for the control structure selection in a mixing tank problem

```
* Mixing tank Regulatory Control Structure Selection problem
SETS j nodes   /1*301/
     i states /h,T,ISE/
     m inputs /FH,FC/
     n(i) measurements /h,T/;
PARAMETER NP; NP=card(j)-1;
PARAMETER tF; tF=5;
PARAMETER dt; dt=tF/NP;
PARAMETER time(j); time(j)=(ord(j) - 1)*dt;
PARAMETER Ac    cross sectional area of the tank m^2      /1/
          TH    temparature of the hot stream degC        /65/
          TC    temperature of the cold stream degC       /15/
          k     constant for the outflow model            /3/
          hsps  initial sp for liquid level in m           /6.25/
          Tsps  initial sp for temperature in deg C       /25/
          Fds   ss value for process stream flowrate      /1/
          Tds   ss value for process stream temperature   /40/
          hsp   sp for liquid level for t>0               /6.25/
          Tsp   sp for temperature for t>0                /30/
          Fd    flowrate for process stream for t>0       /1/
          Td    temperature for process stream for t>0    /40/;
SCALAR    BM  /1/;
VARIABLES f(i,j), x(i,j), e(n,j), u(m,j), us(m), KC(m,n), Obj;
BINARY VARIABLES y(m,n);
EQUATIONS eq1(i,j),eq2(j),eq3(j),eq4(j),eq5(j),eq5a,eq6(j),eq6a,
eq7(m,j),eq7a,eq7b,eq8(m,n),eq9(m,n),eq10(m),eq11(n),ObjFun;
* ODE INTEGRATION
eq1(i,j)$( ord(j) gt 1 ).. x(i,j)=E=x(i,j-1)+0.5*dt*(f(i,j)+f(i,j-1));
* SYSTEM DYNAMICS
eq2(j).. Ac*f('h',j) =E= u('FH',j)+u('FC',j)+Fd-k*sqrt(x('h',j));
eq3(j).. Ac*x('h',j)*f('T',j)=E=u('FH',j)*(TH-x('T',j))+u('FC',j)*(TC-
x('T',j))+Fd*(Td-x('T',j)));
eq4(j)..
f('ISE',j)=E=1000*((e('h',j)/hsp)**2+(e('T',j)/Tsp)**2);
* ERROR DEFINITION
eq5(j)$( ord(j) gt 1 ).. e('h',j)   =E=  -x('h',j)   + hsp;
eq5a..                   e('h','1') =E=  -x('h','1') + hsps;
eq6(j)$( ord(j) gt 1 ).. e('T',j)   =E=  -x('T',j)   + Tsp;
eq6a..                   e('T','1') =E=  -x('T','1') + Tsps;
* CONTROL LAW
eq7(m,j).. u(m,j) =E= us(m) + sum( n, KC(m,n)*e(n,j) ) ;
* CALCULATE u_bias ASSUMING SYSTEM AT SS AT t=0
eq7a.. us('FH')+us('FC')+Fds-k*sqrt(hsps) =E= 0;
eq7b.. us('FH')*(TH-Tsps)+us('FC')*(TC-Tsps)+Fds*(Tds-Tsps)=E=0;
* DECENTRALISED CONTROL STRUCTURE SELECTION
eq8(m,n) .. KC(m,n) =L= BM*y(m,n);
eq9(m,n) .. KC(m,n) =G= -BM*y(m,n);
eq10(m) ..  sum(n,y(m,n))=L=1;
eq11(n) ..  sum(m,y(m,n))=L=1;
* OBJECTIVE FUNCTION
ObjFun.. Obj=e= x('ISE','301');
* INITIAL GUESSES and INITIAL CONDITIONS
x.lo('h',j)=0.1; x.up('h',j)=5; x.l('h',j)=1;
x.lo('T',j)=TC; x.up('h',j)=TH; x.l('T',j)=25;
u.lo(m,j) =0; u.up(m,j)=10;
us.lo(m)=0 ; us.up(m)=10;
* Initial Conditions and states at set points at t=0
x.fx('ISE','1')=0; e.fx('h','1')=0; e.fx('T','1')=0;

MODEL mixing_tank /ALL/ ;
OPTIONS MINLP=SBB;
SOLVE mixing_tank USING miNLP MinIMIZING Obj;
```

Problems

7.1 Investigate the effect that the minimum pressure has on the economics of the single evaporator process (assume that the minimum pressure can be varied in the range 0.1–1 bar). Explain why the minimum always corresponds to the minimum pressure.

7.2 The single-effect evaporator has been optimized using the CONOPT4 solver. Compare the performance of CONOPT4 with other NLP solvers (SNOPT, MINOS, BARON, KNITRO, etc.). Generate a table to compare their performance.

7.3 Increase the number of evaporators in the program of Table 7.3. What is the optimum number of effects? Prepare a short report.

7.4 Use the GAMS program of Table 7.6 to determine the optimal design for the case where the top purity varies from 0.9 to 0.999. Make a plot of the main process variables as a function of the purity of the top product. What do you observe?

7.5 For the batch reactor optimization problem and its implementation in GAMS (Table 7.7), study the cases of a linear and then a quadratic profile for the manipulated variable, and compare your results with the results obtained using the exponentially decreasing profile.

7.6 Start with the GAMS program given in Table 7.8, and increase the number of elements and nodes. What is the best objective that you can achieve? How does it compare with the optimal value of 0.610071345? Was the improvement obtained worth the effort (from the practical implementation point of view)?

7.7 Solve the following optimal control problem (Goh & Teo, *Automatica*, 24(1), pp. 3–18, 1988), and compare your results with the optimal solution reported in the literature.

$$\min_{u(t)} J = \int_0^1 \left(x_1^2(t) + x_2^2(t) + 0.005u^2(t) \right) dt$$

s.t.

$$\dot{x}_1(t) = x_2(t)$$
$$\dot{x}_2(t) = x_2(t) + u(t)$$
$$x_2(t) - 8(t - 0.5)^2 + 0.5 \leq 0, \ \forall t$$
$$x(0) = [0 \ -1]^{\mathrm{T}}$$

7.8 Solve also Example 1 of the abovementioned publication, and prepare a short report.

$$\min_{u(t)} J = \int_0^1 \left(x^2(t) + u^2(t) \right) dt$$

s.t.

$$\frac{dx(t)}{dt} = u(t), x(0) = 1$$

7.9 Several investigators have proposed to use monetary values for the CO_2 emission so as to account for the financial and social impact of pollution on the human health and on the environment. The values that have been suggested vary over several orders of magnitude and range from 0.02 \$/kg CO_{2eq} (20 \$/t CO_{2eq}) up to 650 \$/kg CO_{2eq}. Modify the GAMS program of Table 7.9 so as to be able to minimize a combined objective function that incorporates the unit electricity cost and the cost of the CO_{2eq} emissions. Use the indicative value of 0.1 \$/kg CO_{2eq}, and present the solution obtained.

7.10 Ladakis et al. (*Biochemical Engineering Journal*, 137, p. 262, 2018) have studied experimentally and theoretically the metabolic capabilities of *Actinobacillus succinogenes* and *Basfia succiniciproducens* grown on spent sulfite for succinic acid production. The metabolic network of *B. succiniciproducens* was modified in order to obtain the metabolic pathway of *A. succinogenes*. *A. succinogenes* lacks a complete TCA cycle resulting in only one pathway for succinate production (nonexistent reactions, reactions 22, 23, 24, and 33 in Fig. 7.22). Collect all information necessary, and propose modifications of the GAMS model of Table 7.12 to produce the theoretical maximum yields of *A. succinogenes* (see Fig 7.23) when grown on glucose. The equation for biomass production for *A. succinogenes* is the following:

$$1.502 \cdot OAA + 1.370 \cdot PG3 + 0.686 \cdot RIBO5P + 46.930 \cdot ATP$$
$$+ 13.439 \cdot NADPH + 7.813 \cdot NH3 + 3.094 \cdot NAD + 3.006 \cdot AcCoA$$
$$+ 2.764 \cdot PYR + 0.244 \cdot E4P + 0.528 \cdot PEP + 0.126 \cdot F6P$$
$$+ 0.410 \cdot G6P + 0.099 \cdot GAP$$

$$\rightarrow 1 \cdot \textbf{Biomass} + 13.439 \cdot NADP + 3.094 \cdot NADH + 46.930 \cdot ADP$$

7.11 The following reactions are taking place in a batch reactor of constant volume:

$$A \xrightarrow{k_1} B$$

$$B \xrightarrow{k_2} C$$

$$A + B \xrightarrow{k_3} B + B$$

$$A + B \xrightarrow{k_4} B + C$$

$$A + B \xrightarrow{k_5} B + D$$

where k_i, $i = 1, 2, \ldots, 5$ are the reaction rate constants given by:

$$k_i = k_{0i} \exp\left(-\frac{E_i}{RT}\right)$$

k_{0i} are the pre-exponential constants, R the ideal gas constant, and E_i the activation energy. Values for these constants are given in the table that follows:

i	$\ln(k_{0i})$ [−]	E_i/R [1/K]
1	8.86	10 215.4
2	24.25	18 820.5
3	23.67	17 008.9
4	18.75	14 190.8
5	20.70	15 599.8

The absolute temperature is bounded:

$$698.15 \leq T(t) \leq 748.15$$

The aim is to calculate the optimal temperature variation so as to maximize the concentration of B at the end of the reactor operation. Assume that the reactor operates for $t_f = 10$ h and that the following model is a realistic representation of the reactor operation, and determine the optimal temperature profile:

$$\frac{dC_A(t)}{dt} = -k_1 C_A(t) - (k_3 + k_4 + k_5) C_A(t) C_B(t)$$

$$\frac{dC_B(t)}{dt} = +k_1 C_A(t) + k_3 C_A(t) C_B(t) - k_2 C_B(t)$$

$$C_A(0) = 1 \frac{\text{mol}}{\text{L}}, C_B(0) = 0$$

7.12 For the combination of a PI controller and a FOPDT system, studied in Sect. 7.8, the GAMS model of designing an optimal PI controller for a step change in the set point is given in Table 7.14. We will now study the case of a Proportional-Integral-Derivative (or PID) controller. Show initially following the ideas presented in Sect. 7.8 that the discrete form of the "ideal" PID controller:

$$u(t) = k_c \left(e(t) + \frac{1}{\tau_I} \int_0^t e(\theta) d\theta + \tau_D \frac{de(t)}{dt} \right) \qquad (7.65')$$

is given by:

$$u(n) = u(n-1) + \Delta u(n) \qquad (7.66')$$

$$\Delta u(n) = k_c \left[\left(\frac{T_s}{\tau_I} + 1 \right) e(n) - e(n-1) \right] + k_c \left(\frac{\tau_D}{T_s} \right)$$
$$\times [e(n) - 2e(n-1) + e(n-2)] \qquad (7.67')$$

Modify the GAMS model given in Table 7.14 to incorporate the D part of the controller, and also eliminate the rate constraints on the input. Perform the optimization. Compare the results obtained using optimal open-loop, P, PI, and PID controller. Does the PID controller significantly improve the performance for the system as compared to an optimal PI controller (this is a usual argument in the literature)?

Appendix A: Introduction to MATLAB®

Introduction

MATLAB® is a powerful tool for mathematical and engineering calculations. It can be used as a simple scientific calculator, to visualize data in different ways, to perform complex matrix algebra, to perform polynomial manipulations, numerical integration, and solution of differential equations, etc. In Windows environments, you can invoke MATLAB by double clicking on the relative icon on the desktop of your computer. The string $>>$ in the Command Window is the MATLAB prompt, and you can interact with MATLAB by typing commands and pressing the enter ($\hookleftarrow$) on your keyboard. A simple example is to assign the value of 2 to variable x and the value of 5 to variable y:

```
>> x=2

x =
    2

>> y=5

y =
    5
```

I. K. Kookos, *Practical Chemical Process Optimization*, Springer Optimization and Its Applications 197, https://doi.org/10.1007/978-3-031-11298-0

You can then multiply (*), divide (/), add (+), and subtract (−) x and y:

```
>> x*y

ans =
    10

>> x/y

ans =
    0.4000

>> x+y

ans =
    7

>> x-y

ans =
    -3
```

or you can raise x to the power y:

```
>> x^y

ans =
    32
>>
```

Semicolon (;) can be used to suppress the response by MATLAB. Of course, you can multiply 2 by 5, or add them together directly without assigning the values to specific variables:

```
>> 2*5

ans =

    10

>> 2+5;
>>
```

In the same way, we may define vectors and matrices in MATLAB. To define the vector **x** and the table **A**

$$\mathbf{x} = \begin{bmatrix} 1 \\ 3 \\ 7 \end{bmatrix}, \quad \mathbf{A} = \begin{bmatrix} 1 & 2 & 3 \\ 4 & 5 & 6 \\ 7 & 8 & 9 \end{bmatrix}$$

we type the following in the command window:

```
>> x=[1;3;7]

x =
     1
     3
     7

>> A=[1 2 3; 4 5 6; 7 8 9]

A =
     1    2    3
     4    5    6
     7    8    9
>>
```

MATLAB is the ideal environment for performing matrix manipulations:

```
>> A*x

ans =
    28
    61
    94

>> x*A
Error using *
```

To understand why the MATLAB returns an error message, it is reminded that if **A** is an n-by-p matrix and **B** is an p-by-m, then their product $\mathbf{C}_{n\times m} = \mathbf{A}_{n\times p} * \mathbf{B}_{p\times m}$ is an n-by-m matrix:

$$C_{ij} = \sum_k A_{ik} \cdot B_{kj}$$

The multiplication is possible only if the number of columns of matrix **A** is equal to the number of rows of **B**. **x** is a 3-by-1 matrix (vector), and A is a 3-by-3 matrix, and therefore $\mathbf{A}_{3\times 3}$ * $\mathbf{x}_{3\times 1}$ is permissible, while $\mathbf{x}_{3\times 1}$ * $\mathbf{A}_{3\times 3}$ is not. For matrix addition or subtraction, the two matrices must have the same number of rows and the same number of columns:

$$C_{ij} = A_{ij} \pm B_{ij}$$

In MATLAB we can define special type of matrices such as an n-by-m matrix with all elements equal to one or zero:

```
>> n=4;
>> m=3;
>> A=ones (n,m)

A =
        1   1   1
        1   1   1
        1   1   1
        1   1   1

>> B=zeros (n,m)

B =
        0   0   0
        0   0   0
        0   0   0
        0   0   0
```

or the unit matrix **I** which is a square matrix:

```
>> I=eye (n)

I =
        1   0   0   0
        0   1   0   0
        0   0   1   0
        0   0   0   1
```

For any square (n-by-n), invertible matrix **A**, it holds true that

$$\mathbf{A}^{-1}\mathbf{A} = \mathbf{A}\mathbf{A}^{-1} = \mathbf{I}$$

```
>> A=[1 4 3;2 9 6;7 8 5]

A =

     1    4    3
     2    9    6
     7    8    5

>> inv(A)*A

ans =
     1.0000         0   -0.0000
          0    1.0000         0
    -0.0000   -0.0000    1.0000

>> A*inv(A)

ans =
     1.0000         0   -0.0000
          0    1.0000   -0.0000
          0         0    1.0000
```

For any n-by-m matrix $\mathbf{A}$ the transpose of $\mathbf{A}$ is denoted by $\mathbf{A}^{\mathrm{T}}$ and is a m-by-n matrix with columns the rows of $\mathbf{A}$:

```
>> A=[1 2 3;4 5 6]

A =

     1    2    3
     4    5    6

>> B=A'

B =

     1    4
     2    5
     3    6
```

For square matrices, the following MATLAB commands are useful:

`[V,D] = eig(A)`	Computes the eigenvalues (diagonal elements of V) and the eigenvectors (columns of D) of **A**
`d = det(A)`	Computes the determinant of **A**

To solve the linear system of equations:

$$\mathbf{A} \cdot \mathbf{x} = \mathbf{b}$$

we use the command x=A\b:

```
>> A=[1 4 3;2 9 6;7 8 5];
>> b=[1;2;3];
>> x=A\b

x =
    0.2500
         0
    0.2500
```

or

```
>> x=inv(A)*b

x =
    0.2500
         0
    0.2500
```

The same can be performed symbolically using the syms command from the symbolic toolbox:

```
>> clear all
>> syms a b c d A
>> A=[a b;c d]

A =
[ a, b]
[ c, d]

>> inv(A)

ans =
[  d/(a*d - b*c),  -b/(a*d - b*c)]
[ -c/(a*d - b*c),   a/(a*d - b*c)]

>> det(A)
```

(continued)

```
ans =
a*d - b*c

>> eig(A)

ans =
a/2 + d/2 - (a^2 - 2*a*d + d^2 + 4*b*c)^(1/2)/2
a/2 + d/2 + (a^2 - 2*a*d + d^2 + 4*b*c)^(1/2)/2
```

which gives the analytic expression for the inverse, determinant, and eigenvalues of a 2-by-2 matrix.

The `clear` command can be used at any point to clear the variables from the workspace. Other useful commands are the who, whos, and the help commands:

```
>> who
Your variables are:

A  a  ans b  c  d

>> whos
  Name    Size        Bytes    Class    Attributes
  A       2x2             8    sym
  a       1x1             8    sym
  ans     2x1             8    sym
  b       1x1             8    sym
  c       1x1             8    sym
  d       1x1             8    sym

>> clear
>> who
>> whos
>> help clear
 clear Clear variables and functions from memory.
   clear removes all variables from the workspace.
   clear VARIABLES does the same thing.
   clear GLOBAL removes all global variables.
   clear FUNCTIONS removes all compiled MATLAB and MEX-functions.
   Calling clear FUNCTIONS decreases code performance and is usually
   unnecessary.
   For more information, see the clear Reference page.

   clear ALL removes all variables, globals, functions and MEX
 links.
```

The linspace(x0,xf,n) command generates *n* points uniformly distributed in the interval $[x_0, x_f]$:

```
>> clear all
>> x=linspace(0,10,11)

x =
   0   1   2   3   4   5   6   7   8   9   10
```

The command logspace(n1,n2,n) generates points uniformly distributed in the logarithmic scale:

```
>> logspace(-2,2,5)

ans =
     0.0100   0.1000   1.0000  10.0000  100.0000
```

There is a wide set of build in functions available in MATLAB, and the most commonly used are the following:

abs	Absolute value
ceil	Round towards plus infinity
cos	Cosine
cosh	Hyperbolic cosine
exp	Exponential (e to the power)
fix	Round towards zero
floor	Round towards minus infinity
imag	Complex imaginary part
log	Natural logarithm
log10	Common logarithm
max	Largest component
min	Smallest component
round	Round towards nearest integer
sec	Secant
sech	Hyperbolic secant
sin	Sine
sinh	Hyperbolic sine
sign	Signum function
sqrt	Square root
tan	Tangent
tanh	Hyperbolic tangent

We may use sin and cos to calculate the values of sin and cos for x=[0 π/6 π/3 π/2]:

```
>> clear
>> x=pi*[0 1/6 1/3 1/2];
>> sin(x)

ans =
       0   0.5000   0.8660   1.0000

>> cos(x)

ans =
   1.0000   0.8660   0.5000   0.0000
```

Note that as x is defined as a row vector, `sin(x)` and `cos(x)` are row vectors as well. It is more informative if we use the plotting facilities in MATLAB to plot sin(x) and cos(x). The basic plotting command in MATLAB is `plot(x,y)`:

```
>> clear
>> x=linspace(0,2*pi,101);
>> plot(x,sin(x))
```

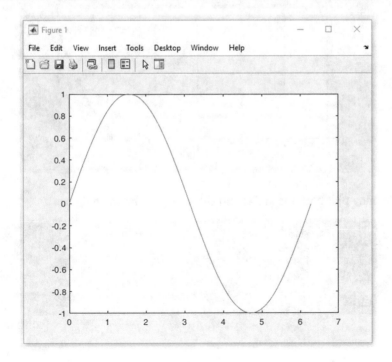

404 Appendix A: Introduction to MATLAB®

The figure generated needs improvement which can be achieved using the following commands (in case their effect is not obvious to the reader, then use the help command to explore them further—observe also their effect on the figure produced when they are applied)

```
>> plot(x,sin(x),'k-','Linewidth',1.2)
>> grid on
>> axis tight
>> xlabel('x')
>> ylabel('sin(x)')
>> title('Sin(x)')
>> hold on
>> plot(x(1:10:length(x)),sin(x(1:10:length(x))),'ko')
```

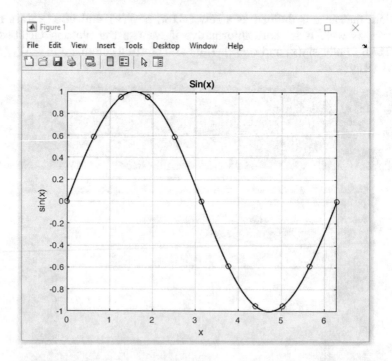

We may plot both the sin(x) and the cos(x) in the same figure:

```
>> figure(2)
>> plot(x,sin(x),'k-','Linewidth',1.2)
>> hold on
>> plot(x,cos(x),'k--','Linewidth',1.2)
>> xlabel('x')
```

(continued)

```
>> ylabel('sin(x) or cos(x)')
>> axis([0 2*pi -1 +1])
>> grid on
```

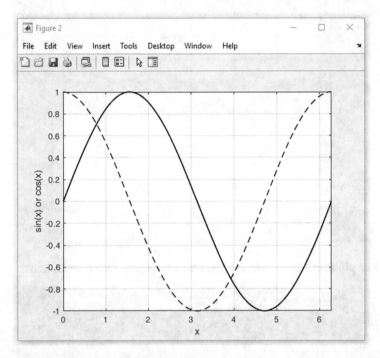

or we may divide the plot to subplots:

```
>> figure(3)
>> subplot(2,1,1)
>> plot(x,sin(x),'k-','Linewidth',1.2)
>> xlabel('x')
>> ylabel('sin(x)')
>> grid on
>> axis tight
>> subplot(2,1,2)
>> plot(x,cos(x),'k-','Linewidth',1.2)
>> grid on
>> xlabel('x')
>> ylabel('cos(x)')
>> axis tight
```

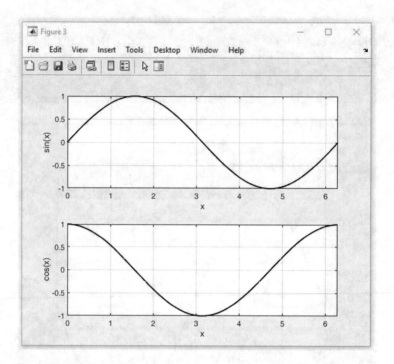

Other plotting commands are the bar, stems, and stairs commands:

```
>> figure(4)
>> x=linspace(0,2*pi,21);
>> subplot(3,1,1)
>> bar(x,sin(x))
>> title('Subplot 1-1-bar')
>> subplot(3,1,2)
>> stem(x,sin(x))
>> title('Subplot 2-1-stem')
>> subplot(3,1,3)
>> stairs(x,sin(x))
>> title('Subplot 3-1-stairs')
```

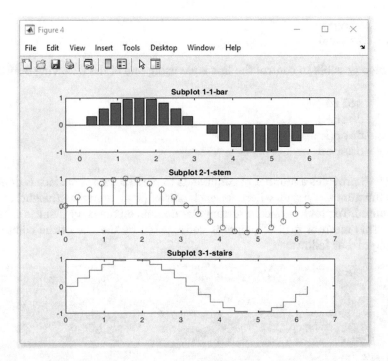

The color options are:

y   yellow
m   magenta
c   cyan
r   red
g   green
b   blue
w   white
k   black

and the choice of symbols are:

.   point
v   triangle (down)
o   circle
^   triangle (up)
x   x-mark
<   triangle (left)
+   plus
>   triangle (right)
*   star
p   pentagram
s   square

    h   hexagram
    d   diamond

It is also possible to control the line style using one of the following options:

    -     solid
    :   dotted
    -.   dashdot
    - -   dashed

MATLAB provides a number of commands to plot 3D data. A surface is defined by a function $z = f(x, y)$, where for each pair of (x, y), the value (height) of z is computed. To plot a surface, a rectangular domain on the (x, y)-plane is created first. This mesh or grid of points is constructed by the use of the command meshgrid as follows:

```
>> clear
>> x=linspace(-8,8,21);
>> y=linspace(-8,8,21);
>> [X,Y]=meshgrid(x,y);
>> plot(X,Y,'ko')
>> xlabel('x');
>> ylabel('y')
>> grid on
```

(note that MATLAB is case sensitive, i.e., x and X are different variables). Then the surface of the function:

$$z = f(x,y) = \frac{\sin\left(\sqrt{x^2 + y^2}\right)}{\sqrt{x^2 + y^2}}$$

can be plotted as follows:

```
>> r=sqrt(X.^2+Y.^2);
>> r=r+(r==0)*eps;
>> z = sin(r)./r;
>> surf(X,Y,z)
```

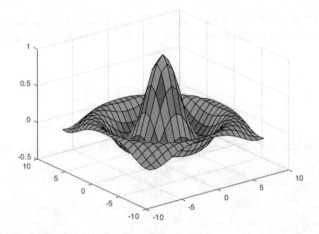

```
>> mesh(X,Y,z)
```

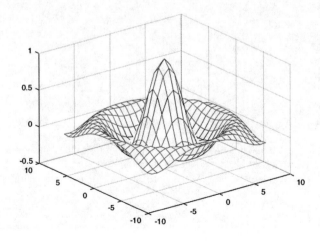

The reader should note that in calculating the intermediate variable r, we performed the operation `r=r+(r==0)*eps`, to avoid division by error. The condition `(r==0)` is first checked (as expressions in parenthesis are performed first in MATLAB followed by multiplications/divisions), and `eps` is added to `r` if `r` is equal to zero. `eps` is the smallest number that can be detected in your system as being different than zero. To use control flow commands, it is necessary to perform operations that result in logical values: 1 (TRUE) or 0 (FALSE). The relational operators:

<	less than
<=	less or equal to
>	greater that
>=	greater or equal to
==	equal to
~=	not equal to

can be used to compare two arrays of the same size or an array to a scalar. Combinations of logical expressions can be performed using the logical operators:

&	and
|	or
~	not.

The example that follows demonstrates the use of logical operations:

```
>> x=[1 2 3 5 11]

x =
     1    2    3    5   11

>> y=(x<=2) | (x>10)

y =
     1  1  0  0  1

>> z=(x>2) & (x<=10)

z =
     0  0  1  1  0

>> (y==~z)

ans =

     1  1  1  1  1
```

Another new feature that was introduced is the use of the dot (.) operator:

```
>> r=sqrt (X.^2+Y.^2);
```

When the dot operator is used, then the intended mathematical operation is performed on an element-by-element basis:

```
>> A=[1 2;3 4]

A =
        1    2
        3    4

>> A^2

ans =
        7   10
       15   22

>> A.^2

ans =
        1    4
        9   16
```

Note that while A^2 is the usual operation of raising a matrix to a power of 2:

$$\mathbf{A}^2 = \mathbf{A} \cdot \mathbf{A} = \begin{bmatrix} 1 & 2 \\ 3 & 4 \end{bmatrix} \cdot \begin{bmatrix} 1 & 2 \\ 3 & 4 \end{bmatrix} = \begin{bmatrix} 1 \cdot 1 + 2 \cdot 3 & 1 \cdot 2 + 2 \cdot 4 \\ 3 \cdot 1 + 4 \cdot 3 & 3 \cdot 2 + 4 \cdot 4 \end{bmatrix} = \begin{bmatrix} 7 & 10 \\ 15 & 22 \end{bmatrix}$$

the A.^2 causes each element of A to be raised to the power of 2:

$$\mathbf{A}.^2 = \begin{bmatrix} 1^2 & 2^2 \\ 3^2 & 4^2 \end{bmatrix} = \begin{bmatrix} 1 & 4 \\ 9 & 16 \end{bmatrix}$$

Controlling the Flow

Like all computer programming languages, MATLAB features elements that allow us to control the flow of command execution. There are three main control flow elements:

The `if/else` structure:

```
if condition 1;
% commands if condition 1 is true
    statements 1;
elseif condition 2;
% commands if condition 2 is true
    statements 2;
else;
% commands if condition 1 and 2 are both false
    statements 3;
end;
```

The `for` structure:

```
for variable=start:step:end;
    statements;
end;
```

The `while` structure:

```
while condition;
    statements executed while condition is true;
end;
```

We will now demonstrate the use of flow control through some simple examples.

We will first demonstrate the use of the if element. Let assume that you have been asked to determine the drag coefficient for a spherical object, given the Reynolds number which must be greater that 0.0001 ($N_{Re} > 0.0001$). It is known that the Drag coefficient of a sphere decreases with the Reynolds number and drag coefficient becomes almost a constant ($C_D = 0.44$) for a Reynolds number between 10^3 and 2×10^5. More specifically, you have been given the following information:

$$C_D = \begin{cases} \dfrac{24}{N_{Re}}, & 0.0001 \le N_{Re} < 0.2 \\[2mm] \dfrac{24}{N_{Re}} + 0.44, & 0.2 \le N_{Re} < 1000 \\[2mm] 0.44, & 10^3 \le N_{Re} \le 2 \cdot 10^5 \\[2mm] 0.05, & N_{Re} > 2 \cdot 10^5 \end{cases}$$

The calculation for a given value of the Reynolds number can be performed as follows:

```
if (NRe<1e-4)
    disp('Equation not available for N_{Re}<1e-4')
    CD = NaN;
```

(continued)

```
elseif ((Nre>=1e-4)&(Nre<0.2))
    CD = 24/Nre;
elseif ((Nre>=0.2)&(Nre<1e3))
    CD = 24/Nre+0.44;
elseif ((Nre>=1e3)&(Nre<2*1e5))
    CD = 0.44;
else
    CD = 0.05;
end
```

The best way to implement this sequence of commands is to create a subroutine or `function` in MATLAB that can be built as an `m-file`. Function can be defined in MATLAB by selecting New/Function:

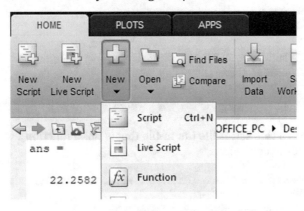

The text editor is invoked, and we may create our function:

```
1    function CD = DragSphere(NRe)
2
3 -      if (NRe<1e-4)
4 -          disp('Equation not available for N_{Re}<30')
5 -          CD = NaN;
6 -      elseif ((NRe>=1e-4)&(NRe<0.2))
7 -          CD = 24/NRe;
8 -      elseif ((NRe>=0.2)&(NRe<1e3))
9 -          CD = 24/NRe+0.44;
10 -     elseif ((NRe>=1e3)&(NRe<2*1e5))
11 -         CD = 0.45;
12 -     else
13 -         CD = 0.05;
14 -     end
```

We then save our file by selecting Save/Save As. Be careful to save the file in your current directory of MATLAB.

After this small but important diversion, we return to demonstrate now the `for` element. Let now assume that we want to prepare a plot of the drug coefficient. We first create a vector with values of the Reynolds number in the range $[10^{-3}, 10^6]$ and then use the for element to evaluate the drug coefficient for each element of the vector of the Reynolds numbers (use help to find information about the `length`, `semilog`, and `loglog` commands in MATLAB):

```
DragSphere.m   runDrugSphere.m*   +
1 -     clear
2
3 -     NRe=logspace(-3,6,100);
4
5 -     for i=1:length(NRe)
6 -         CD(i)=DragSphere(NRe(i));
7 -     end
8
9 -     loglog(NRe,CD,'k','Linewidth',1)
10 -    grid on
11 -    axis([1E-3 1E6 1E-2 1E5])
12 -    xlabel('N_{Re}')
13 -    ylabel('C_{D}')
```

When we run this script file (an m-file that is not a function), we obtain the graph of the drag coefficient:

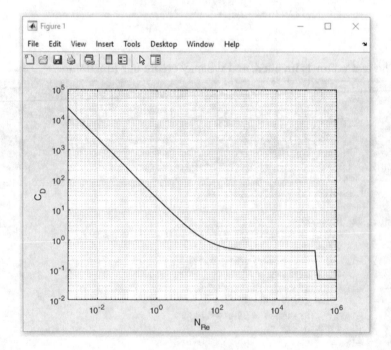

To demonstrate the use of `while` element, we will use the analytic solution for the temperature distribution at the center of a slab, with uniform initial temperature T_0, in which the surface temperature changes suddenly to T_1 and remains constant. If the thermal diffusivity is a and the thickness of the slab is $2b$, then (see Bird et al., Introductory Transport Phenomena, Wiley, 2015, p. 350):

$$\Theta(\tau) = \frac{T(0, \tau) - T_1}{T_1 - T_0} = 2 \sum_{n=0}^{\infty} (-1)^n \frac{\exp\left\{-\lambda_n^2 \cdot \tau\right\}}{\lambda_n}, \quad \lambda_n = \left(n + \frac{1}{2}\right)\pi, \quad \tau = \frac{at}{b^2}$$

The problem in the calculation is that it involves adding an infinite number of terms. However, as the absolute value of the function inside the summation is a decreasing function of n, we normally stop adding terms when:

$$f(n, \tau) = \frac{\exp\left\{-\lambda_n^2 \cdot \tau\right\}}{\lambda_n} \leq \text{tol}$$

The `while` element is the most appropriate for performing this calculation:

```
Slab.m  ×   runSlab.m  ×  +
1    function [Theta,n]=Slab(tau,tol)
2
3    n=0;
4    lamda=(n+1/2)*pi;
5    f=exp(-lamda^2*tau)/lamda;
6    Theta=2*(-1)^n*f;
7
8    while (f>tol)
9        n=n+1;
10       lamda=(n+1/2)*pi;
11       f=exp(-lamda^2*tau)/lamda;
12       Theta=Theta+2*(-1)^n*f;
13   end
14
15   return
```

```
Slab.m  ×   runSlab.m  ×  +
1    clear
2
3    tau=logspace(-2,1,101)
4    tol=1e-6;
5
6    for i=1:length(tau)
7        [Theta(i),n(i)]=Slab(tau(i),
8    end
9
10   semilogx(tau,Theta)
11   axis tight
12   grid on
13
14   xlabel('Dimensionless time \tau'
15   ylabel('Temperature \Theta')
```

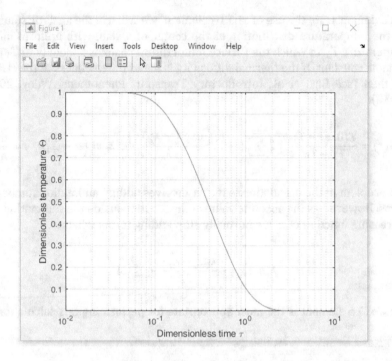

Vectorization

It is particularly advantageous in MATLAB to use matrix operation instead of using for loops. Let's take a specific example to demonstrate the reduction in computational time when we use a matrix operation to perform calculations. A widely acceptable approximation to the Moody friction coefficient for fluid flow in rough pipes, as determined by the Colebrook equation, is given by the Eck equation:

$$\frac{1}{\sqrt{f}} = -2 \cdot \log_{10}\left(\frac{e/D}{3.715} + \frac{15}{N_{Re}}\right)$$

The advantage of the Eck approximation is that the friction coefficient is calculated directly, while the Colebrook equation requires a numerical solution. If we select e/D (the ratio of the roughness e over the pipe diameter D) to belong to the range [0.00005,0.05] and the Reynolds number to belong to the interval [$3 \cdot 10^3$, 10^8], then we can generate the Moody diagram using the Eck equation using nested for loops or matrix operations (use help to collect information for the tic/toc commands):

Nested for loops:

```
clear

NRe=logspace(log10(3000),8,20000);
e_D=[0.5 0.75 1 2.5 5 7.5 10 25 50 75 100 250 500]*1e-4;

tstart1 = tic;
for i=1:length(NRe)
    for j=1:length(e_D)
        X=-2*log10(e_D(j)/3.715+15/NRe(i));
        f(i,j)=1/X^2;
    end
end
telapsed1 = toc(tstart1);

loglog(NRe,f)
grid on
axis([1e3 1e8 1e-2 1e-1])
xlabel('Reynolds Number')
ylabel('Moody friction factor f')
```

Vectorized:

```
tstart2 = tic;
f1=1./(-2*log10(e_D'/3.715+15./NRe)).^2;
telapsed2 = toc(tstart2);
```

and we finally compare them:

```
ratio=telapsed1/telapsed2
```

```
>> EckLoop

ratio =

 100.8723
```

We therefore observe that the vectorized form "runs" 100 times faster (!) to produce the following graph:

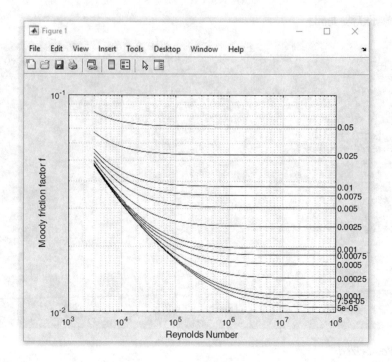

Basic Numerical Calculations in MATLAB®

We will demonstrate how to perform three basic numerical computations in MATLAB:

- Solving nonlinear algebraic equations
- Solving differential and algebraic equations
- Calculating integrals

To demonstrate the first, we consider the flash tank of Fig. A.1. An equimolar mixture of component A (more volatile) and component B, with known molar flowrate (F) and composition (z, mole fraction of more volatile component), is fed to the flash tank that operates at constant and known pressure (P) and temperature (T). Our aim is to determine the molar flow rate (V) and composition (y, mole fraction of the more volatile component) of the vapor product and of the liquid product (L, x). The mathematical model of the process consists of:

Fig A.1 The case study of the flash tank

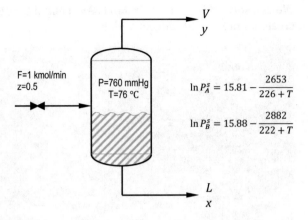

The overall material balance:

$$f_1(L, V) = F - L - V = 0$$

The material balance of the more volatile component:

$$f_2(L, V, x, y) = Fz - Lx - Vy = 0$$

The equilibrium for the more volatile component:

$$f_3(x, y) = xP_A^s - yP = 0$$

The equilibrium for the less volatile component:

$$f_4(x, y) = (1 - x)P_B^s - (1 - y)P = 0$$

The component material balance is nonlinear and thus this is a nonlinear problem of four equations in four unknowns (L, V, x, y). The problem is to find **x**:

$$\mathbf{x} = \begin{bmatrix} L \\ V \\ x \\ y \end{bmatrix}$$

so that:

$$\mathbf{f(x)} = \begin{bmatrix} f_1(\mathbf{x}) \\ f_2(\mathbf{x}) \\ f_3(\mathbf{x}) \\ f_4(\mathbf{x}) \end{bmatrix} = \mathbf{0}$$

We can solve this problem in MATLAB using the `fsolve` command. We first create an m-file with the equalities:

```
function f=flash(X)
% flash tank problem
L=X(1); V=X(2); x=X(3); y=X(4);
F = 1 ; % feed molar flowrate kmol/min
z = 0.5; % feed composition
P = 760; % pressure mmHg
T = 75; % temperature deg C
Avp=[15.81 15.88];
Bvp=[ 2653 2882];
Cvp=[ 226  222];
Ps = exp(Avp-Bvp./(Cvp+T));
f(1) = F  - L  - V;
f(2) = F*z - L*x -V*y;
f(3) =    x*Ps(1) -    y*P;
f(4) = (1-x)*Ps(2) - (1-y)*P;
f=f(:);
return
```

Note that the `%` is used to introduce comments in MATLAB. We then use `fsolve` to solve the problem of:

```
>> fsolve(@flash, [0.5;0.5;0.4;0.6])

Equation solved.

fsolve completed because the vector of function values is near zero
as measured by the default value of the function tolerance, and the
problem appears regular as measured by the gradient.

ans =
    0.7805
    0.2195
    0.4562
    0.6557
```

To demonstrate the solution of differential equations in MATLAB, we consider the batch bioreactor of Fig. A.2. Fresh material is loaded into the batch reactor at time $t = 0$. The volume of the reactor $V = 1$ m^3 is constant, and substrate (carbon source) with concentration $S(t)$ is consumed to produce cells with concentration $X(t)$. The cell growth follows a Monod kinetics with $\mu_{max} = 1$ 1/h and $K_s = 1$ g/L. The initial concentration of the substrate is $S(0) = 60$ g/L and of the cells 0.1 g/L. If the yield coefficient is $Y_{x/s} = 0.1$ g/g, we want to simulate the reactor operation and calculate how the concentrations change with time.

Fig A.2 A batch bioreactor

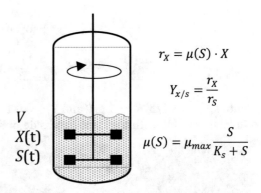

$$r_X = \mu(S) \cdot X$$

$$Y_{X/S} = \frac{r_X}{r_S}$$

$$\mu(S) = \mu_{max} \frac{S}{K_S + S}$$

The mathematical model of the process consists of the material balances of biomass and substrate:

$$\frac{dX(t)}{dt} = r_X(S, X), \quad X(0) = 0.1$$

$$\frac{dS(t)}{dt} = -r_S(S, X), \quad S(0) = 100$$

There are also two algebraic equations for calculating the rates of biomass formation and substrate consumption:

$$\mu - \mu_{max} \frac{S}{K_S + S} = 0$$

$$r_X - \mu \cdot X = 0$$

$$r_S - \frac{r_X}{Y_{\frac{x}{s}}} = 0$$

We may solve directly the system of differential and algebraic equations (DAE), as given above, which has the following general mathematical description:

$$\frac{d\mathbf{x}}{dt} = \mathbf{f}(t, \mathbf{x}, \mathbf{z})$$

$$\mathbf{0} = \mathbf{g}(t, \mathbf{x}, \mathbf{z})$$

where $\mathbf{x}$ is the vector of differential variables ($\mathbf{x} = [X\ S]^T$) and $\mathbf{z}$ is the vector of algebraic variables ($\mathbf{z} = [\mu\ r_X\ r_S]^T$). We will, however, first solve the problem as an ordinary differential equations (ODE) problem by substituting all algebraic equations into the differential equations:

$$\frac{dX(t)}{dt} = \mu_{\max} \frac{S(t)}{K_S + S(t)} X(t), \ X(0) = 0.1$$

$$\frac{dS(t)}{dt} = -\frac{\mu_{\max}}{Y_{x/s}} \frac{S(t)}{K_S + S(t)} X(t), \ S(0) = 100$$

with $\mu_{\max} = 1 \ 1/h$, $K_s = 1 \ g/L$ and $Y_{x/s} = 0.1 \ g/g$. To achieve that we create the following m-file with the right-hand side of the ODE system:

```
function dxdt=BatchBio(t,x)

X = x(1); % biomass concentration in g/L
S = x(2); % substrate concentration in g/L
miu_max=1; % maximum spc growh rate in 1/h
Ks = 1;   % saturation constant in g/L
Yxs= 1/10; % yield coeficient in g/g


rX = miu_max*S*X/(Ks+S);

dxdt= [1; -1/Yxs]*rX;
dxdt=dxdt(:);
```

We then solve the ODE using the ODE solver `ode15s` available in MATLAB and plot the results:

```
clear
x0=[0.1;100]
[t,y]=ode15s(@BatchBio,[0 5],x0)

figure(1)
subplot(2,1,1)
plot(t,y(:,2),'k','Linewidth',1)
xlabel('Time (h)')
ylabel('S(t) (g/L)')
axis([0 5 0 100])
grid on

subplot(2,1,2)
plot(t,y(:,1),'k','Linewidth',1)
xlabel('Time (h)')
ylabel('X(t) (g/L)')
axis([0 5 0 12])
grid on
```

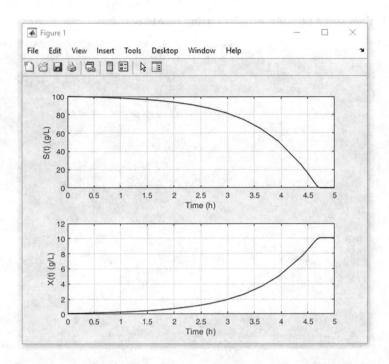

Other commonly used ODE solvers in MATLAB are the ode113 and ode45. We will now solve the problem as a DAE problem. We first create the rhs of the differential and algebraic equations in an m-file:

```
function dxdt=BatchBioDAE(t,x)

X  = x(1);      % dif: biomass concentration in g/L
S  = x(2);      % dif: substrate concentration in g/L
miu= x(3);      % alg: specific growth rate in 1/h
rX = x(4);      % alg: rate of biomass form. in g/L h
rS = x(5);      % alg: rate of substrate cons. in g/L h
miu_max=1;      % maximum spc growh rate in 1/h
Ks = 1;         % saturation constant in g/L
Yxs= 1/10;      % yield coeficient in g/g
% dif eqs
f(1) = +rX;
f(2) = -rS
% alg eqs
g(1) = miu - miu_max*S/(Ks+S);
g(2) = rX - miu*X;
g(3) = rS - rX/Yxs;
%
dxdt= [f(:);g(:)];
```

We then solve the DAE problem using `ode15s`:

```
clear
X0 =0.1;
S0 =100;
miu0=1*S0/(1+S0);
rX0 = miu0*X0;
rS0 = rX0/0.1;
x0=[X0;S0;miu0;rX0;rS0];
M=diag([1;1;0;0;0])
options=odeset('Mass',M);
[t,y]=ode15s(@BatchBioDAE,[0 5],x0,options)

figure(2)
clf
subplot(3,1,1)
plot(t,y(:,3),'k','Linewidth',1)
xlabel('Time (h)')
ylabel('\mu(t) (1/L)')
axis([0 5 0 1])
grid on

subplot(3,1,2)
plot(t,y(:,4),'k','Linewidth',1)
xlabel('Time (h)')
ylabel('r_{X}(t) (g/(L h))')
axis([0 5 0 10])
grid on

subplot(3,1,3)
plot(t,-y(:,5),'k','Linewidth',1)
xlabel('Time (h)')
ylabel('r_{S}(t) (g/(L h))')
axis([0 5 -100 0])
grid on
```

Note how Tex Markup is used (\mu) to introduce letters of the Greek alphabet or create subscripts (r_{X}). Special symbols of character formatting can be used through Tex Markup to improve the quality of the figures created in MATLAB.

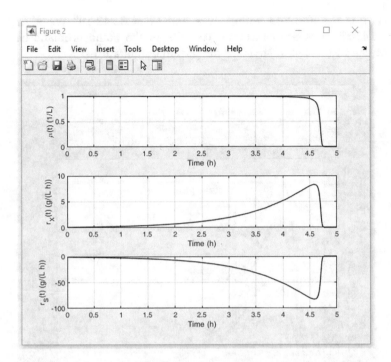

Note that important information concerning the algebraic variables of the problem are made available through the solution of the DAE systems of equations. Furthermore, the solution of the DAE system can be more robust compared to the solution of the ODE system obtained using variable substitution or, unfortunately, variable elimination (a common practice in chemical engineering literature which can be the source of important mistakes and erroneous conclusions).

We will conclude this short introduction on numerical calculations in MATLAB by demonstrating how numerical integration can be performed. In radiative heat transfer, we learn that for a blackbody, the total emissive power is given by:

$$E_b = \int_{\lambda=0}^{\lambda=\infty} E(\lambda)d\lambda$$

where λ is wavelength of radiation per unit area and unit time and $E(\lambda)$ is the "monochromatic" emissive power given by the notorious Plank's equation:

$$E(\lambda) = \frac{1}{\lambda^5}\frac{C_1}{\exp\left(\frac{C_2}{\lambda T}\right)-1}$$

where T is the absolute temperature and $C_1 = 3.742 \cdot 10^8$ W.μm^4/m^2 and $C_2 = 14{,}388$ μm.K are constants. The plot of the monochromatic emissive power can be created in MATLAB for certain temperatures:

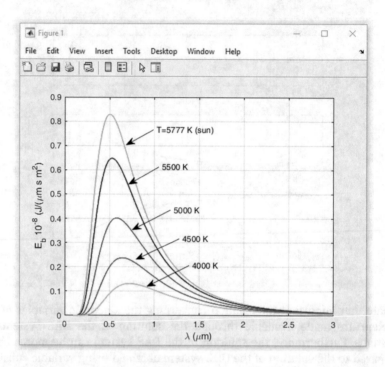

```
clear all
C1=3.742E8; C2=14388;
T=[5777 5500 5000 4500 4000];
lamda=linspace(0.1,3,1000);
for i=1:length(T)
  Eb=(C1./lamda.^5)./(exp(C2./(lamda*T(i)))-1);
  plot(lamda,Eb/1e8,'Linewidth',1.2)
  hold on
end
grid on
xlabel('\lambda (\mum)')
ylabel('E_{b} 10^{-8} (J/(\mum s m^{2})')
```

We will prove numerically that:

$$E_b = \sigma T^4 = \int\limits_{\lambda=0}^{\lambda=\infty} E(\lambda)d\lambda$$

where $\sigma = 5.67 \cdot 10^{-8}$ W/(m^2 K^4) is the Stefan-Boltzmann constant. To this end we will use the `trapz` command available in MATLAB for performing numerical integration using the trapezoidal rule:

```
clear all
C1=3.742E8; C2=14388;
T=linspace(100,6000,51);
lamda=linspace(0.01,1000,100000)';
Eb=[]
for i=1:length(T)
  E=(C1./lamda.^5)./(exp(C2./(lamda*T(i)))-1);
  Eb=[Eb E];
end
dlamda=lamda(2)-lamda(1);
Integral=dlamda*trapz(Eb);
sigma=mean(Integral./T.^4)
```

```
>> plank
```

```
sigma =

   5.6703e-08
```

The result validates the value of σ reported in the literature.

We will present another case study on numerical integration performed in MATLAB. Fogler (*Elements of Chemical Reaction Engineering*, Prentice Hall, 4th ed., p. 46) gives the following "experimental" data for the reaction rate of the reaction $A \rightarrow B$:

i	Conversion X_i [−]	Reaction rate $-r_i$ (mol/(m^3 s)
1	0.0	0.450
2	0.1	0.370
3	0.2	0.300
4	0.4	0.195
5	0.6	0.113
6	0.7	0.079
7	0.8	0.050

For the case of a plug flow reactor (PFR), the volume for achieving a specific conversion X is given by the following integral equation (page 50 of the same reference):

$$\frac{V_{PFR}}{F_0} = \int_0^X \left[\frac{1}{-r(X)} \right] \cdot dX$$

We have been asked to determine the volume of a PFR necessary that will achieve 80% conversion of the reactant A. We will first fit the "experimental data to a polynomial function:

```
clear
X= [0.0  0.1  0.2  0.4  0.6  0.7  0.8];
r=-[0.450 0.370 0.300 0.195 0.113 0.079 0.050];

c=polyfit(X,r,2);

figure(1)
plot(X,-r,'o')
hold on
plot(X,-(c(1)*X.^2+c(2)*X+c(3)),'k-')
axis([0 0.8 0 0.5])
grid on
xlabel('Conversion X [-]')
ylabel('-r (mol/(m^{3} s))')
```

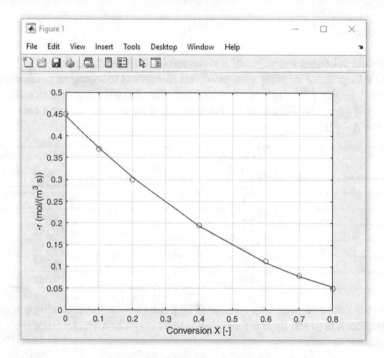

Using the polynomial function for the rate, we may determine the volume of the PFR reactor using the integral command:

```
% Integrate 1/[-r(X)] from 0 to 0.8:
f = @(X) -1./(c(1)*X.^2+c(2)*X+c(3))
Vpfr_F0 = integral(f,0,0.8)
```

```
>> pfr_volume

f =

  function_handle with value:

    @(X)-1./(c(1)*X.^2+c(2)*X+c(3))

Vpfr_F0 =

    5.3936
```

Literature and Notes for Further Study

The following books can be used for a formal and comprehensive introduction to the general theory of nonlinear programming (all books with the symbol a following the book are strongly recommended):

[1]. Arora J.A. *Introduction to Optimum Design*, Academic Press, London, UK, 4th ed., 2017.[a]

[2]. Bazaraa M. A., H. D. Sherali, and C. M. Shetty. *Nonlinear programming – Theory and algorithms.* Wiley, Hoboken, 3rd ed., 2006.[a]

[3]. Beck A. *Nonlinear optimization Theory, Algorithms, and Applications with MATLAB.* SIAM, Philadelphia, PA, 2014.

[4]. Betts J. T. (2001), *Practical Method for Optimal Control using Nonlinear Programming*, SIAM, 2nd ed., Philadelphia, 2010.

[5]. Biegler L. T. *Nonlinear Programming – Concepts, Algorithms and Applications to Chemical Processes.* SIAM, Philadelphia, 2010.

[6]. Boyd S. and L. Vandenberghe. *Convex optimization.* Cambridge University Press, New York, 2009.

[7]. Butenko S. and Pardalos P.M. *Numerical Methods and Optimization: Ana Introduction.* CRC Press, Boca Raton, FL, 2014.[a]

[8]. Eiselt H.A. and C. L. Sandblom. *Nonlinear Optimization, Methods and Applications.* Springer Nature Switzerland AG, Cham, 2019.

[9]. Fletcher R. *Practical Methods of Optimization.* Wiley, Chichester, 1991.

[10]. Floudas C.A., *Nonlinear and Mixed-Integer Optimization, Fundamentals and Applications*, Oxford University Press, USA, 1995.[a]

[11]. Gill P. E., W. Murray, and M. H. Wright. *Practical optimization.* Academic Press, London, 1981.

[12]. Griva I., Nash S.G. and Sofer A. *Linear and Nonlinear Optimization*, SIAM, 2nd ed., Philadelphia, 2009.[a]

[13]. Luenberger D. G. and Y. Ye. *Linear and Nonlinear Programming.* Springer, Heidelberg, 4th ed., 2016.

© The Editor(s) (if applicable) and The Author(s), under exclusive license to Springer Nature Switzerland AG 2022
I. K. Kookos, *Practical Chemical Process Optimization*, Springer Optimization and Its Applications 197, https://doi.org/10.1007/978-3-031-11298-0

[14]. Nocedal J. and S. J. Wright. *Numerical optimization*. Springer, New York, 2nd edition, 2006.

[15]. Rao S.S. *Engineering Optimization – Theory and Practice*. John Wiley, 5th edition, Hoboken, NJ, 2020.[a]

Books [2], [12], and [13] are strongly recommended for those readers with strong mathematical background (book [12] is easier to follow). Books [1] and [15] are extensively used by engineers and have a balanced exposition to both theory and applications. They are more appropriate for readers with a mechanical engineering background.

The following books are more appropriate for readers with a background in chemical engineering (the same holds true for book [10]):

[16]. Biegler L.T., I. Grossmann, and A. Westerberg, *Systematic Methods of Chemical Process Design*, Prentice-Hall, Englewood Cliffs, NJ.[a]

[17]. Edgar T.F., D. M. Himmelblau and Lasdon L.S., *Optimization of Chemical Processes*. McGraw-Hill, 2nd ed., NY, 2001.

[18]. Grossmann I.E., *Advanced Optimization for Process Systems Engineering*, Cambridge University Press, NY, 2021.

[19]. Maranas C.D. and A.R. Zomorrodi, *Optimization Methods in Metabolic Networks*, John Wiley & Sons, NJ, 2016.

[20]. Ray W. H. and J. Szekely. *Process Optimization with applications in metallurgy and chemical engineering*. John Wiley, New York, 1973.

[21]. Ravindran A, K. M. Ragsdell, Reklaitis G. V. *Engineering Optimization: Methods and Applications*. John Wiley & Sons, New York, 2nd ed., 2006.[a]

[22]. Vasiliadis V., Kahn W., Chanona E. A., Yuan Y. *Optimization for Chemical Engineering and Biochemical Engineering*, Cambridge University Press, NY, 2021.

There are many excellent books available that are specific to linear programming and mixed-integer linear programming:

[23]. Dantzig G.B. and M. N. Thapa. *Linear programming 1: Introduction*. Springer, New York, 1997.

[24]. Dantzig G.B. and M. N. Thapa. *Linear programming 2: Theories and extensions*. Springer, New York, 2003.

[25]. Eiselt H.A. and C. L. Sandblom. *Operations Research*. Springer-Verlag Berlin Heidelberg, 2012.[a]

[26]. Hillier F.S. and G.J. Lieberman. *Introduction to Operations Research*, McGraw Hill Eduacation, NY, 10th ed., 2015.

[27]. Pochet Y. and L.A. Wolsey. *Production Planning by Integer Programming*, Springer, NY, 2006.

[28]. Taha H.A. *Operations Research: An Introduction*. Pearson Education Limited, 10th ed., Harlow, 2017.[a]

[29]. Taha H.A., *Integer Programming, Theory, Applications and Computations*, Academic Press, NY, 1975.

[30]. Williams H.P. *Model Building in Mathematical Programming*, John Wiley & Sons, Chichester, 5th ed., 2013.[a]

[31]. Williams, H.P. *Model Solving in Mathematical Programming*, John Wiley & Sons, Chichester, 1993.[a]

[32]. Williams, H.P. *Logic and Integer Programming*, Springer, New York, 2009.[a]

[33]. Wolsey, L.A. *Integer Programming*, John Wiley & Sons, NJ, 2nd ed., 2021.[a]

Book [10] remains the best and most comprehensive introduction to mixed-integer nonlinear programming theory and problems but contains an unusual number of typos. Excellent (but relatively short) introductions are also presented in books [16], [18], [19], and [22].

Index

Printed in the United States
by Baker & Taylor Publisher Services